Ferdinand Hucho

Neurochemistry

©VCH Verlagsgesellschaft mbH, D-6940 Weinheim (Federal Republic of Germany), 1986

> Distribution:
> VCH Verlagsgesellschaft, P. O. Box 1260/1280, D-6940 Weinheim (Federal Republic of Germany)
> USA and Canada: VCH Publishers, 303 N. W. 12th Avenue, Deerfield Beach, FL 33442-1705 (USA)

ISBN 3-527-25989-9 (VCH Verlagsgesellschaft)
ISBN 3-89573-225-4 (VCH Publishers)

Ferdinand Hucho

Neurochemistry

Fundamentals and Concepts

Translated by Margaret Dickins

Prof. Dr. Ferdinand Hucho
Freie Universität Berlin
Institut für Biochemie
Thielallee 63
1000 Berlin 33

QP
356.3
.H8313
1986

Editorial Director: Dr. Hans-Joachim Kraus
Production Manager: Dipl.-Ing. (FH) Hans Jörg Maier

Library of Congress Card No.86-9066

Deutsche Bibliothek Cataloguing-in-Publication Data

Hucho, Ferdinand:
Neurochemistry: fundamentals and concepts /
Ferdinand Hucho. Transl. by Margaret Dickins. –
Weinheim; Deerfield Beach, FL: VCH, 1986.
  Dt. Ausg. u.d.T.: Hucho, Ferdinand:
Einführung in die Neurochemie
  ISBN 3-527-25989-9 (Weinheim)
  ISBN 0-89573-225-4 (Deerfield Beach)

©VCH Verlagsgesellschaft mbH, D-6940 Weinheim (Federal Republic of Germany), 1986
All rights reserved (including those of translation into other languages). No part of this book may be reproduced in any form – by photoprint, microfilm, or any other means – nor transmitted or translated into a machine language without written permission from the publishers. Registered names, trademarks etc. used in this book, even when not specifically marked as such, are not to be considered unprotected by law.
Composition: Hans Richarz Publikations-Service, D-5205 St. Augustin 1
Printing: betz-druck gmbh, D-6100 Darmstadt 12
Bookbinding: Josef Spinner, D-7583 Ottersweier
Printed in the Federal Republic of Germany

For Barbara
with gratitude for 25 years

Preface

Neurochemistry is probably one of the most rapidly advancing areas of the life sciences. This book was initially planned as a translation of a text that I wrote six years ago in German. But soon it became obvious that much more than translating and slightly updating was required to do justice to the many new developments. Recombinant DNA techniques have come into play, new concepts of signal transduction through membranes have emerged, and the characterization of several neurotransmitter receptors have clarified hitherto vague hypotheses. As I have taken all this into account, "Neurochemistry, principles and concepts" has become a new book. However, one aspect has remained unchanged: I wanted again to present a readable and enjoyable compilation of the highlights of neurochemistry, rather than a comprehensive reference work. This book is intended for the biochemists, biologists, chemists, biophysicists, pharmacologists and medical students who wish to familiarize themselves with the exciting interdisciplinary topics of neurochemistry. First of all I would like to thank Dr. Margaret Dickins, who contributed much more than just the translation to this work. I thank also my coworkers, who kept the laboratory work going while I was distracted by collecting and writing down the facts. Further thanks are due to my students, who provided the testing ground for the didactic and factual concepts that underlie the following twelve chapters. Finally, I wish to thank the publisher, and especially Dr. Kraus, who patiently accepted the many last-minute additions which are necessary to keep a book up-to-date during the process of its production.

Berlin, March 1986

Ferdinand Hucho

Contents

Chapter 1

Neurochemistry: Definition Exemplified by the Visual Process 1

The mechanism of information transfer, not the information itself, can be analysed biochemically 1
The neurochemistry of the retina does not describe the seeing of pictures . . 3
First step: vitamin A aldehyde absorbs light 3
Second step: a nerve impulse is triggered (transduction) 7
Rhodopsin is photophosphorylated 10
The cones and colour vision 11
Invertebrates see differently 11
Third step: nerve impulses are integrated 11
Neurochemistry as an integrating science 14
The neuron: functional elements as subjects of neurochemistry 15
Summary 22
References 23

Chapter 2

Membrane Molecules 25

Neurochemistry is more than the chemistry of "neuro-molecules" 25
Lipids, proteins and carbohydrates are the building blocks of the nerve membrane 26
The construction of phospholipids allows a great variety of molecular structure . . 27
Phospholipases initiate the degradation of phospholipids 32
Sphingolipids are not only present in nerves, but are especially important there . 34
Speculations about the function of sphingolipids 38
Gangliosides are receptors for bacterial toxins 39
Lipidoses are caused by enzyme defects in glycolipid metabolism 41
Carbohydrate confers multiple specificities on cell surfaces: glycoproteins . . 44
Glycoproteins may be important for the specificity of neural connections . . 46
Summary 47
References 48

Chapter 3

Membranes 51

The neuronal membrane as a plasma membrane 51
From Gorter and Grendel to Singer and Nicolson: membrane models . . . 52
Models are not copies of reality 53
Membrane properties: the lipid phase 53
Bio-membranes are "fluid-crystalline" 56
Lysolecithin damages membranes 57
Cholesterol-containing membranes are neither crystalline nor fluid crystalline . 58
Drugs affect membrane fluidity 58
Effect of ions on lipid membranes 59
Asymmetry of biological membranes 60
Carbohydrates and proteins are also distributed asymmetrically in the bilayer . 60
Protein-lipid interactions cause phase separation and asymmetry of the membrane 62
Lipid-exchange proteins 63
Mild detergents can replace lipids 63
Artificial lipid membranes are models of biological membranes 66
Summary 70
References 70

Chapter 4

Myelin 73

Functions of myelin: (I) insulation; acceleration of conduction velocity . . . 73
Functions of myelin: (II) space and energy saving 75
Myelin is a compact spiral made from plasma membranes 76
The bilayer structure of myelin 76
Chemical composition of myelin 80
Myelin is not metabolized as a whole 82
Proteins of unknown function 82
Experimental autoimmune disease: a model for multiple sclerosis? 85
Diseases caused by myelin defects 86
Summary 87
References 88

Chapter 5

Electrophysiology 91

The resting potential 91
Excitation of the neuron: local potential and action potential 95
Synaptic impulse transmission 99
Single channels and noise analysis: electrophysiology at the molecular level . 101

Summary	105
References	107

Chapter 6

Ion Channels — 109

Active an passive ion transport are independent of each other	109
Chemical and electrical regulation of passive ion currents	111
Passive Na^+ and K^+ transport are independent of each other	111
Gate and selectivity filter are functional elements of ion channels	112
The sodium channel: the gating mechanism	112
Calcium ions affect the treshold of excitation, but not m^3	114
The sodium channel: the selectivity filter	114
Channel or carrier?	115
Some physical properties of the sodium channel	117
Biochemical characterization of the sodium channel	117
Drugs influence the action potential	121
Neurotoxins as tools for the investigation of ion channels	121
Anaesthesia	125
The potassium channel	129
Some physical properties of the potassium channel	132
Biochemical characterization of the potassium channel	132
Structure of the axonal membrane: biochemistry, electron microscopy, spectroscopy	133
Summary	136
References	136

Chapter 7

Active Ion Transport — 139

Three examples of ATP-driven ion pumps	142
The sodium/potassium pump	143
Pumps which transport charges: electrogenic pumps	146
Heart glycosides inhibit the sodium/potassium pump	147
Ca^{2+} transport mechanisms	148
The proton pump: ATP synthesis is the opposite of active transport	149
Models of proton pumps	150
What an ion pump looks like	152
Summary	154
References	154

Chapter 8

The Synapse, Part 1 — 157

Neuron and synapse, historical remarks	157

Contents

Why is the synapse so interesting?	157
Electrical and chemical synapses	158
Properties of the electrical synapse	158
The chemical synapse, control site of the nervous system	159
The cholinergic synapse, peripheral and central	162
Two classes of cholinergic synapses, muscarinic and nicotinic	162
Individual stages of chemical synaptic transmission	163
The nicotinic cholinergic synapse	164
Choline reacts with acetyl-CoA forming acetylcholine	166
Acetylcholine is packaged in vesicles	167
How does acetylcholine get into the synaptic cleft?	167
Exocytosis – endocytosis	168
The role of calcium	168
Acetylcholine is bound by the postsynaptic membrane	169
Transmitter inactivation by enzymatic hydrolysis	172
Inhibitors of the individual steps in synaptic transmission	175
Neurotoxins from snake venoms	176
Other transmitters: criteria and classification	178
Neuropeptides: transmitters and hormones	180
A transmitter can have several functions	181
Catecholamines	182
Synthesis is tightly regulated	182
Catecholamines are also packed into vesicles	185
Release and binding	185
Multiple receptors assure variability of transmitter effects	186
Various types of adrenergic effects	186
Inactivation of catecholamines by uptake and degradation	187
Serotonin	190
Synthesis from tryptophan, degradation by MAO	190
The serotonin cycle is analogous to other transmitter cycles	191
Amino acids as transmitters: GABA, glycine and others	192
GABA, an inhibitory transmitter	192
GABA regulates chloride channels	194
GABA antagonists are convulsants	194
Glycine, another inhibitory transmitter	195
Glutamate and aspartate: excitatory transmitters	195
Kainic acid and the method of chemical lesions	195
Enkephalins and other neuropeptides: neuromodulators and suspected neuro transmitters	196
Substance P, the longest-known neuropeptide	199
Summary	200
References	200

Chapter 9

The Synapse, Part 2: Receptors 203

Receptors, defined as physiological sites of action 203
Three criteria define a binding site as a receptor 203
Three levels of receptor research: the data must be correlated 204
Receptor models 204
Binding studies: binding does not equal effect 208
Methodological remarks: irreversible binding, affinity labelling 210
Mobile receptors: the "floating receptor" hypothesis 212
Acetylcholine receptors 214
The nicotinic acetylcholine receptor – First level: intact cells 214
Second level: receptor membrane 215
Third level: receptor molecules 217
Pharmacological desensitization: model for synapse modulation 218
Acetylcholine receptors from muscle tissue 219
Hypersensitization: another model for receptor modulation? 219
Phosphorylation of acetylcholine receptors 220
Myasthenia gravis, an antoimmune disease of the nicotinic cholinergic synapse 220
The muscarinic acetylcholine receptor 222
The catecholamine receptors 222
β-adrenergic receptors: interaction of receptor, cyclase and regulator (R, C and N) 223
Greengard's hypothesis: cAMP regulates via protein phosphorylation . . . 226
Synapsin I, substrate of various protein kinases 228
Cyclic nucleotides and Ca^{2+}, two classical "second messengers" 228
α-adrenergic receptors (α-adrenoceptors) 229
Toxins as tools: cholera toxin and pertussis toxin cause ADP-ribosylation of
 N-proteins 231
Dopamine receptors 231
Dopamine and schizophrenia 236
Parkinson's disease 236
Opiate receptors 237
Presynaptic inhibition by opiates 240
Addiction and tolerance – a molecular model 240
GABA receptors, inhibitory and allosteric protein complexes 242
Benzodiazepines and barbiturates enhance GABA effects allosterically . . 242
Glycine receptors are also inhibitory; the first central receptor isolated . . 243
Glutamate receptors, serotonin receptors and the many other receptors of general
 interest have not yet been isolated 244
Autoreceptors and some remarks concerning the fine tuning of nervous activity . 245
Receptors are under regulatory control 246
Summary 247
References 248

Chapter 10

Neuronal Cytoplasm and Neuron-Specific Proteins 251

Axonal transport, the intracellular communication system 251
Everything is transported: proteins, lipids, transmitters, mitochondria etc. . . 253
Kinetics of transport: different rates for different components 254
Mechanism: energy dependent, membrane-bound and carried by filamentous structures 254
Tubulin and associated proteins 257
Neurofilaments 258
Actin, myosin: a role in mechanical work? 258
Calmodulin, a mediator protein for calcium regulation 258
Neuron-specific proteins 260
Summary 261
References 262

Chapter 11

Development, Stabilization and Plasticity of the Nervous System 263

Differentiation of transmitters 265
Transsynaptic regulation: orthograde and retrograde 265
Differentiation of ion channels and excitability 266
There are various differentiation programmes 267
Trophic factors 267
Nerve growth factor (NGF) 268
The molecule NGF 269
The mechanism of NGF action is unknown 269
NGF regulates differentiation, survival and target-oriented growth of nerve cells 270
Are there more nerve growth factors? 271
Synaptogenesis 271
"Selective stabilization" of synapses, a plausible hypothesis 273
Plasticity 274
Protein structure and activity are dependent on the chemical environment . . 274
Biochemical basis of learning 275
Learning depends on the plasticity of the nervous system 275
Learning paradigms 276
The site of learning 278
Post-tetanic facilitation, a synaptic "memory"? 279
Intervention experiments for the localization of memory 281
Without protein synthesis there is no long-term memory 281
The search for memory-specific proteins: S-100 and others 283
General or specific effect? 284
Aplysia sensitization – a learning model completely described from behaviour to the molecular events 285

Do catecholamines, acetylcholine and pituitary hormones take part in learning? . 288
Summary 288
References 289

Chapter 12

Experimental Model Systems 291

Chemotaxis 294
Behavioural model: *Paramecium* 297
Developmental biology model: *Hydra* 298
Drosophila 298
Mouse mutants: genetics as a method for analysis of motor behaviour . . . 299
Electroplaques of electric fish: synapse model 301
E.coli of neurobiology: cell cultures 304
Genetic engineering revolutionizes neurochemistry 305
Hybridoma cloning for monoclonal antibodies 306
References 307

Index 309

Chapter 1

Neurochemistry: Definition Exemplified by the Visual Process

Modesty should be in the forefront of neurochemistry. A quarter of a century of molecular biology has so increased biochemists' self-confidence that some have assumed that all problems of living nature can be solved by biochemical methods or through the molecular approach. In investigating the mechanism of heredity there has been a temptation to think of the human brain as just another molecular puzzle. As more fundamental problems in molecular genetics have been solved and there are fewer break-throughs to be made, prominent molecular biologists have tended to turn to neurobiology. Here, however, there are limits to the molecular approach. I do not want to advocate a new vitalism but a definition of the limits and possibilities of the subject can be a good way of introducing it, and a comparison between molecular genetics and molecular biology may serve to illustrate it.

The key to molecular genetics is the DNA molecule. In it the genetic information is contained in a code which we modestly call "universal". This implies that when the transmission of hereditary information in one cell is understood the molecular mechanism in all cells, not only in the same organ or organism but in principle in all living things is also known. There are certainly differences between the regulation of heredity in prokaryotes and eukaryotes, but the situation in general is well summarized in Monod's famous saying, "what applies to *E. coli* applies also to the elephant". Heredity is the biochemistry of DNA.

The mechanism of information transfer, not the information itself, can be analysed biochemically

The subject of neurochemistry is the fundamental unit of all nervous systems, the nerve cell or neuron. Why don't we say that nerve function is the biochemistry of neurons? Because we cannot include in this definition the vital higher functions of nervous systems: perception, sensation, thought, learning or consciousness. All these are certainly not simple biochemical properties of neurons but the result of the integration of hundreds of millions of neurons into complex networks. The biochemical analysis of the human brain alone contributes no more to an understanding of brain function than the analysis of pigments to the appreciation of a painting. In spite of the infinite multiplicity and complexity of its functions the biochemistry of the neuron is quite stereotyped; the activity of a single neuron is almost primitive compared to its function in the nervous system.

Let me illustrate this. The true function of the neuron is to transmit signals. We will see however (Chapter 5) that in the nervous system there are only two types of signal: electrical and chemical. It is important to note that the signal itself contains very little information. Its specificity depends on its origins and endpoint – what it links together. Thus, for example, the reason we hear a sound rather than see it is not in the electrical or chemical code of the nerve impulse but in the connections which it makes between the neural cells of the retina and those of the visual cortex in the occipital lobe of the brain. If we stimulate the retina electrically or mechanically instead of optically we also "see". Anyone who has seen stars after a punch in the eye can confirm this. The quality of information transmitted by the neuron therefore depends on the specificity of the connection – the quantity alone appears to be contained in the signal; a strong stimulus sends more nerve impulses from the receptor to the end organ than a weaker one. Again: nerve impulses in, say, the optic or the acoustic area of our nervous system are practically indistinguishable from those of totally different systems in, for example, much more primitive life forms. In isolation, they contain little information even for the expert. So the neurochemist investigating the biochemistry of neurons can only explain the mechanism of the initiation and conduction of the signals; the specific content of the signal is not accessible by his methods. He can investigate the general molecular reactions underlying the processing of the signals, but not the result of this processing, the "information".

By contrast, the molecular geneticist can not only record biochemically the mechanism of transcription and translation of genetic information, but also, simultaneously, the information itself – for example the enzyme pattern which makes a liver cell a liver cell or a neuron a neuron. This, however, needs qualifying: in the nervous system there is not only the biochemistry of single cells, there is the biochemistry of embryonic development of the neural network, the metabolism of the nervous system, the integration of hormonal action, the interaction between parts of the nervous system and between the nerve cells and their environment e.g. the glial cells etc. There are thus many neurobiological subjects with neurochemical aspects which go well beyond the biochemistry of the single neuron.

Neurons have such varied functions that it must be assumed that there are adaptations by single neurons to specific functions. The central nervous system, for example, is a fundamentally more differentiated organ than the liver where the various functions are divided up among relatively few cell types. Although neurons with their stereotyped signals have many of their basic functions in common, there are certainly specialized neurons with their own peculiar neurochemistry. As always, at the beginning of a new subject, we will deal at first with what neurons have in common.

The neurochemistry of the retina does not describe the seeing of pictures

Instead of discussing in general terms the possibilities and limits of neurochemistry, it is better to take a specific example. Let us take "seeing" as a typical complex nervous function. Seeing is a multi-stage process: light energy must be captured, converted into a nerve impulse and passed on in this form. Nerve impulses must be processed, i.e. they must be integrated to yield a special information. Only the first two steps can be described as exclusively molecular neurochemical processes and these do not represent "seeing" in the sense of an image perception. We do not see with the eyes, but with the brain.

First step: vitamin A aldehyde absorbs light

In 1877 Franz Boll discovered visual purple, retina rhodopsin, and in the century that has followed a complete picture of its function has been developed. Signals from our environment are received through special structures in our sense organs – the light receptors – and transmitted to the central nervous system. The photoreceptors for the reception of visual signals are the rods and cones of the retina at the back of our eyes. Hecht showed in 1931 that a single photon there can trigger a nerve impulse and in 1933 one of Hecht's pupils, G. Wald, clarified the decisive stage of this molecular process: he discovered vitamin A in the retina. The structure of this vitamin had already been established by Karrer and in 1934 Wald formulated his cyclical theory of photoreception (Fig. 1.1A) which is still fundamentally accepted today.

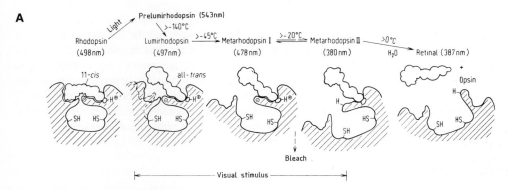

Fig. 1.1A. Bleaching of rhodopsin. Following the photoisomerization of retinal (11-*cis*-retinal → all-*trans*-retinal) which induces a sequence of conformational changes in opsin, the complex of protein (opsin) and chromophore dissociates. The individual conformations are stabilized at very low temperatures and can then be characterized by their absorption spectra. The nerve impulses are released before the dissociation (bleaching), probably at the metarhodopsin II stage (after G. Wald).

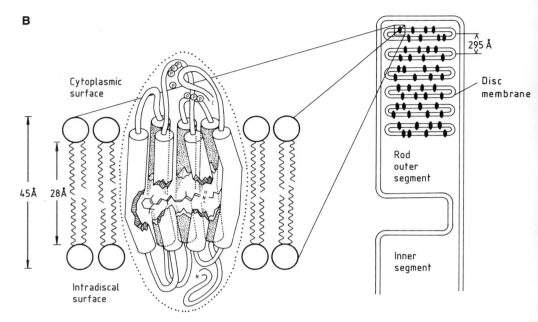

Fig. 1.1B. Diagram of a cross-section of the rod outer segment. The drawing is not to scale as there are 600-2000 discs/rod and 2×10^4–8×10^3 rhodopsins/disc depending on species. A magnified portion of the rod outer segment is shown on the left, which is a schematic model of the general structure of rhodopsin and its association with the lipid bilayer of the rod outer segment disc membrane. Rhodopsin is represented as an elongated bundle of somewhat irregular helices embedded in the lipid bilayer. The 11-*cis*-retinal binding site, oriented nearly parallel to the membrane plane, is schematically shown. The seven known phosphorylation sites are marked with the letter P. Reproduced, with kind permission, by Elsevier Science Publishers, from TIBS 8, p. 128 (1983).

Rhodopsin consists of a protein, *opsin*, (Fig. 1.1B), bound in a 1:1 molecular ratio to a chromophore, *vitamin A aldehyde (retinal)*. There are different chromophores, retinal$_1$ and retinal$_2$ (Fig. 1.2) in land and sea vertebrates, which are adapted to the different light conditions in their environments. Rhodopsin of both rods and cones, however, contains the same retinal in spite of its different absorption spectrum (Fig. 1.3); the difference is due to the protein component. Retinal forms a Schiff's base with its aldehyde group and an amino group of an opsin-lysine residue and is thus covalently linked to the protein. The fundamental step in photoreception is the utilization of the absorbed light energy through the isomerization of retinal: 11-*cis*-retinal (Fig. 1.2) is transformed to all-*trans*-retinal. This is followed by the destabilization of the rhodopsin complex and its hydrolysis to retinal and opsin. Spectroscopically more intermediate stages can be identified (Fig. 1.4) and it can also be shown biochemically that the protein goes through a series of conformational changes. Thus in native rhodopsin the Schiff's base cannot be reduced in the normal way with sodium borohydride (NaBH$_4$); only at the metarhodopsin II stage does it become accessible to this reagent. An amino acid of the protein, whose isoelectric constant of 6.4

suggests that an imidazole ring of a histidine residue is involved, can be titrated following irradiation; the thiol groups of opsin also show conformation-dependent changes in reactivity. It thus appears that light causes changes in the structure of the protein via the isomerization of retinal.

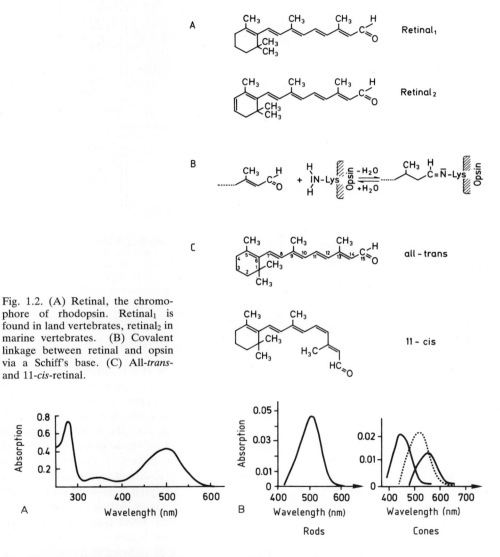

Fig. 1.2. (A) Retinal, the chromophore of rhodopsin. Retinal$_1$ is found in land vertebrates, retinal$_2$ in marine vertebrates. (B) Covalent linkage between retinal and opsin via a Schiff's base. (C) All-*trans*- and 11-*cis*-retinal.

Fig. 1.3. Absorption spectrum of rhodopsin of rods (A). Although the chromophore is the same in all cases, the spectrum of rhodopsin is different in the rods and the three types of cones responsible for colour vision (B).

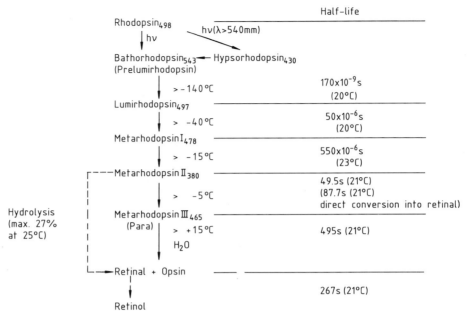

Fig. 1.4. Sequence of intermediate stages and their half-lifes during the bleaching of rhodopsin of frog retina [2].

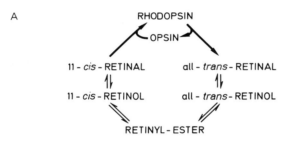

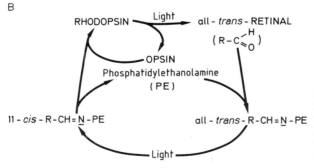

Fig. 1.5. Regeneration of 11-*cis*-retinal. (A) Enzymatic regeneration. (B) Possible mechanism of light-induced regeneration with participation of phosphatidylethanolamine (PE).

We shall return to the dissociation of rhodopsin which is, as we have seen, the decisive event for the second stage in the visual process - the conversion of light into a nerve impulse. The Wald's cycle must now be completed by the regeneration of native rhodopsin from the bleached form. This is shown in Fig. 1.5. It may be brought about in part by an enzymatic reduction of retinal to retinol (vitamin A) followed by the isomerization of the latter (all-*trans*-retinol to 11-*cis*-retinol) and oxidation to 11-*cis*-retinal followed by recombination to rhodopsin. In parallel to this process, which takes place in the dark, light may itself promote the regeneration of 11-*cis*-retinal (Fig. 1.5). This second process is possibly catalysed by a component of the rod membrane, phosphatidylethanolamine (PE). However these two routes are not fully understood and here, as in other aspects of photoreception, we are still as it were in the dark!

Second step: a nerve impulse is triggered (transduction)

Everything up to now has been biochemical and the next step in the primary process of light perception can also be described in molecular terms. Here we shall be mainly concerned with one of the two vertebrate photoreceptors, the rods. The cones, which are responsible for colour vision, are less well understood and we shall put them on one side for the moment. The rods subserve the recognition of brightness contrast and are used primarily for seeing in the dark. The human eye contains about 120×10^6 rods and 6.5×10^6 cones. A rod (Fig. 1.6) consists of an outer and an inner segment which are connected together by a cilium. Inside the inner segment are the mitochondria responsible for the metabolic activity of the cell, and the cell nucleus. The outer segment contains many stacked bilayer membranes (1000-2000 per rod) known as discs into which numerous rhodopsin molecules are inserted. Rhodopsin is a glycoprotein i.e. it contains 14 covalently bound sugar residues (5 N-acetyl-glucosamine and 9 mannose residues). It is insoluble in water but can be released by mild detergents from its lipophilic environment in the membrane without too much damage to its structure. In humans each segment contains 10^7 rhodopsin molecules of relative molecular mass 41 000, (in the frog, 10^9 so that the concentration of rhodopsin in the frog's rods is amazingly high – about 2 mmol/l). Its primary structure has been elucidated [1] and it has been shown to be an intrinsic membrane protein spanning the membrane. This implies that it has an extended configuration *in situ*, because as a globular molecule its diameter would amount to only 4 nm, less than the thickness of the membrane. Anyway, by its high concentration alone it would affect the physical characteristics of the bilayer, particularly its electrical properties.

Following the absorption of light by rhodopsin – the first stage of the visual process – there is a change of membrane potential of the rod cells; as a result of a difference in the distribution of ions across the plasma (cell) membrane (see Chapter 5) a difference of potential of about 40 mV (inside negative) is created between the inner and outer sides of the rod. This so-called resting potential is raised by the action of light. The new temporarily

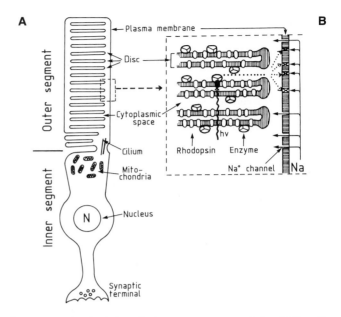

Fig. 1.6. Structure of the rod. (A) Structure as a whole (schematic). (B) Outer segment with details of the rhodopsin-containing disc membranes (see also Fig. 1.1B.). Kindly supplied by H. Kühn, Jülich.

increased membrane potential (hyperpolarization) is called a *receptor potential* and like all changes of membrane potential it is brought about by a change in the ionic permeability of the membrane (Chapter 5). It is thought that it is specifically the sodium permeability of the membrane that is affected through stage 1 of the visual process – the interaction of light with rhodopsin – about 1000 sodium channels being closed. The hyperpolarization of the membrane is only seen in vertebrates; in invertebrates depolarization occurs, i.e. the interior of the cells becomes less negative on irradiation.

Yet how can rhodopsin regulate the sodium channels of the plasma membrane when it is itself located in the disc membranes? It is thought that calcium ions and cGMP function as messengers between the two membrane systems. A calcium-activated ATPase has been identified as a constituent of the disc membranes; this acts as a calcium pump utilizing metabolic energy to increase the calcium concentration in the spaces between the disc membranes. Possibly, as postulated in one of Hagins' models (Fig. 1.7A) [3], the incident light opens the calcium channels. The question remains as to the role of rhodopsin in the opening of the calcium channels in the disc membrane. Is it an enzyme which catalyses changes in the membrane structure? Or does it constitute the channel itself which opens or closes as a result of conformational changes?

The cyclic nucleotide cGMP may play the most important role in this connection (Fig. 1.7B). It has been found in the rod outer segment in high concentrations and appears to keep the sodium channels of the rod membrane open. Furthermore, a light-activated phosphodiesterase which cleaves cGMP has been discovered. According to a hypothesis proposed by Stryer this enzyme is activated by a protein called "transducin". In the dark transducin contains bound GDP which is replaced by GTP under the influence of bleached rhodopsin. It is the transducin/GTP complex which activates the phosphodiesterase. One

Fig. 1.7. (A) Hagins' model of the release of the nerve impulse in vertebrate rods and cones. The light-induced conformational change of rhodopsin sets free calcium ions from the intramembranous space between the bilayer cells; this then closes the sodium channels in the plasma membrane, thus bringing about a hyperpolarization. Calcium ions thus act as a messenger between the disc membrane, where the primary photoreaction takes place, and the cell membrane which generates the nerve impulse. (Right half of the diagram: in cones the disc membranes are simply invaginations of the plasma membrane. The same model applies with this modification) [3].

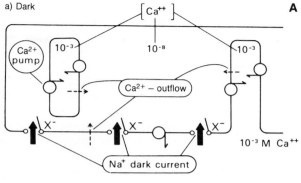

(B) An alternative model, the amplification cascade of light transduction (after Stryer), postulating cGMP to be the second messenger regulating the Na^+-channels of the rod outer membrane. Rh, rhodopsin; Rh*, light-activated rhodopsin; T, transducin (also called G-protein); PDE, phosphodiesterase; PDE*, phosphodiesterase activated by the transducin-GTP complex.

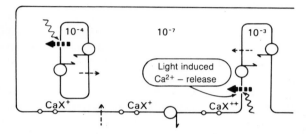

Amplification   1   :   500   :   500000

molecule of bleached rhodopsin can cause GDP-GTP exchange in several transducin molecules which activate by complex formation an equal number of molecules of phosphodiesterase. These in turn cleave many molecules of cGMP, an impressive cascade amplifying the bleaching effect of a single photon and closing in its final stage the sodium channels of the rod plasma membrane [4].

It has recently been shown by Kühn, that light-activated rhodopsin (which is identical with metarhodopsin II) directly interacts with transducin by forming a transient complex. Thus the triggering of the enzyme cascade is a direct action of light-activated rhodopsin. Moreover, microinjection of activated transducin into rod cells mimics the effect of light, i.e. it hyperpolarizes the cell. This shows that the activation of the cGMP-degrading enzyme cascade plays a major role in phototransduction. It is much less clear how the light-dependent changes in cGMP metabolism and in Ca^{2+} concentration are interrelated and what the mechanism or substance is that finally regulates the light-dependent permeability of the rod cell membrane.

Transducin, also called G-protein, is strikingly similar to the N-proteins which we shall discuss in Chapter 9 as the membrane components transducing hormonal or neurotransmitter signals from certain receptors to the enzyme adenylate cyclase. Both transducin and N-proteins consist of three polypeptide chains, α, β and γ and their mechanism of action appears to be very similar. The similarity between the rhodopsin/transducin/phosphodiesterase system and that of the β-adrenergic receptor/N-protein/adenylate cyclase, for example, is great enough to allow cross-recombinations between isolated components. In one such reconstitution experiment transducin has been shown to transmit signals from β-receptors to adenylate cyclase in cells deficient in N-protein.

Rhodopsin is photophosphorylated

Another intriguing enzymatic reaction is activated by light: in 1972 H. Kühn discovered a Mg^{2+} requiring, cAMP-independent protein kinase that specifically phosphorylates photobleached rhodopsin at serine and threonine residues [5]. The kinase itself seems to be active in the dark as well as in the light; light appears to convert rhodopsin from a conformation which cannot be phosphorylated into one which is recognized by the kinase as a substrate. 7–9 phosphate groups can be incorporated into each bleached rhodopsin molecule. *In vitro*, phosphorylation of rhodopsin seems to be a slow reaction, taking many minutes, if much rhodopsin is bleached. At low bleaching levels, however, the reaction becomes much faster.

Phosphorylation is probably involved in the inactivation process following light activation of rhodopsin. In other words: light is the "turn-on signal" and phosphorylation may be the "turn-off signal" for rhodopsin to activate the enzyme cascade [6].

The cones and colour vision

The cones which are the receptors for colour vision seem rather more complicated though in principle they probably operate in much the same way. We have already mentioned that their chromophore is the same as in the rods. Differences in their absorption spectra (Fig. 1.3) depend on differences in the opsin to which the retinal is bound. Even less is known about the structure of these proteins than about the opsin of the rods. They are thought to be products of different genes and could thus have different amino acid sequences. The genetic origin of the cone opsins is supported by the fact that colour blindness has a recessive sex-linked inherited character. About 1% of men are red-blind and 2% are green-blind, whereas in women this defect occurs much less often. The three types of cones are also distinguished morphologically from the rods. Besides being conical in form, their disc membranes are not separate organelles but are invaginations of the plasma membrane; the plasma and disc membranes form a continuum. This is included in Hagins' model of photoreception (Fig. 1.7 A, right half) and the link between the light absorption and the closing of the sodium channels is again brought about by calcium which flows into the cytoplasm from the extracellular photoreceptor (disc) membrane space.

Invertebrates see differently

The visual process is quite different in *invertebrates*. Light absorption does not lead to hyperpolarization but to depolarization of the receptor cell; i.e. the cell interior becomes less negative with respect to the outside, because ionic conductivity of the membrane increases. Calcium ions are probably not involved in the coupling between rhodopsin bleaching and the change in membrane conductivity. There is not yet, however, a coherent picture of the biochemistry of light reception in invertebrates.

Third step: nerve impulses are integrated

In the next stage of the visual process the receptor potential must now be converted into a nerve impulse which, by a series of further steps, will eventually reach the occipital lobe of the cerebrum. As we will see in later chapters, the total phenomenon of triggering the release and conduction of nerve impulses can be described neurochemically, i.e. in molecular terms. But on its way from the rhodopsin of the receptor cell to its target the light signal is transformed by successively taking up information from a flash of light in the retina to a picture formed in the cerebrum. Here the process can only be outlined; for a comprehensive account the book of Kuffler, Nicholls, and Martin [7] should be consulted.

Mere detection of light is not seeing; the photoreceptor is a "light measurer" and not a photographic plate. The visual system consists of a hierarchy of steps in which the simple light stimulus is increasingly supplemented with information. The photoreceptor (rods or cones) reacts quite simply in proportion to the amount of incident light. But in the ganglion cells of the retina it is already hardly possible to establish a proportionality between light intensity and response. Such cells react primarily to light contrast. The processing of the signal leaving the photoreceptor is called *integration*. It depends on two properties of the *retina*: firstly its three-layered structure (Fig. 1.8) of *receptor, bipolar* and *ganglion cells* which are interlayered with further cell types (*amacrine* and *horizontal* cells), and secondly its high degree of *convergence*: on average there is only one ganglion cell for 100 receptors, i.e. numerous light signals are funnelled via two intermediate steps into one ganglion cell. It is not clear how this happens without loss of the high degree of image resolution, which is directly dependent on the high density of receptors. 'Contrast' now comes into play as each ganglion cell receives its signal (through the intermediate stages mentioned above) from a definite group of receptors. Such a group, called a *receptive field*, consists of a centre whose light signal stimulates the ganglion cell, and a concentric ring of cells which inhibit this stimulation (Fig. 1.9). There are also reversed receptive fields which consist of an inhibitory centre and an excitatory ring. When the whole receptive field is irradiated with diffuse light, the excitation and inhibition cancel each other out; the ganglion cells "see nothing", although rhodopsin is being bleached in the receptors. Only a small point of light cast on the centre of the receptive field or a halo of light around the centre is seen (also an edge or corner which forms a contrast against its background).

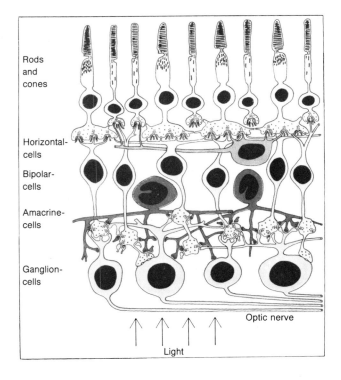

Fig. 1.8. Construction of the retina showing three layers of cells: rods and cones, bipolar cells, and ganglion cells. There is an interlacing network of horizontal and amacrine cells. This schematic diagram does not show convergence; for every 100 rods or cones there is only one ganglion cell. At this level there is already a substantial integration and processing of light impulses. Reproduced, with kind permission, from the Proceedings of the Royal Society [8].

Fig. 1.9. Diagram of two types of receptive field. A group of receptor cells transmit their signals to a ganglion cell; those stimulating the ganglion cell are surrounded by a ring of inhibitory receptors (a), the reverse configuration is observed in (b).

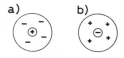

+ Excitatory
− Inhibitory

From the ganglion cells nerve fibres carry the impulse via a relay station in the thalamus, the lateral geniculate body, to the visual cortex (Fig. 1.10). The nerve fibres from the right half of the visual field of both eyes are conducted to the left cerebral hemisphere and those from the left half to the right cerebral hemisphere. Note that it is not the nerve fibres of the left or right eyes but the left and right sectors of each eye that are represented in the contralateral cortex. The receptive field of the retina corresponds to a receptive field in the geniculate body and contrast is enhanced here. The processing of information is continued in the cortex. Here again the principle of receptive fields is utilized but now the interaction of stimulatory and inhibitory impulses generates a new kind of specific contrast. The form and direction of the light source now decide whether a particular cortical cell is stimulated or not. Other cells react only to moving light sources and can signal the direction of the movement. The pioneers in this fascinating field, the Americans Hubel and Wiesel, have distinguished according to the specificity of the information and to the complexity of the connections with other cells "simple", "complex", "hypercomplex cells" and "hypercomplex cells of a higher order", and there would seem to be no end to this hierarchy. As a result of it, simple photochemical reactions in the retina form a picture in the cerebral cortex. The whole process is based on the principle that not every photon or receptor potential results in a nerve impulse and direct excitation of a cortical nerve cell, but rather that numerous receptor potentials are integrated first in the retina, again in the thalamus and yet again in various layers of the cortex.

How then can integration be investigated? By electrophysiological techniques in which fine capillary electrodes, inserted into single nerve cells, record which of these cells are

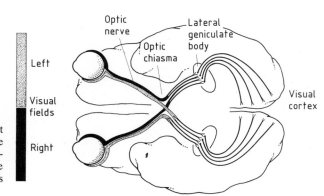

Fig. 1.10. The path of the light impulse from the retina to the cerebral visual cortex. Reproduced, with kind permission, by the authors and Sinauer Associates Inc. [7].

being excited by a specific light stimulus. In simple cases the network of neurons and their connections can be probed and reconstructed, but this is obviously not a field for biochemists!

Neurochemistry as an integrating science

These limitations of neurochemistry – the clarification of higher nervous functions – must now be followed by an account of what it can achieve. This is the subject of later chapters of this book, here we will confine ourselves to definitions: neurochemistry is the science of the molecular basis of all branches of the neurosciences. The neurosciences consist of neuroanatomy and neurophysiology (structural and functional anatomy, metabolism, and electrophysiology of the nervous system), neurology (the science of nervous diseases), neuropharmacology, neurotoxicology, behavioural physiology (including psychopharmacology) and the cellular and developmental biology of the nervous system. Biochemistry, biophysics, cybernetics and mathematics contribute as supporting sciences. The position of neurochemistry is not alongside the other disciplines; it acts rather as an integrating science representing the molecular and mechanistic aspects of all areas of the neurosciences (Fig. 1.11).

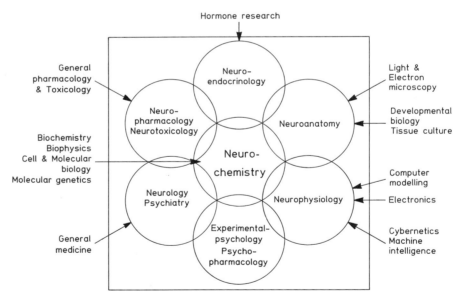

Fig. 1.11. Neurochemistry and its interrelationship with the other neurosciences and their supporting subjects. Kindly supplied by V.P. Whittaker, Göttingen.

In a narrower sense neurochemistry is the biochemistry of neurons. Therefore we will next consider the nerve cell with its characteristic functional elements.

The neuron: functional elements as subjects of neurochemistry

The nervous system is the organ of communication and integration in a complex organism. The nerve cell, also called the neuron, is adapted to this function through its polar structure (Fig. 1.12). It consists of a cell body (*soma*) and fibre-like processes. The polarity arises from the fact that one kind of process, the *dendrites*, conduct the nerve impulse to the cell body, whereas another, the *axon*, conducts it away from the cell body. Dendrites and axons are thus responsible for *input* and *output* respectively. While a cell body may have numerous dendrites, it has only one axon which however can divide into several branches (*collaterals*). Nerve fibres have two functions: they either convey an impulse from a sense receptor to the central nervous system or conversely from the central nervous system to the target organ. The first type is called *afferent* and the second *efferent*. Typical functions are summarized in Fig. 1.13.

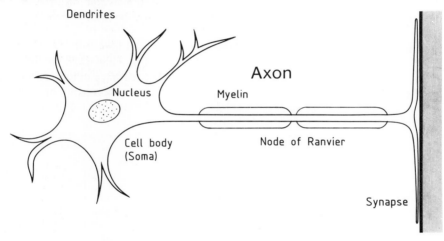

Fig. 1.12. The nerve cell (schematic diagram).

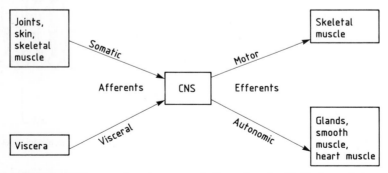

Fig. 1.13. Typical nerve functions (CNS: central nervous system). Reproduced, with kind permission, by R. F. Schmidt and the Springer-Verlag [9].

The following functional elements are of particular interest to the biochemist:

The plasma membrane. The neuron, like every other cell, is surrounded and delimited by a membrane. Its properties are, however, more intimately bound up with the function of the neuron than is the case with other cells because the nerve impulses are transmitted as changes in electrical membrane potential.

Synapses are functional contact points between neuronal plasma membranes. Substances modifying nervous activity, such as endogenous transmitter substances and numerous exogenous drugs (e.g. local anaesthetics, neurotoxins) act at the level of the membrane. Biological or pathological changes in the nervous system are frequently consequences of changes in the neural membrane. Thus basic neurochemistry must include a knowledge of the formation and properties of biological membranes. In Chapters 2 and 3 we will deal with membrane molecules and membrane models, together with the function and metabolism of specific molecules of the nerve membrane and disturbances of their metabolism. Relevant methods of membrane research will be briefly presented.

Myelin, a specialized membrane, will be considered in Chapter 4. Most nerve fibres, especially in higher organisms, are surrounded by a coat of many layers in order to isolate and accelerate the conduction of the nerve impulse. Due perhaps to its specialization, this membrane has a particularly simple structure and has been thoroughly investigated from several aspects. Diseases such as multiple sclerosis are due to a defect in the myelin sheath.

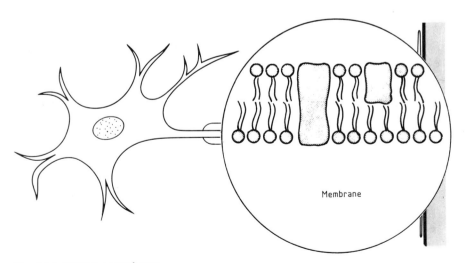

Fig. 1.14. Neuron : membrane.

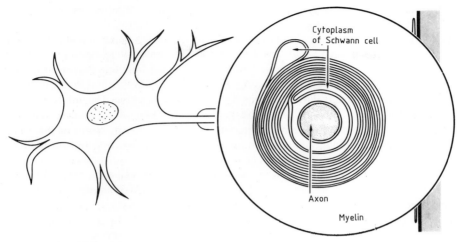

Fig. 1.15. Neuron : myelin.

The node of Ranvier. A nerve fibre is not continuously covered by a myelin sheath; the cover is interrupted by short and regularly spaced lengths of myelin-free membrane – the *nodes of Ranvier*. The myelinated area between two nodes is called the *internode*. It is electrically passive in contrast to the node membrane which is involved in the onward conduction of the nerve impulse. The properties, ion transport systems and ion channels of the neuronal membrane will be covered in Chapters 6 and 7.

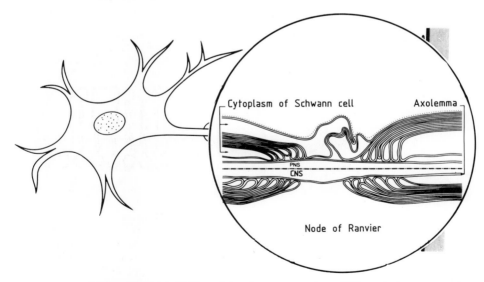

Fig. 1.16. Neuron : node of Ranvier. PNS, peripheral nervous system; CNS, central nervous system.

Axon and dendrites. The function of these processes is to transmit information from one part of the neuron to another, which, as we have already learnt, depends on the electrical excitation of the neuronal membrane. In Chapter 5 we will give a short summary of electrophysiology, which will serve to identify the most important phenomena, the molecular basis of which will be described in the following Chapters (6 and 7). Chapter 6 is concerned with the structure and function of the axonal ionic channels through which there is a passive ionic current during the passage of a nerve impulse. The subtle ionic balance between the inside and outside of the cell must be maintained or restored after the transient ionic current accompanying a nerve impulse. This energy-requiring process, called *active transport*, is brought about by an ion pump. We will deal with this in Chapter 7.

The synapse. Nerve impulses must be transmitted from one cell to another. At the site of this transmission there are specialized contact areas, called *synapses*. An axon can be connected to the soma of a second cell. This is called an *axo-somatic* synapse. Correspondingly there are *axo-dendritic* and *axo-axonic* synapses.

The synapse between the axon and the muscle fibre has a special form known as the *neuromuscular end plate*. There are approximately 10^{14} synaptic connections between more than 10^{10} neurons in our central nervous system. Synapses are the regulation points of the nervous system. Their morphology and biochemistry are very well adapted to this function. In Chapters 8 and 9 we will be concerned with their structure and function. We will particularly emphasize their ontogenesis and possible mechanisms of synaptic modification and modulation, how they react to strong stimuli, the action of drugs on them, and diseases such as Parkinson's disease and myasthenia gravis which involve disturbances of synaptic function. Their possible role in behavioural adaptation through training, learning, drug addiction or aging are the subjects of Chapters 9 and 11.

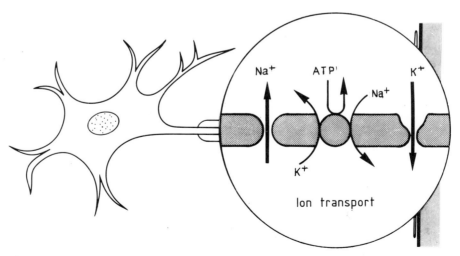

Fig. 1.17. Neuron : ion transport mechanisms.

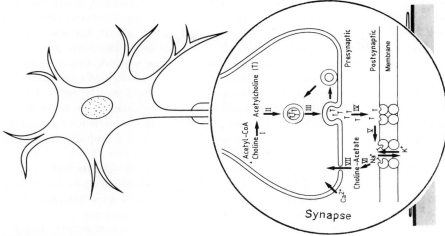

Fig. 1.18. Neuron : synapse.

The neuronal cytoplasm is sometimes extended over a metre. Thus the synapse as an especially important point of neuronal activity is often a long way from the cell body, which contains the nucleus, the regulator of the protein biosynthesis necessary for metabolism. There must therefore be a cytoplasmic communication and metabolic exchange between the peripheral synapse and that part of the cell body, the *perikaryon*, surrounding the nucleus. In Chapter 10 we shall be concerned with this cytoplasmic communication system, called *axoplasmic transport*. Further subjects of neuronal metabolism, energy needs, synthesis of special proteins, etc. will also be covered in Chapter 10.

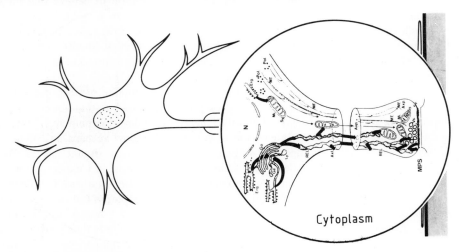

Fig. 1.19. Neuron : cytoplasm (Explanation of abbreviations see legend of Fig. 10.4.).

Glial cells. In 1846 Rudolf Virchow described a substance which he took to be the cement which holds the nervous system together. He called this *neuroglia* (from Greek: *glia*, glue) but we know today that the glue consists of cells which are not nerve cells and do not themselves conduct nerve impulses though they are clearly a vitally important constituent of the nervous system. Their function is still uncertain and we shall limit ourselves here to a short description. Glial cells make up about 50% of the volume and more than 90% of the cells of our central nervous system. They correspond to the *Schwann cells* of the peripheral nervous system. They differ fundamentally from nerve cells: unlike the latter they retain as differentiated cells the ability to divide and proliferate. Most brain tumours are thus *gliomas*, that is, malignant growths of glial cells. A further difference is their "non-excitability", that is, they do not generate action potentials (see Chapter 5) and therefore take no direct part in the conduction of nerve impulses. They possess a somewhat higher membrane potential than nerve cells. It is a pure potassium potential, that is, its magnitude is determined by the concentration gradient of K^+ ions between the intracellular and extracellular spaces, as defined by the Nernst equation. In neurons, sodium ions also contribute to the membrane potential (see Chapters 5 and 6); during the excitation of a nerve fibre there is an increase in permeability of the membrane both to sodium and potassium; potassium ions flow out transitorily influencing the membrane potential of the surrounding glial cells. The outflowing K^+ ions are in part taken up by the glial cells, so that they act as a potassium buffer, keeping the extracellular potassium concentration constant. For, if the latter increases, then the threshold of excitation of the nerve fibre decreases (Chapter 5), possibly to a value at which spontaneous activity of the fibre occurs. It has been suggested that this could be a trigger for epileptiform convulsions.

Glial cells are sometimes connected to each other by contact zones, through which metabolic exchange may take place. By contrast they are always divided from the nerve cells by a cleft at least 20 nm wide. However, there is reason to believe that metabolic exchange can also take place between glia and axons, and this may be mediated by the extracellular potassium concentration [10]. In the giant axon of the squid, the model system for so much basic neurobiological research, it has been shown that 20-40% of glial proteins, of a relative molecular mass between 20 000 and 200 000 is transferred to the axon by an as yet unknown mechanism [10].

There are numerous observations, though as yet no complete picture, suggesting that glial cells are not simply "cement", i.e. supporting tissue, but that they also play an important active role. They may determine the extracellular environment of the neuron and contribute directly to the integration of groups of neurons. Perhaps, in addition, they supply the nerve cell with important substances, metabolites and trophic factors. We shall explore their role in greater detail, for example in ontogenesis, in Chapter 11, where we shall see that at least in cell culture, expression of transmitter synthesis is influenced by these "non-nerve cells" of a ganglion. To take another example, in cell culture, whereas a *neuroblastoma* cell line is capable of true neurite outgrowth, though not of functional synapse formation, *neuroblastoma x glioma*-hybrids form synapses thus providing further evidence of the supporting function of glial cells. Glial cells of the periphery, the Schwann cells, are involved in the repair of damaged nerves; it has even been shown that after denervation a Schwann cell can take the place of the degenerating nerve ending in the muscle and even release quanta of transmitter.

One function of the glial cells is certain: *oligodendroglial cells* in the central nervous system, and *Schwann cells* in the periphery (Fig. 1.20) wrap themselves round the axon and form the *myelin* that insulates it and accelerates the conduction of the impulse (see Chapter 4). The various projections of one oligodendroglial cell can simultaneously enwrap several axons. Perhaps in this way they can bring about the coordination and integration of whole groups of neurons. In the central nervous system one must distinguish the oligodendroglial cells morphologically from the smaller, many branched *astrocytes* and the *microglia*. The latter act as macrophages; the former, through their numerous processes making contact with the walls of microcapillaries, act as selective filters allowing some substances to pass from the blood into the extracellular environment of the neurons, and preventing others, thus constituting a selective *blood-brain-barrier*.

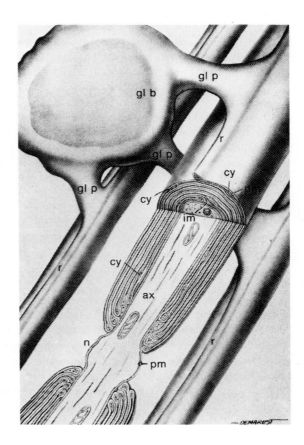

Fig. 1.20. Oligodendroglial cell, whose processes can form myelin sheaths of several axons simultaneously. glb, glial cell body; glp, glial process; ax, axon; n, node; pm, plasma membrane; cy, glial cell cytoplasm; im, inner mesaxon; r, ridge of cytoplasm. Reproduced, with kind permission, from Bunge et al., *J. Cell Biol.* **10**, 67-94 (1961).

Model systems. In many cases "nervous tissue" is too complex a research material to use in biochemical investigations of the nervous system. There are a number of simpler preparations which possess basic functions of higher systems and some of these are described in Chapter 12. Nerve cell differentiation has been investigated in cell culture. There are also model systems for metabolic and pharmacological investigations. In the biology of behaviour and its molecular correlates an increasing number of organisms are used that are easy to breed and have genetically determined mutant characteristics, e.g. *Paramecium,* nematodes, *Drosophila* and mice. For investigations into the biochemistry of stimulation and conduction chemotactic unicellular organisms such as ciliated bacteria and slime moulds are suitable. In the study of the molecular basis of behaviour and its modification during conditioning and learning organisms such as the sea-snail *Aplysia* with its identifiable neurons have proved useful models (Chapter 11). In Chapter 12 we shall also summarize a few principles and applications of the vast field of recombinant DNA techniques in molecular neurobiology.

Summary

Neurochemistry is concerned with the molecular basis of neurobiology. As opposed to molecular genetics molecular neurobiology can investigate, by biochemical methods, mechanisms of transmitting information though not the content of this information. This is exemplified by the visual process. From the stimulation of the retina by light to the conscious perception of a picture, there are many stages of information processing, of which only the first – light absorption, release of the nerve impulse and its transmission – can be described biochemically. The impulse is then, through convergence and integration, i.e. through the specific wiring diagram of the optic tract, processed, coded and passed on to the central nervous system. Perception and consciousness are emergent properties of such wiring diagrams. Neurochemistry in this context can describe mainly the stereotyped basic function of the single nerve cell.

"Seeing" starts with the absorption of light by chromophores of the retinal rods and cones. The molecular events are described by the Wald-cycle, a sequence of bleaching and regeneration of rhodopsin. Rhodopsin consists of 11-*cis*-retinal which forms a Schiff's base with opsin, a protein of relative molecular mass 41 000. Light causes the isomerization of 11-*cis*- to all-*trans*-retinal, a dissociation of the retinal-opsin complex and the release of a nerve impulse. For this a chemical second messenger, perhaps Ca^{2+} and/or cGMP, must reach the plasma membrane of the rod cells from the rhodopsin-containing disc membrane. Rhodopsin phosphorylation is stimulated by light, a process which may have something to do with the adaption of the retina to varying light conditions.

The basic element of all nervous systems is the neuron. The propagator of the nerve impulse is the neuronal membrane. The nerve impulse propagation is accelerated by a special membrane system, the myelin sheath. Further structural parts of the neuron are the afferent dendrites, the efferent axon, the soma which integrates and regulates the cell metabolism by means of its nucleus, and the synapse which constitutes the communication point between nerve cells or between neurons and their non-neuronal target cells.

References

Cited:

[1] Ovchinnikov, Y.A., "Rhodopsin and bacteriorhodopsin: Structure-function relationship", *FEBS Lett.* **148**, 179-191 (1982).
[2] Ostroy, S.E., "Rhodopsin and the visual process", *Biochem. Biophys. Acta* **463**, 91-125 (1977).
[3] Hagins, A.A., "The visual process: Excitatory mechanism in the primary receptor cells", *Ann. Rev. Biophys. Bioeng.* **1**, 131-158 (1972).
[4] Liebman, P.A., Sitaramayya, A., Parkes, J.H., and Buzdygon, B., "Mechanism of cGMP control in retinal rod outer segments", *TIPS* **5**, 293-296 (1984).
[5] Kühn, H., and Dreyer, W.J., "Light-dependent phosphorylation of rhodopsin by ATP", *FEBS Lett.* **20**, 1–6 (1972).
[6] Kühn, H.: "Interactions between photoexcited rhodopsin and light-activated enzymes in rods". In: *Progress in Retinal Research.* Osborne, N.N., and Chader, G.J., (eds.), Vol. 3, p. 123–156. Pergamon Press 1984.
[7] Kuffler, S.W., Nicholls, J.G., and Martin, A.R.: *From Neuron to Brain,* 2nd edition. Sinauer Associates, Inc. Sunderland, MA 1984.
[8] Dowling, J.E., and Boycott, B.N., "Organisation of the primate retina: electron microscopy", *Proc. R. Soc.* London, Ser. B **166**, 80-111 (1966).
[9] Schmidt, R.F.: *Grundriß der Neurophysiologie,* 3. Auflage. Springer-Verlag, Berlin-Heidelberg-New York 1974.
[10] Lasek, R.J., Gainer, H., and Backer, J.L., "Cell-to-cell transfer of glial proteins to the squid giant axon", *J. Cell Biol.* **74**, 501-523 (1977).

Further reading:

Kuffler, S.W., Nicholls, J.G., and Martin, A.R.: *From Neuron to Brain.* 2nd edition. Sinauer Associates, Inc., Sunderland, MA 1984.
"The Brain". *Sci. Am.* **241**, the whole issue No. 3, September 1979.
Kandel, E., and Schwartz, J. (eds.): *Principles of Neural Science.* Arnold, London 1981.
Hoppe, W., Lohmann, W., Markl, H., and Ziegler H. (eds.): *Biophysics.* Springer Verlag, Berlin-Heidelberg-New York 1983.
Shichi, H.: *Biochemistry of Vision,* Academic Press, New York 1984.
See also ref. 1. and 6.

Chapter 2

Membrane Molecules

Neurochemistry must not be defined as the chemistry of the "neuro-molecules", but nevertheless we will first consider the important "building blocks" of which the nerve membrane is made up. In principle these are the same as are found in the plasma membranes of other types of cell, although their relative proportions in the dry weight of membrane are fundamentally different (Table 2.1).

Table 2.1. Composition of typical membranes [1].

Component	Myelin (Ox)	Erythrocyte	Mitochondrion	Bacterium (B. Megaterium)	Chloroplast (Spinach)
Lipid : Protein (Mass ratio)	3 : 1	1 : 2	1 : 3	1 : 3	1 : 1
Phospholipids (% of total lipid)	43	61	90	48	12
Glycolipids (% of total lipid)	42	11	–	52	80
Sterols (% of total lipid)	17	28	Trace	–	Trace

Values may not sum to 100% due to analytical errors.

Neurochemistry is more than the chemistry of "neuro-molecules"

The term "building block" could be misleading if it gives the impression that the function of membrane molecules is predominantly structural. The two following chapters will show that most, perhaps all molecules have additional functions. Among many others they can act as barriers or gates, antigens or receptors, enzymes or ion pumps, as translocases (proteins serving as a carrier to transport metabolites across a membrane) or as specific recognition sites. The individual molecule should not be considered in isolation. Its characteristic properties are expressed by its interaction with other membrane molecules. The rapid developments in recent years in the area of immunology, cell biology and neurobiology have only been possible because the cell membrane has been seen not just as an interesting structure but as a highly active and cooperative system. Removed from this system, the membrane molecule loses, by definition, an essential part of its function, and its structure is only preserved under limited conditions. A biochemist who isolates an ion channel or "pore" of the nerve membrane is like a gourmet trying to pick out the holes from a Swiss cheese!

However, a careful biochemical analysis is a necessary preliminary to the understanding of the functional mechanisms of the nerve membrane. After this, ideally an artificial functional membrane system should be assembled in which the role of single types of molecules in the whole system can be investigated under well defined conditions. This kind of reconstitution experiment could be as important for the "proof" of the mechanism of the molecular membrane, as, for instance, the total synthesis of the molecule for the correct structure of a vitamin.

Lipids, proteins and carbohydrates are the building blocks of the nerve membrane

Lipids, proteins and carbohydrates are, in various combinations, the constituents of the cell membrane. In addition metal ions and not least water are essential components. Proteins predominate (Table 2.1) but, as can be seen in the case of myelin (see Chapter 4) three-quarters of the membrane weight can consist of lipid. According to their structure the lipids are classified as phospholipids, glycolipids and sterols; Tables 2.1 and 2.3 show that they occur in very different quantities in membranes. Today it is beginning to be understood that the property of a membrane is significantly dependent on the proportion of the individual lipids it contains (see Chapter 3). How difficult it is to generalize is shown in Table 2.2: the erythrocyte membrane, which in principle performs the same function in all organisms, shows considerable differences of phospholipid composition in different species. Similar species variations are also found in nervous system membranes [3]. There is therefore not just one "plasma membrane", "axon membrane" or "myelin".

Table 2.2. Phospholipid composition of the erythrocyte membrane of various species [1]. PC, phosphatidylcholine; PE, phosphatidylethanolamine; PS, phosphatidylserine; PI, phosphatidylinositol.

	PC (%)	*Sphingo-myelin* (%)	PE + PS + PI (%)
Sheep	1	63	36
Ox	7	61	32
Pig	29	36	35
Man	39	37	24
Rabbit	44	29	27
Rat	56	26	18

Table 2.3. Composition of the membranes of the central nervous system (ox) [2].

Component	Oligoden-droglia	Axon	Myelin	White matter
Total lipid[a])	29.5	13.4	75.3	55.0
Cholesterol (% of total lipid)	14.1	20.1	28.1	23.6
Cerebroside (% of total lipid)	7.3	12.9	23.2	22.5
Sulphatide (% of total lipid)	1.5	7.2	4.1	5.0
Phospholipid (% of total lipid)	62.2	60.0	43.0	46.3
Ganglioside (% of total lipid)	0.75	0.74	0.26	0.54
Phospholipids (mol/100 mol lipid-P)				
PC	48	31	24	27
PE	24	24	43	32
PS	8	9	14	24
Sphingomyelin	9	16	16	14
Unidentified	6	12	–	–
Plasmalogens	16	17	35	28

[a]) Value refers to % of dry weight of tissue.
Values may not sum to 100 % due to analytical errors.

Before we look at the membrane structure more closely (Chapter 3) we need to review the structure of the membrane constituents that are common to all membranes: the *lipids*. We will only deal briefly with their metabolism but will discuss in greater detail the nerve membrane's most important lipid class, the *gangliosides* and their possible function. Then after a review of the molecular mechanisms involved in pathological disorders of lipid metabolism, the carbohydrate-containing *glycoproteins* will be discussed. Special proteins of the nerve membrane will be introduced in later chapters.

The construction of phospholipids allows a great variety of molecular structure

The basic structure of the *phospholipids* is the monoester of phosphoric acid, glycerol-3-phosphate (Fig. 2.1). It results from the reduction of the product of glycolysis, dihydroxyacetone phosphate, by means of the cytosolic NAD-dependent glycerol-3-phosphate dehydrogenase (Fig. 2.2).

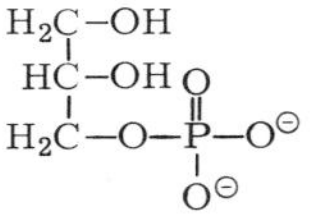

2 Membrane Molecules

```
H2C—OH
HC—OH  O
       ‖
H2C—O—P—O⁻
       |
       O⁻
```

Fig. 2.1. Glycerol-3-phosphate.

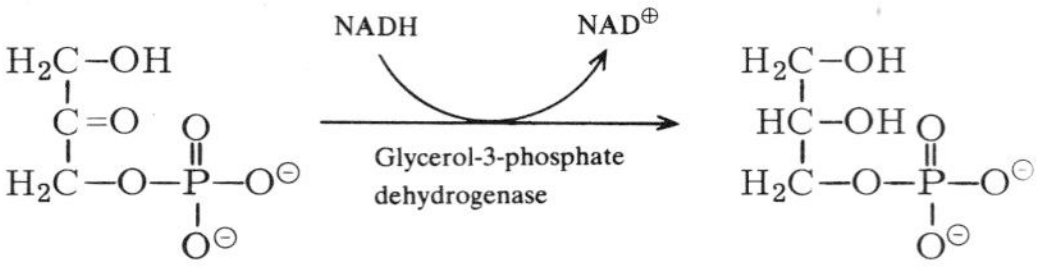

```
H2C—OH                          H2C—OH
|         NADH    NAD⊕          |
C=O   O                         HC—OH  O
|     ‖       Glycerol-3-phosphate     ‖
H2C—O—P—O⁻   dehydrogenase      H2C—O—P—O⁻
      |                               |
      O⁻                              O⁻
```

Dihydroxyacetone phosphate   Glycerol-3-phosphate

Fig. 2.2. Formation of glycerol-3-phosphate from dihydroxyacetonephosphate.

Glycerol-3-phosphate can also result from the phosphorylation of glycerol which is catalysed by a MgATP-dependent kinase. Glycerol-3-phosphate is then acylated in two stages by the corresponding acylcoenzyme A derivatives, first to lysophosphatidic acid and then to phosphatidic acid (Fig. 2.3).

```
                                                         O
                                                         ‖
                                                   H2C—O—C—R¹
                  O                                      |    O
H2C—OH  Acyl-CoA  CoA-SH  ‖                              |    ‖
|                   H2C—O—C—R1    Acyl-CoA  CoA-SH  HC—O—C—R²
HC—OH  O                 |                               |    O
|      ‖            HC—OH  O                             |    ‖
H2C—O—P—O⁻               |     ‖                   H2C—O—P—O⁻
       |            H2C—O—P—O⁻                           |
       O⁻                  |                             O⁻
                           O⁻
```

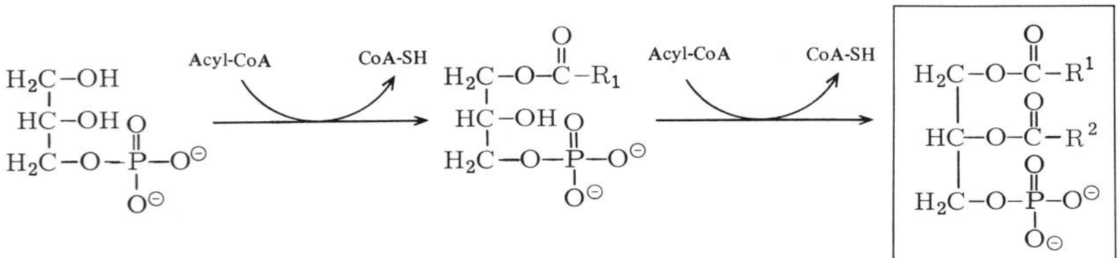

Glycerol-3-phosphate   Lysophosphatidic acid   Phosphatidic acid

Fig. 2.3. Synthesis of phosphatidic acid.

The following derivatives of phosphatidic acid are essential membrane building blocks (Fig. 2.4): phosphatidylcholine (lecithin, abbreviation PC), phosphatidylethanolamine (PE), phosphatidylserine (PS), phosphatidylinositol (PI) and plasmalogen.

Each formula (Fig. 2.4) represents a whole series of phospholipids, since the acyl group R can consist of saturated or unsaturated hydrocarbon chains of different lengths. (Of these, those with 16 or 18 carbon atoms predominate). In the case of phosphatidylinositol one or more of the hydroxy groups of inositol can be esterified by phosphoric acid. This variability results in a diversity of membrane building blocks, the biological significance of which is not clear. The structure of the acyl group influences the properties of the membrane (see Chapter 3) and the membranes from different regions of one organism differ in composition. It can well be imagined that there is a connection between phospholipid structure and the biological function of a membrane.

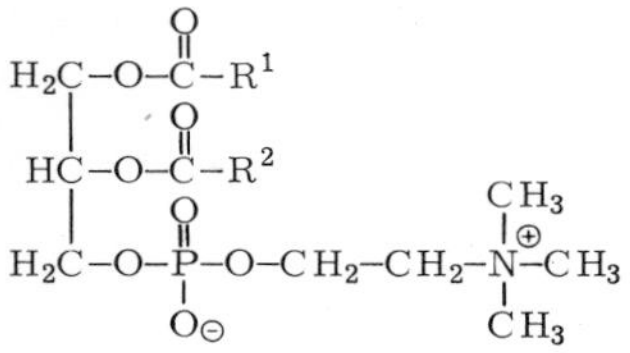

Phosphatidylcholine (PC)
(Lecithin)

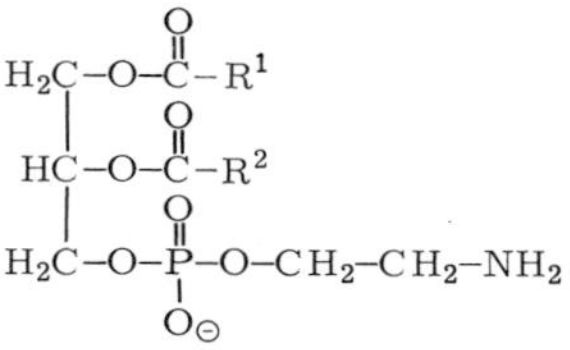

$$H_2C-O-\overset{O}{\underset{\|}{C}}-R^1$$
$$HC-O-\overset{O}{\underset{\|}{C}}-R^2$$
$$H_2C-O-\underset{\underset{O^\ominus}{|}}{\overset{O}{\underset{\|}{P}}}-O-CH_2-CH_2-\overset{CH_3}{\underset{CH_3}{\overset{|\oplus}{N}}}-CH_3$$

Phosphatidylethanolamine (PE)

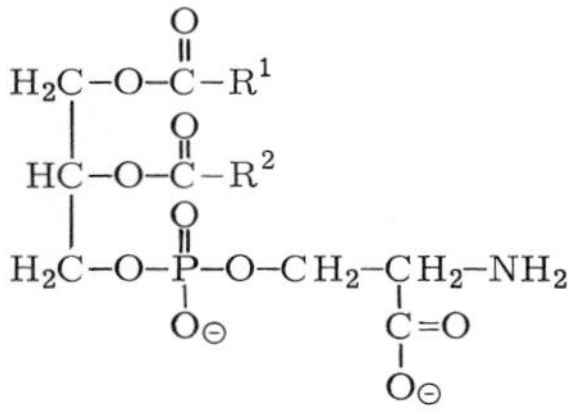

$$H_2C-O-\overset{O}{\underset{\|}{C}}-R^1$$
$$HC-O-\overset{O}{\underset{\|}{C}}-R^2$$
$$H_2C-O-\underset{\underset{O^\ominus}{|}}{\overset{O}{\underset{\|}{P}}}-O-CH_2-CH_2-NH_2$$

Phosphatidylserine (PS)

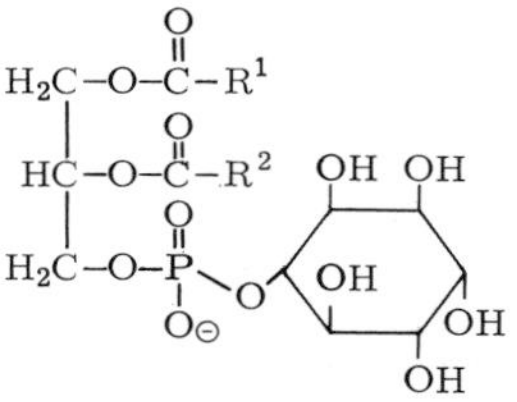

$$H_2C-O-\overset{O}{\underset{\|}{C}}-R^1$$
$$HC-O-\overset{O}{\underset{\|}{C}}-R^2$$
$$H_2C-O-\underset{\underset{O^\ominus}{|}}{\overset{O}{\underset{\|}{P}}}-O-CH_2-\underset{\underset{O^\ominus}{\underset{\|}{C=O}}}{CH}-NH_2$$

Phosphatidylinositol (PI)

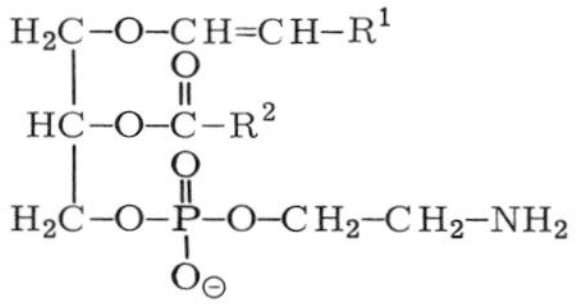

(inositol ring with OH groups)

Plasmalogen (Plasmene)

$$H_2C-O-CH=CH-R^1$$
$$HC-O-\overset{O}{\underset{\|}{C}}-R^2$$
$$H_2C-O-\underset{\underset{O^\ominus}{|}}{\overset{O}{\underset{\|}{P}}}-O-CH_2-CH_2-NH_2$$

Fig. 2.4. Phosphatidic acid derivatives: The most abundant phospholipids of the cell membrane.

The synthesis of phosphatidylglycerol starts with the activation of phosphatidic acid by cytidine triphosphate (CTP). The resulting CMP-phosphatidic acid reacts with inositol to form *phosphatidylinositol* or with the amino acid serine to form *phosphatidylserine*.

Phosphatidylserine is decarboxylated enzymatically (with pyridoxalphosphate, PLP, as coenzyme) to *phosphatidylethanolamine* which with methyl groups transferred from S-adenosylmethionine (SAM) is methylated to *phosphatidylcholine*. PS, PC and PE on the other hand can also be formed from diacylglycerol and the activated cytidine phosphate compounds, CDP-serine, CDP-ethanolamine and CDP-choline (Fig. 2.5).

Phospholipid methylation. Methylation of PE appears to be involved in signal transmission through cell membranes: a methyltransferase is located on the inside of many cell membranes methylating PE to phosphatidyl-N-monomethylethanolamine. A second methyltransferase located on the outside methylates further to PC. The methyl donor in each case is again SAM. Methylation of PE influences membrane fluidity and is stimulated by catecholamine neurotransmitters, e.g. adrenaline, resulting in Ca^{2+} influx, generation of cAMP, histamine release etc. [4].

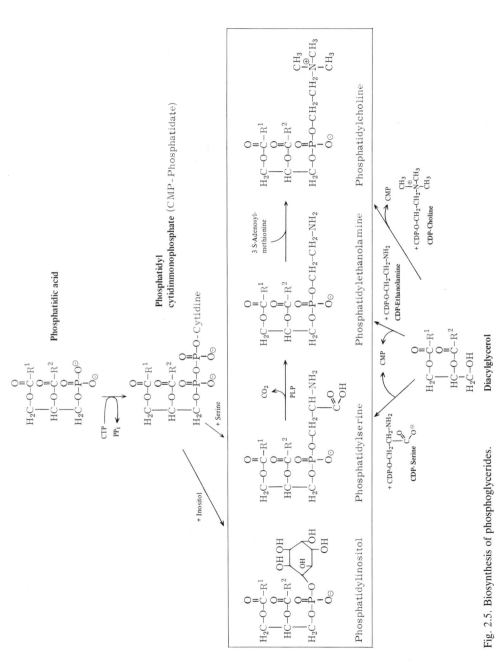

Fig. 2.5. Biosynthesis of phosphoglycerides.

Plasmalogens. Plasmalogens in which an acyl-group is replaced by an enol-ether group, originate by reduction of the acyl group and removal of the elements of water. In myelin it makes up more than 14% of the phosphoglyceride. The ethanolamine-plasmalogen (see Fig. 2.4) is the main compound of this class. Recently plasmalogens have been renamed to plasmenes.

Phosphatidylinositol. This compound is particularly interesting for neurochemists. The OH group of inositol can be esterified with one or more phosphate groups. For the lipid molecule as a membrane building block this means an increase in negative charge which can affect the physical properties of the membrane. In addition, the ability to bind divalent ions (Ca^{2+}, Mg^{2+}) increases with the number of phosphate groups. It has been observed that a higher turnover of phosphate groups is associated with nervous activity [4–6].

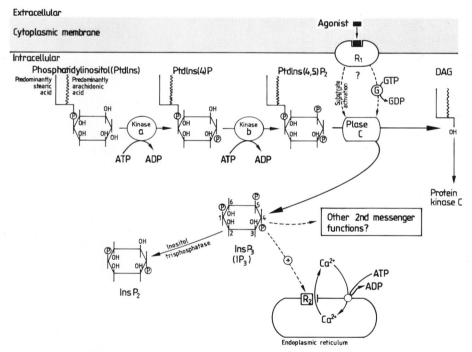

Fig. 2.6. Phosphatidylinositol in signal transduction through cell membranes. A signal (agonist) binds to a receptor (R_1) on the external cell surface. The agonist-receptor complex activates via a coupling protein (G or N) or directly phospholipase C (plase C) which cleaves trisphosphoinositol ($InsP_3$, IP_3) from phosphatidylinositol 4,5-bisphosphate. IP_3 then triggers Ca^{2+} release from the endoplasmic reticulum (receptor R_2). The other cleavage product produced by plase C is diacylglycerol (DAG), another second messenger activating protein kinase C. – (a) and (b) are phosphorylation steps leading to $InsP_3$. After [6] with modifications.

In 1953 Hokin and Hokin had already observed that the neurotransmitter acetylcholine (see Chapter 8) stimulated ^{32}P incorporation into phospholipids in the pancreas, and similar effects have since been detected in a variety of cells in response to a wide range of external signals. Michell proposed that phosphatidylinositol (PI) turnover is involved in those instances where agonists affect the intracellular Ca^{2+} level [5]. The following picture is now emerging [6]: certain external stimuli promote through cell surface receptors the mobilisation of Ca^{2+} from intracellular stores. The second messenger triggering this mobilisation is inositol trisphosphate (inositol 1,4,5-trisphosphate, IP$_3$). It is formed by the action of a Ca^{2+}-dependent phospholipase C, activated in the membrane by agonist-stimulated receptors such as e.g. muscarinic cholinergic, α_1-adrenergic, serotonergic, and peptidergic (substance P, vasopressin, angiotensin, bradykinin). In most, if not in all cases, a coupling (G or N-) protein also located in the membrane (see Chapter 9) may be involved [7]. The other cleavage product of the phospholipase C is diacyl glycerol (DAG), another important second messenger which activates proteinkinase C [8] (see Chapter 9). This hypothesis as summarized by Berridge and Irvine is shown in Fig. 2.6. It may not be fully complete but it is introduced here, because it effectively coordinates the many recently discovered facts in a very important and rapidly developing field of research.

Phospholipases initiate the degradation of phospholipids

The degradation of the phosphatidylglycerols is initiated by the splitting off of the acyl groups. This is catalysed by lipases which differ in their specificity according to the group to be split off. Phospholipase A$_2$, for example, hydrolyses the ester link of the middle C-atom of glycerol with the formation of lysolecithin; phospholipase A$_1$, on the other hand, splits off the other acyl residue; further degradation is performed by phospholipases C and D (Fig. 2.7). The resulting fatty acids are degraded by β-oxidation [9].

The unsaturated fatty acid *arachidonic acid* deserves special mention. It is released by phospholipase A$_2$ and plays a multiple role as a precursor of prostaglandins, thromboxan-

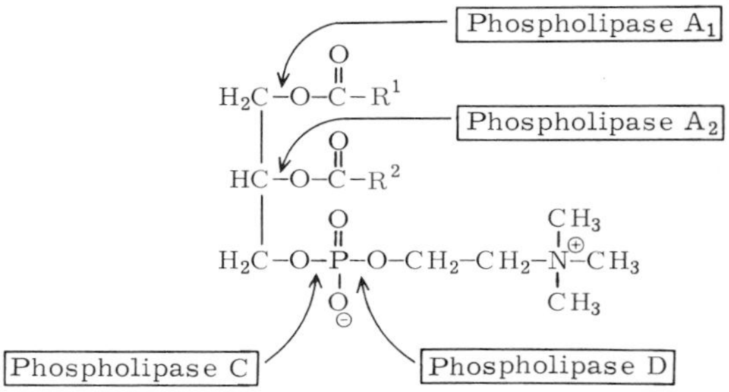

Fig. 2.7. Specifity of phospholipases.

es and leukotrienes and also as an important cellular regulator (either itself or after oxidation by lipoxidase); e.g. it has been shown to stimulate guanylate cyclase. Neurotransmitter-mediated cGMP formation (see Chapter 9) appears to be regulated via receptor-mediated release of arachidonic acid; cAMP formation on the other hand has been shown in some tissues to be regulated by certain prostaglandins.

Lysolecithin, the product of the action of phospholipase A_2 on lecithin (phosphatidylcholine) can be reacylated to form the same or another lecithin (Fig. 2.8).

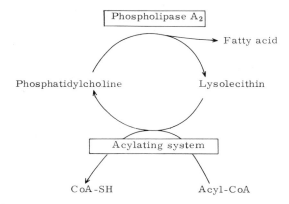

Fig. 2.8. Deacylation – reacylation cycle of phospholipids.

As this cycle proceeds with such a rapid turnover it is assumed that it plays a role in the maintenance of the concentration of specific lipids. It is important to prevent too high a concentration of lysolecithin, as this can cause lysis of the membrane.

Cholesterol is particularly important for its ability to affect the fluidity of the lipid membrane (see Chapter 3). The complex biosynthesis of this molecule is summarized in Fig. 2.9; textbooks of biochemistry should be consulted for a detailed description [9] of its metabolic pathways.

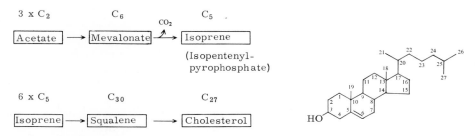

Fig. 2.9. Biosynthesis of cholesterol.

All of the 27 C-atoms of cholesterol originate from acetyl coenzyme A via isoprene or its activated form. Six of these isoprene units combine to form squalene which is finally cyclized to form cholesterol.

Sphingolipids are not only present in nerves, but are especially important there

Sphingolipids, particularly the carbohydrate derivatives, the gangliosides and cerebrosides, are attracting great interest because of their special function in the communication of the nerve cell with its environment [10].

There is a lot of evidence that their main function is to participate in the transmission of signals from the exterior cell surface to the interior of the cell. Their structure and localization subserve this function: they consist of a lipophilic part, *ceramide*, and a hydrophilic carbohydrate portion (Fig. 2.12). This allows them to be firmly anchored to the lipid phase of the cell membrane and also to interact with the surrounding polar medium. They are directed exclusively towards the outside; the cytoplasmic side of the membrane seems to be free of carbohydrate. They can be bearers of both specificity and information through the variability of their carbohydrate structure. Gangliosides contain N-acetylneuraminic acid groups, which, through their negative charge, possess binding properties for ions and other ligands. As a result of these properties the following roles for gangliosides have been proposed: 1. They are receptors for signals from outside. Later in this chapter the example of bacterial neurotoxins will be described. 2. Together with glycoproteins, they are concerned with cell surface specificity, cell recognition and cell adhesion. An illustration of this is the change in the ganglioside composition of virus-transformed cells. 3. They may play an important role during the development of the nervous system in the formation of "correct" intercellular connections. Evidence for this is the temporary increase in the ganglioside content of the central nervous system shortly before the onset of myelination. Thus they may be concerned with the communication between an axonal membrane and its enveloping oligodendroglial cell. 4. They may also take part in a similar way in the functional adaptation of the mature nervous system. Phenomena like conditioning and learning (see Chapter 11) could involve an increased concentration of certain glycolipids and glycoproteins at synapses [11].

The sphingolipids are derived from the amino alcohol sphingosine which itself originates from palmitoyl-coenzyme A and the amino acid serine in several stages (Fig. 2.10).

The amino group of sphingosine can react with acyl-coenzyme A to form an amide link and the terminal OH-group reacts with CDP choline to form the phosphorylcholine ester, sphingomyelin. The N-acylsphingosine (ceramide) is the product of the first step and the starting point for further reactions which lead to the formation of cerebrosides and gangliosides (Fig. 2.11).

Cerebrosides. In the central nervous system the galactose derivatives predominate among the cerebrosides and their sulphate esters, the sulphatides. Galactosylceramide is found in higher concentrations in the white matter than in the grey matter of the brain; it is a

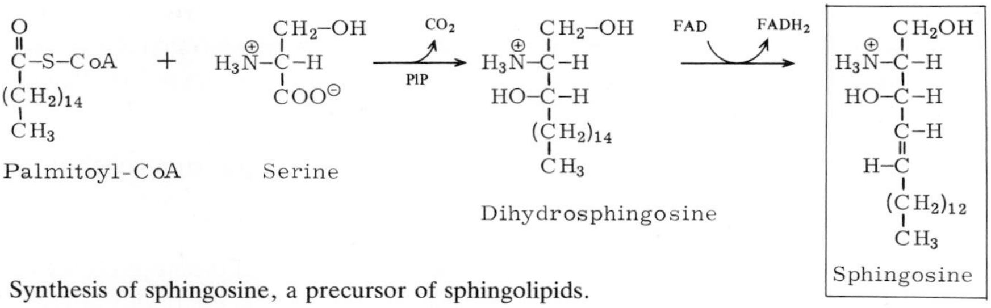

Fig. 2.10. Synthesis of sphingosine, a precursor of sphingolipids.

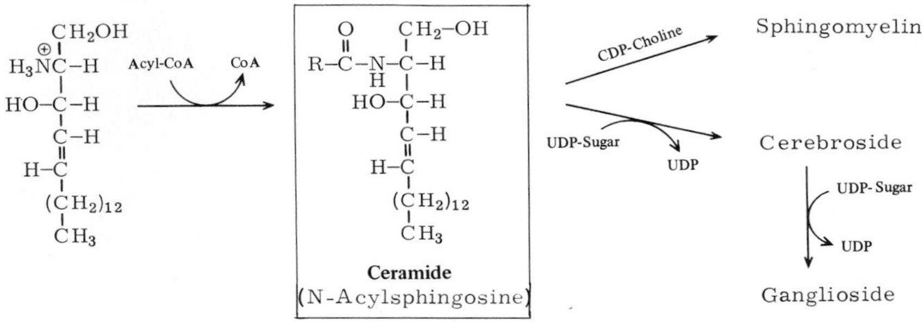

Fig. 2.11. Ceramide, the parent substance for gangliosides, cerebrosides and sphingomyelin.

characteristic constituent of myelin. Galactosylcerebroside however is found only in small quantities in the neuronal perikarya, but occurs in considerable concentration in the axon membrane. The significance of this similarity between axon and myelin membranes needs further investigation. Since glucosylceramide is found in astrocytes as well as in neurons the original assumption that this glycolipid was specific for neurons, and galactosyl-ceramide was confined to oligodendroglial cells and thus to myelin, is clearly not correct.

Gangliosides. In cerebrosides the terminal hydroxy-group of ceramide forms a link with one glucose or galactose residue; with gangliosides it is linked to an oligosaccharide. In contrast to cerebrosides gangliosides are found primarily in the grey matter of the brain and in the cell bodies of neurons; their concentration in the white matter is only 10% of that in the grey matter. The variability of the oligosaccharide part theoretically permits a very large number of gangliosides but in fact, only twelve have been found so far. The four main ones are: G_{M1}, G_{D1a}, G_{D1b}, G_{T1}. In the nomenclature proposed by Svennerholm, who has contributed substantially to the characterization of these compounds [12], G stands for ganglioside, M for monosialo- (i.e. with *one* N-acetylneuraminic acid), D for disialo-, T for trisialo- etc. "1" indicates the complete carbohydrate structure with the terminal galactose, "2" the first degradation product i.e. without the galactose etc. (see Fig. 2.12, Fig. 2.13). Gangliosides are found in almost every tissue; the highest concentration is in nervous tissue. If this is fractionated the synaptosome fraction is richest in gangliosides.

Gangliosides are ubiquitous in the nervous system; the distribution of the different classes shows considerable regional variations: thus the visual cortex is relatively rich in G_{T1}

and G_{D1b}; the cerebellar cortex contains mainly G_{T1}. White matter is rich in G_{M1} and, in man, also contains sialogalactosylceramide. In the peripheral nervous system the same four gangliosides are found except that in G_{M1} N-acetylgalactosamine is replaced by N-acetylglucosamine.

Metabolism of gangliosides. Gangliosides are built up from ceramide in stages by the addition of single monosaccharides (Fig. 2.14). These are first activated by being coupled to uridine diphosphate (forming UDP glucose etc.). The transfer of the sugar residue is catalysed by a group of glycosyltransferases, each of which is specific for one of the steps shown in Fig. 2.14. The glycosyltransferases are membrane-bound enzymes which are found in especially high activity in synaptosome membranes. This is thought to be further evidence that gangliosides are involved in synaptic transmission. Probably the glycosyltransferases exist as multienzyme complexes and this may be one of the rare examples in nature in which a series of building blocks is synthesized in a specific sequence into an oligomeric molecule without the help of a matrix, unlike protein synthesis; thus, as with some peptide antibiotics, the sequence of enzymes codes for the product instead of a nucleic acid sequence.

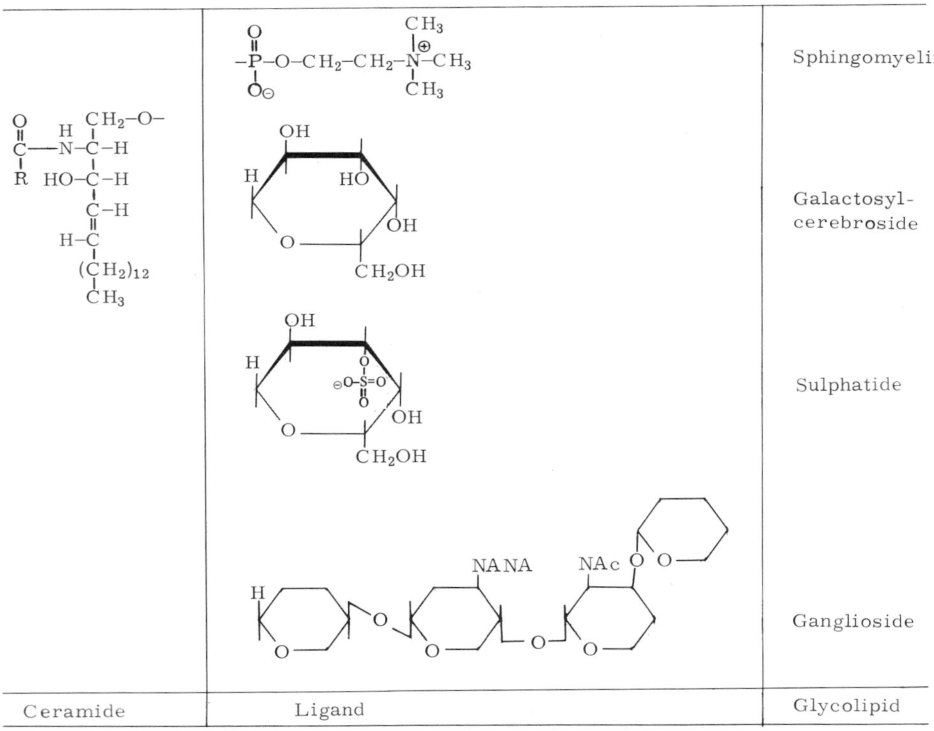

Sphingosine or its N-acyl derivative, ceramide, gives rise to four classes of sphingolipids. This diagram shows a representative of each.

Fig. 2.12. The four classes of sphingolipids. They all consist of a hydrophobic ceramide and hydrophilic substitutents.

The degradation of gangliosides also takes place in stages by specific glycosidases and a neuraminidase. Defects in these hydrolases cause a series of lipid storage diseases (see below). The glycosidases and the neuraminidases are lysosomal enzymes. The membrane bound neuraminidase of mammals is a protein, which hydrolyses the terminal N-acetylneuraminic acid group from gangliosides.

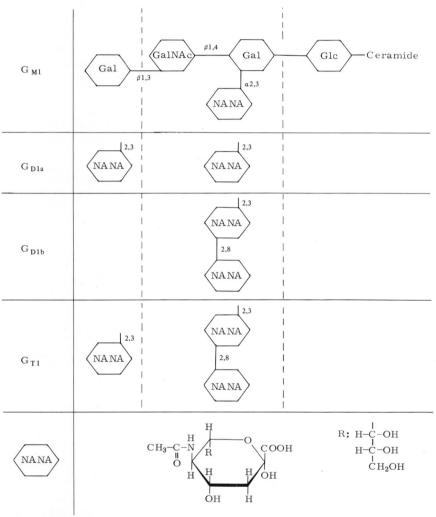

Fig. 2.13. Scheme showing nomenclature of gangliosides. The most important gangliosides are differentiated by the number of their N-acetylneuraminic acid (sialic acid, NANA) groups: mono- (M), di- (D), tri- (T) sialoganglioside. When the terminal NANA group of the ganglioside G_{DIa} is hydrolysed by neuraminidase G_{M1} is formed and this is converted to G_{M2} by the hydrolysis of its terminal galactose.

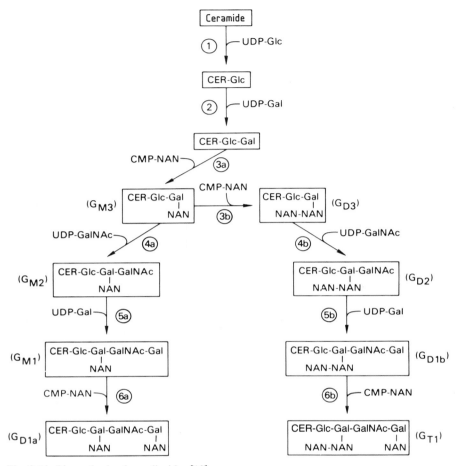

Fig. 2.14. Biosynthesis of gangliosides [10].

Speculations about the function of sphingolipids

If the synthesis of sphingolipid is traced during the development of the central nervous system three groups can be clearly distinguished: galactosylceramide and galactosylsulphatide increase significantly with myelinization, while the synthesis of gangliosides appears to be connected with differentiation of neurons. The third group of sphingolipids, glucosylceramide and lactosylceramide, remain relatively constant during development. These observations correspond to the results obtained with differentiated cell types. The activity of the enzymes involved shows parallel changes to the lipid concentrations.

There is not yet a clear conception of the function of sphingolipids; there are however a series of observations and a number of promising clues [13]. As already mentioned they have been associated with cell adhesion, cell recognition, immunological specificity*⁾, synaptic transmission and the receptor properties of cell surfaces. We will now review some pertinent observations that might provide a model for ganglioside function.

Gangliosides are receptors for bacterial toxins

Tetanus toxin. The receptor property of gangliosides for bacterial toxins is well documented [14]. Tetanus toxin, a product of *Clostridium tetani*, is a protein with a relative molecular mass of 160 000. Reduction with SH reagents cleaves it into two polypeptides, A (relative molecular mass 55 000) and B (relative molecular mass 105 000) which apparently originate from a common polypeptide chain by proteolysis. Van Heyningen showed that the toxin binds specifically to the gangliosides G_{D1a} and G_{T1} and he considered these to be the natural receptors for the neurotoxin [15]. The toxin-ganglioside-complex is no longer toxic, while incubation of gangliosides with neuraminidase results in products which no longer bind to the toxin. The binding site for ganglioside is found on the B-chain of the toxin, yet neither B or A are toxic by themselves.

Botulinum toxin. There is evidence of a similar neurotoxic mode of action for botulinum toxin, the exotoxin of *Clostridium botulinum*, which can transform bad canned meat into a deadly foodstuff.

Various bacterial strains produce serologically different toxins, but they conform to a common structural pattern [16]: they are converted from a pro-toxin, a protein with a relative molecular mass of ca. 145 000, into the active toxin by proteolytic cleavage of a peptide link. The resulting subunit peptides with relative molecular masses of 50 000 and about 100 000 are linked by a disulphide bridge; the reductive cleavage of the bridge leads to loss of toxicity [17]. Botulinum toxin binds specifically to gangliosides but not to cerebroside or other lipids. The strength of the link increases with the number of sialic acid residues in the ganglioside (from G_{M1} to G_{T1}). Possibly the toxin can also react with glycoproteins. *In vitro*, botulinum toxin is selectively bound to synaptosomes. *In vivo* it blocks chemical synapses by inhibiting the presynaptic release of transmitter molecules.

Cholera toxin. *Vibrio cholerae* produces a protein which is not a neurotoxin but is nevertheless interesting because of its ability to bind ganglioside G_{M1} and its mode of action. The toxin is bound to the intestinal mucosa and stimulates the secretory cells of the small intestine to such an extent that the organism loses dangerously large amounts of water and electrolytes. Cholera toxin is a protein of relative molecular mass 82 000; it does

---

*) Immunologically sphingolipids are *haptenes* rather than *antigens*, i.e. the isolated ganglioside molecule elicits only a weak antibody response and cerebroside none at all. But they will bind to preformed antibodies.

not contain carbohydrate or lipid; it interacts selectively with ganglioside G_{M1}. The protein consists of several polypeptide chains: the five B-chains (relative molecular mass ca. 10 000) carry the binding site for the ganglioside, but do not give rise to the cholera reaction. When the cell is disrupted the subunit A (relative molecular mass 28 000) stimulates the adenylatecyclase in the presence of NAD. It does this by a covalent modification and permanent activation of the enzyme's regulatory subunit, the G- or N-protein which we shall discuss in detail in connection with the catecholamine receptors (Chapter 9). The modification consists of an ADP-ribosylation i.e. a transfer of ADP-ribose from NAD to the α-polypeptide chain of the G-protein. With the intact cell the subunit A is only active in the presence of B which is apparently necessary for binding to and penetration of the cell membrane. A can be further split into A_1 (relative molecular mass 22 000) and A_2 (relative molecular mass 5 000) by mercaptoethanol. A_1 retains its cyclase activating action. The primary structure of the B-chain has been worked out [18] and shows significant homologies with the β-chain of the thyrotropic hormone (TSH) which explains its ability to react with TSH receptors in the thyroid gland.

Tetanus toxin and cholera toxin are similar in structure having two polypeptide chains, one interacting with the membrane and one toxic. Diphtheria toxin from *Corynebacterium diphtheriae* has a similar division of labour – although its mode of action is quite different and gangliosides are not involved – so this may point to a broader concept of protein toxicity. All three toxins carry, in addition to a binding and a catalytic moiety, a peptide sequence enabling them to cross the membrane, a tunnel through which the catalytically active A chain can slip into the cell (Fig. 2.15 [19]).

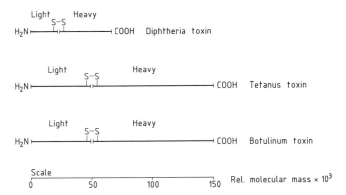

Fig. 2.15. Scheme of the peptide structure of three bacterial toxins. They all consist of a light and a heavy chain connected by a disulphide bond. The heavy chains contain the sites for binding to the membrane and they form channels in membranes which may function as "tunnels" translocating the active fragment (located on the light chain) into the interior of infected cells. As in cholera toxin the active fragment of diphtheria toxin (neither neurotoxins!) catalyses ADP ribosylation of certain proteins. The activity of the light chains of tetanus and botulinum toxin have not yet been identified.

Lipidoses are caused by enzyme defects in glycolipid metabolism

At least ten genetic disturbances of glycolipid metabolism are known; these are the basis for a group of lipid storage diseases or lipidoses, the *sphingolipidoses* [20,21]. They produce incurable symptoms and usually result in early death.

As the name lipid storage disease indicates, the cause of the sphingolipidoses is an enzyme defect in the catabolism of the sphingolipid, which produces an accumulation of the lipid or an intermediate product of its degradation. As an inherited disease the fault lies in the genetically controlled synthesis of a degrading enzyme. At least in one case, the defect is not found in the enzyme itself but in a protein factor, activated by a hydrolase which helps to bind the enzyme to its lipophilic substrate.

Tay-Sachs syndrome. a) Symptoms: development of the newborn child occurs apparently normally until the 6th month of life; then disturbances of the motor system and the mental faculties appear which lead to convulsions, blindness and death within three years. A characteristic of the disease is a cherry red spot on the retina.

b) Molecular causes: the marked concentration of ganglioside G_{M2} in the brain is attributed to a defect in hexosaminidase A which splits off N-acetylgalactosamine (GalNAc) from the so-called Tay-Sachs ganglioside.

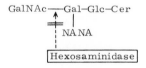

Tay-Sachs disease is the so-called B-variant of a group of G_{M2} gangliosidoses. The defective hexosaminidase A is one of two isozymes of hexosaminidase present in human tissue. In another variant, the O-variant (Sandhoff-Jatzkewitz disease) both isozymes A and B are defective. There is a particularly interesting AB-variant of G_{M2} gangliosidoses in which, in spite of the normal or even increased activity of both isozymes, there is an accumulation of G_{M2} as in the other two variants. Here the defect is not in the degradative enzymes themselves, but in the activator-protein, which promotes the binding of the enzyme to its lipophilic substrate [22].

c) Genetics: Tay-Sachs syndrome is dependent on an autosomal recessive mutation which is carried by one in three hundred Americans but is much commoner in Jewish Americans (one in thirty).

d) Therapy: there is no effective therapy. An injection of a large dose of hexosaminidase A does not decrease the symptoms, as the enzyme does not pass through the blood-brain barrier.

Gaucher syndrome. a) Symptoms: there is a marked enlargement of the liver and spleen. Pain occurs in the bones of the hips and spine. There is a progressive osteoporosis in the spinal column and there are large lipid-containing cells in the bone marrow. The patient is increasingly anaemic. The disease is divided into various types according to the age group which it affects. Children die within two years.

b) Molecular causes: this was the first example of the fact that in certain storage diseases the accumulation of lipids is not due to an increased synthesis but to a defect in catabolism. In this case there is an increase in glucocerebroside because glucocerebroside-β-glucosidase (also called glucocerebrosidase) is inactive or, in other types of the disease, is of reduced activity.

c) Genetics: the disease is autosomal recessive. From two carriers of the disease, one out of four offspring will have Gaucher syndrome. Heterozygous carriers are free of symptoms.

d) Therapy: in some cases an injection of cerebrosidase alleviates symptoms, but, in general, this disease is just as incurable as many other genetically caused metabolic diseases. A preventive approach is more promising. The tendency to the disease can be detected in heterozygous carriers by determining the level of activity of the affected enzyme in the leucocytes or skin fibroblasts.

Niemann-Pick syndrome. a) Symptoms: this syndrome is characterized by a marked enlargement of the liver and spleen and progressive deterioration of the brain, and leads to death within two years of birth.

b) Molecular causes: the main accumulating lipid is sphingomyelin. The defective enzyme is sphingomyelinase which splits sphingomyelin into ceramide and phosphorylcholine.

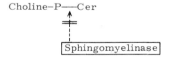

c) Genetics: in this case too, the heterozygotic carrier of the disease is free of symptoms, as it behaves as a recessive mutant.

d) Therapy: there is as yet no form of therapy. Although sphingomyelinase has been isolated, treatment by injection of the enzyme would demand extreme care for as sphingomyelin is a widespread cell component, it would be difficult to avoid serious side effects, such as haemolysis.

Fabry syndrome. a) Symptoms: male patients show severer symptoms of this syndrome than females: burning sensations in hands and feet, augmented by the inability to sweat, arteriosclerosis and increasing kidney damage. Death occurs by the 50th year mainly due to kidney failure.

b) Molecular causes: ceramide-trihexoside (Cer-Glc-Gal-Gal) accumulates, as there is a defect of the α-galactosidase needed to split off the terminal galactose group.

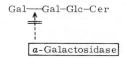

c) Genetics: the enzyme defect is an autosomal recessive localized on the X-chromosome.

d) Therapy: as death occurs from kidney failure, kidney transplantation achieves a certain success. There is also considerable promise in treatment with intravenous injections

Table 2.4. Sphingolipidoses (Lipid storage diseases).

Metabolic disorder	Accumulating intermediate	Enzyme defect	Reaction block (point of attack of enzyme)
Tay-Sachs	Ganglioside G_{M2} (Tay-Sachs Ganglioside)	Hexosaminidase A (one of the two isoenzymes)	GalNAc-Gal-Glc-Cer ↑ │ NANA
Gaucher	Glucocerebroside	Glucocerebrosidase	Glc-Cer ↑
Niemann-Pick	Spingomyelin	Sphingomyelinase	Choline-P-Cer ↑
Fabry	Ceramide trihexoside	α-Galactosidase	Gal-Gal-Glc-Cer ↑
Sandhoff-Jatzkewitz	Ganglioside G_{M2} and Globoside	Hexosaminidase A and B (both isoenzymes)	GalNAc-Gal-Glc-Cer ↑ │ NANA
General gangliosidosis	Ganglioside G_{M1}	Ganglioside G_{M1}-β-galactosidase	GalNAc-Gal-Glc-Cer │← │ Gal NANA
Krabbe	not known	Galactocerebroside-β-galactosidase	Gal-Cer ↑
Metachromatic leucodystrophy	Sulphatide	Galactocerebroside-3-O-sulphatase	Gal-Cer │← OSO_3
Fucosidosis	H-Isoantigen	α-Fucosidase	thought to be: GalNAc-Gal-Glc-Cer │← Fucose
Farber	Ceramide	Ceramidase	Cer → Sphingosine ↑ + fatty acid

of α-galactosidase which has been isolated from human placenta. It is rapidly absorbed, markedly reduces the ceramide-trihexoside level in the blood and, in a majority of the injected patients, the enzyme activity of the liver increases.

Further diseases caused by defects in one or other of the enzymes involved in the catabolism of sphingolipid are summarized in Table 2.4. They are all relatively rare and are due to recessive mutations, which apparently do not usually lead to a complete absence of the enzyme concerned, but to less active form of it. In some cases antibodies against the active enzyme have been successfully used to detect the inactive mutant.

Carbohydrate confers multiple specificities on cell surfaces: glycoproteins

2% to 10% of the mass of the plasma membrane can consist of carbohydrate most of which is bound to protein; a small part is a component of glycolipids. As has been mentioned they are completely localized to the outer surface of the plasma membrane and they impart to it its individuality and specificity [23].

Glycoproteins contain relatively short oligosaccharide chains of 4 to 15 carbohydrate residues. The following carbohydrates are found: D-galactose, D-mannose, D-glucosamine, D-glucose (rarely), D-galactosamine (always as N-acetyl derivative), L-fucose and N-acetylneuraminic acid. In the same way that protein primary structures are constructed from the twenty most frequent amino acids, an almost unlimited number of oligosaccharide structures can be made according to whether the chains are linear or branching.

The possible permutations from one oligosaccharide with 13 carbohydrate residues have been calculated. For three residues each of mannose, N-acetylglucosamine, galactose and N-acetylneuraminic acid bound to a protein via a further molecule of N-acetylglucosamine, the number of possible structures is of the order of 10^{24}!

The carbohydrate chains are either N-glycosidic bound to an asparagine residue or O-glycosidic bound to a serine or threonine (Fig. 2.16).

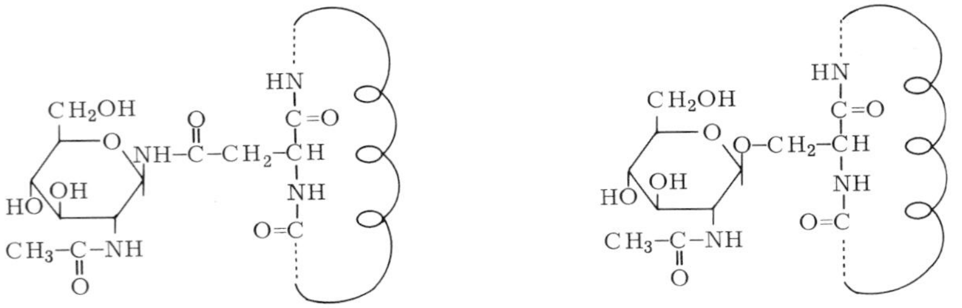

Fig. 2.16. Glycoproteins: (a) N-glycosidic linkage of N-acetylglucosamine with an asparagine residue of the protein; (b) O-glycosidic linkage of N-acetylglucosamine with a serine residue of the protein.

Biosynthesis is different for the two types: O-glycosidic coupling occurs by the direct transfer of the carbohydrate residue from the sugar-dinucleotide (e.g. UDP-glucose) to an OH-group of the protein (Fig. 2.17).

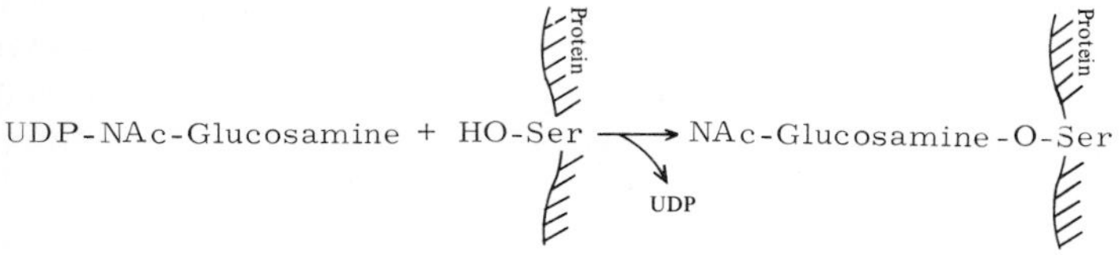

$$\text{UDP-NAc-Glucosamine} + \text{HO-Ser-Protein} \longrightarrow \text{NAc-Glucosamine-O-Ser-Protein} + \text{UDP}$$

Fig. 2.17. Transfer of a UDP activated sugar to a serine-OH-group of a protein.

N-glycosidic coupling of a sugar with asparagine is via an intermediate step in which the sugar is bound to a lipid, the polyprenol dolichol, the structure of which is given in Fig. 2.18 [24,25], (n is 11 for bacteria and 17-21 for mammals). Glycoprotein synthesis takes place as shown in the scheme in Fig. 2.19.

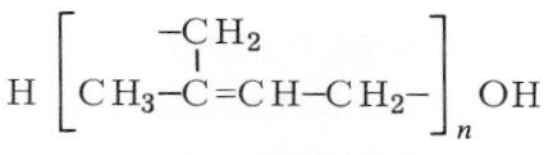

$$H\left[CH_3-\underset{\underset{CH_2}{|}}{C}=CH-CH_2-\right]_n OH$$

Fig. 2.18. Dolichol, a polyprenol, an intermediate in the biosynthesis of the N-glycosidic linkage of glycoproteins.

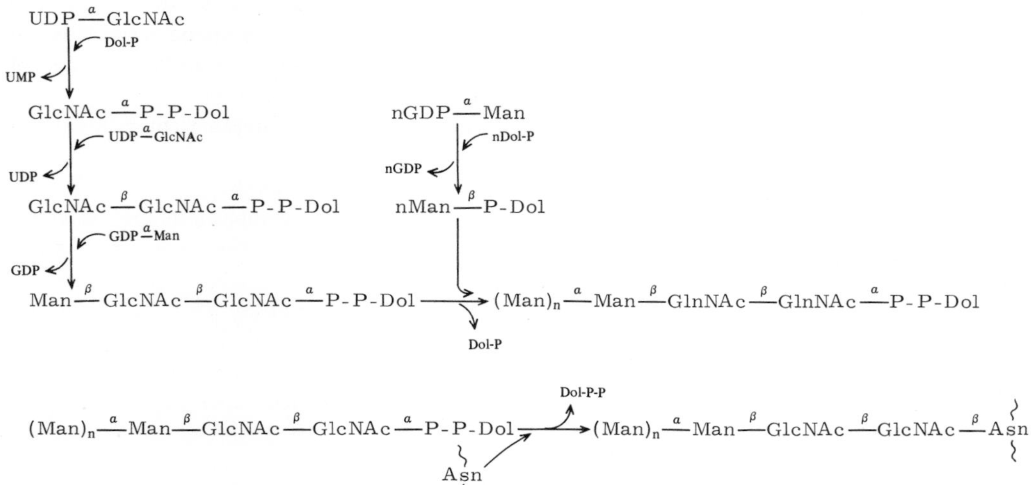

UDP $\xrightarrow{a}$ GlcNAc
↓ ← Dol-P
UMP ↙
GlcNAc $\xrightarrow{a}$ P-P-Dol   nGDP $\xrightarrow{a}$ Man
↓ ← UDP $\xrightarrow{a}$ GlcNAc   ↓ ← nDol-P
UDP ↙   nGDP ↙
GlcNAc $\xrightarrow{\beta}$ GlcNAc $\xrightarrow{a}$ P-P-Dol   nMan $\xrightarrow{\beta}$ P-Dol
↓ ← GDP $\xrightarrow{a}$ Man
GDP ↙
Man $\xrightarrow{\beta}$ GlcNAc $\xrightarrow{\beta}$ GlcNAc $\xrightarrow{a}$ P-P-Dol $\longrightarrow$ (Man)$_n$ $\xrightarrow{a}$ Man $\xrightarrow{\beta}$ GlnNAc $\xrightarrow{\beta}$ GlnNAc $\xrightarrow{a}$ P-P-Dol
↓ Dol-P

(Man)$_n$ $\xrightarrow{a}$ Man $\xrightarrow{\beta}$ GlcNAc $\xrightarrow{\beta}$ GlcNAc $\xrightarrow{a}$ P-P-Dol $\xrightarrow{\text{Dol-P-P}}$ (Man)$_n$ $\xrightarrow{a}$ Man $\xrightarrow{\beta}$ GlcNAc $\xrightarrow{\beta}$ GlcNAc $\xrightarrow{\beta}$ Asn
                                                            ↑ Asn

Fig. 2.19. Biosynthesis of a glycoprotein [24]: postulated sequence with dolichol derivatives (Dol) as intermediates.

It is not clear why glycosylation occurs sometimes with and sometimes without a dolichol intermediate step. Because dolichols are lipophilic their participation may indicate which steps are localized close to the lipophilic part of the membrane.

Each glycoprotein molecule can incorporate several identical oligosaccharide chains which are specific to the protein involved. However the oligosaccharide sequences need not be complete – one or more monosaccharides may be missing. The structure of oligosaccharides is determined by specific glycosyltransferases. Blood group glycoproteins exemplify this type of specific oligosaccharide structure: blood group A differs from blood group B in having terminal N-acetylgalactosamine residues in place of galactose residues in two branches of the oligosaccharide; the glycoprotein of blood group O contains neither carbohydrate. The surface properties of nerve membranes may be similarly specified.

Glycoproteins may be important for the specificity of neural connections

What function can glycoproteins have in biological mechanisms? One hypothesis links them to the intracellular sorting of proteins which are synthesized at the rough endoplasmic reticulum and transported to its destination by a hitherto unknown mechanism. The carbohydrate side chains of a newly synthesized glycoprotein may serve as a "flag" which determines into which membrane it will be finally inserted or whether it is to be secreted into the extraplasmic serum. Because of the great variability and specificity of oligosaccharides it is tempting to ascribe to glycoproteins and gangliosides a specific role in cell recognition and contact and consequently in the correct wiring of the neural network in the brain. The following are three suggested mechanisms by which carbohydrate chains of neighbouring cells might specifically interact (Fig. 2.20). There could be:
1. An interaction between the sugar residues of two cells for example through hydrogen bonds.
2. An interaction analogous to the one between antigen and antibody.
3. An interaction between enzyme and substrate, i.e. the membrane-bound glycosyltransferase of one cell could recognise and bind its substrate, the corresponding oligosaccharide of the other cell.

The third mechanism is consistent with many observations. On the other hand cells are known that show normal adhesion (and contact inhibition) in cell culture, although no membrane-bound galactosyltransferase can be detected. This elegant hypothesis thereby loses some of its attraction and the question of cell recognition and adhesion remains open. For the neurochemist here may lie the key to problems like the specificity of the structure of the 10^{14} synapses of the central nervous system, because there is no lack of glycoprotein in the pre- or postsynaptic membranes nor in the intervening basal lamina. We will return in a later chapter to the problem of specificity of cell-cell interaction and its special relevance to the formation of complex neuronal networks.

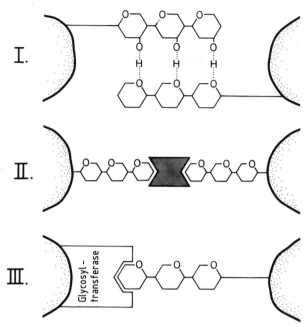

Fig. 2.20. Hypothetic mechanisms for specific intercellular recognition and adhesion with participation of plasma membrane oligosaccharides. (I) Hydrogen bonding; (II) antigen-antibody-like interaction; (III) enzyme-substrate interaction.

Summary

The plasma membrane consists of phospholipids, sterols, glycolipids and glycoproteins plus metal ions and water. They are all more than just building blocks, i.e. structural elements, as they take part in very different ways in cooperative activities. The proportion of different molecules in the membrane is very variable. There is as yet no clear connection between lipid composition and membrane function, for membranes of similar function in different organisms can be of varying composition (for example erythrocytes). There is no one erythrocyte or axon membrane.

Phospholipids are derived from glycerol-3-phosphate via its diacyl ester phosphatidic acid. Phosphatidic acid can be activated to CMP phosphatidic acid by CTP, and then be esterified with serine to form phosphatidylserine (PS) and with inositol to form phosphatidylinositol (PI). Decarboxylation of PS (with a pyridoxalphosphate enzyme) produces phosphatidylethanolamine (PE) the amino group of which is methylated by S-adenosylmethionine to form phosphatidylcholine (PC). Conversely, serine, ethanolamine and choline can be changed into activated derivatives via CTP and then react with diacyl glycerol to form PS, PE and PC. Methylation of PE by two methyltransferases is stimulated by certain neurotransmitters and is an important step in signal transmission through cell membranes.

The parent compound of sphingolipids is sphingosine formed by decarboxylation and oxidation of palmitoyl CoA and serine. Its amino group makes an amide link with a fatty acid producing ceramide. The ceramide-OH group, originating from serine, can react with various ligands: 1. phosphorylcholine forming sphingomyelin; 2. a carbohydrate (galactose or glucose) forming cerebroside which on esterification with a sulphuric acid residue gives a sulphatide; 3. an oligosaccharide coupled with varying numbers of N-acetylneuraminic acid residues forms gangliosides. Galactosylceramide (cerebroside) is found mainly in the white matter of the brain, ganglioside in the grey matter. Gangliosides appear to be concentrated in the synaptic area of the neurons and may be concerned with the function and correct connections of synapses. No clear cut biological function can as yet be assigned to these compounds. Bacterial toxins give some indication of a biological function. Tetanus toxin, botulinum toxin and cholera toxin appear to include as a common structural feature a ganglioside-binding site which is linked with a further polypeptide region via a S-S bridge. The toxic function appears to reside in this second region and in cholera toxin (not a neurotoxin!) involves the activation of an adenylcyclase, in botulinum toxin a presynaptic blocking of transmitter release.

Disturbances in the degradation of sphingolipids cause a group of syndromes, the sphingolipidoses. The ten known types all have recessive characteristics. They originate from the inactivity of a catabolic enzyme which results in the accumulation of the corresponding lipid in the CNS and other organs (liver, spleen). Only occasionally are the symptoms alleviated by replacement of the defective enzyme.

Glycoproteins contain defined oligosaccharides whose structures make possible the great variety of specific properties characteristic of membrane surfaces. They are bound to serine or threonine residues through O-glycosidic links or to asparagine through N-glycosidic links. The link occurs directly via the UDP sugar or, in the case of N-glycoside, via a dolichol intermediate utilizing specific glycosyltransferases. Their function in cell adhesion and recognition remains to be clarified.

References

Cited:

[1] Gurr, M.I., and James, A.T.: *Lipid Biochemistry. An Introduction.* Chapman and Hall, London 1975.
[2] De Vries,G.H., Hadfield, M.G., and Cornbrooks, C., "The isolation and lipid composition of myelin-free axons from rat CNS", *J. Neurochem.* **26**, 725-731 (1976).
[3] Norton, W.T., Abe, T., Poduslo, S.E., and De Vries, G.H., "The lipid composition of isolated brain cells and axons", *J. Neurosci. Res.* **1**, 57-75 (1975).
[4] Hirata, F., and Axelrod, J., "Phospholipid methylation and biological signal transmission", *Science* **209**, 1082-1090 (1980).
[5] Michell, R.H., "Inositol phospholipids and cell surface receptor function", *Biochim. Biophys. Acta* **145**, 81-147 (1975).

[6] Berridge, M.J., and Irvine, R.F., "Inositol trisphosphate, a novel second messenger in cellular signal transduction", *Nature* **312**, 315–321 (1984).
[7] Cockcroft, S., and Gomperts, B.D., "Role of guanine nucleotide binding protein in the activation of polyphosphoinositide phosphodiesterase", *Nature* **314**, 534–536 (1985).
[8] Nishizuka Y., "The role of protein kinase C in cell surface signal transduction and tumour promotion", *Nature* **308**, 693–697 (1984).
[9] Lehninger, A.L.: *Biochemie,* 2. Auflage. Verlag Chemie, Weinheim, New York 1977.
[10] Fishman, P.H., and Brady, R.O., "Biosynthesis and function of gangliosides", *Science* **194**, 906-915 (1976).
[11] Rahmann, H.H., Rösner, H., and Breer, H., "A functional model of sialoglyco-macromolecules in synaptic transmission and memory formation", *J. theor. Biol.* **56**, 231-237 (1976).
[12] Svennerholm, L., "Chromatographic separation of human brain gangliosides", *J. Neurochem.* **10**, 613-623 (1963).
[13] Maggio, B., Cumar, F.A., and Caputto, R., "Molecular behaviour of glycosphingolipids in interfaces. Possible participation in some properties of nerve membranes", *Biochim. Biophys. Acta* **650**, 69-87 (1981).
[14] Van Heyningen, S., "The structure of bacterial toxins", *TIBS* **1**, 114-116 (1976).
[15] Van Heyningen, S., "Binding of ganglioside by the chains of tetanus toxin", *FEBS Lett.* **68**, 5-7 (1976).
[16] Das Gupta, B.R., and Sugiyama, H., "A common subunit structure in *Clostridium botulinum* type A, B and E toxins", *Biochim. Biophys. Res. Commun.* **48**, 108-112 (1972).
[17] Sugiyama, H., Das Gupta, B.R., and Yang, K.H., "Disulfide-toxicity relationship of botulinal toxin types A, E, and F", *Proc. Soc. Exp. Biol. Med.* **143**, 589-591 (1972).
[18] Nakashima, Y., Napiorkowski, P., Schäfer, D.E., and Konigsberg, W.H., "Primary structure of the B subunit of cholera enterotoxin", *FEBS Lett.* **68**, 275-278 (1973).
[19] Hoch, D.H., Romero-Mira, M., Ehrlich, B.E., Finkelstein, A., Das Gupta, B.R., and Simpson, L.L., "Channels formed by botulinum, tetanus, and diphtheria toxins in planar lipid bilayers: Relevance to translocation of proteins across membranes", *Proc. Natl. Acad. Sci. USA* **82**, 1692–1696 (1985).
[20] Sandhoff, K., "Biochemie der Sphingolipidspeicherkrankheiten", *Angew. Chem.* **89**, 283-295 (1977).
[21] Brady, R.O., "Inherited metabolic storage disorders", *Ann. Rev. Neurosci.* **5**, 33-56 (1982).
[22] Conzelmann, E., and Sandhoff, K., "AB variant of infantile G_{M2} gangliosidosis, deficiency of a factor necessary for stimulation of hexosaminidase A-catalyzed degradation of ganglioside G_{M2} and glycolipid G_{a2}", *Proc. Natl. Acad. Sci. USA* **75**, 3979-3983 (1978).
[23] Olden, K., Parent, J.B., and White, S.L., "Carbohydrate moieties of glycoproteins. A re-evaluation of their function", *Biochim. Biophys. Acta* **650**, 209-232 (1982).
[24] Waechter, C.J., and Lennarz, W.J., "The role of polyprenol-linked sugars in glycoprotein synthesis", *Ann. Rev. Biochem.* **45**, 95-112 (1976).
[25] Kornfeld, R., and Kornfeld, S., "Assembly of asparagine-linked oligosaccharides", *Ann. Rev. Biochem.* **54**, 631-664 (1985).

Further reading:

on lipids and plasma membranes: Bretscher, M.S., and Raff, M.C., "Mammalian plasma membranes", *Nature* **258**, 43–49 (1975);
see also ref. 1 and the standard textbooks of biochemistry.
on glycoproteins: Lennarz, W.J. (ed.): *The Biochemistry of Glycoproteins and Proteoglycans.* Plenum Press, New York & London 1980.
Alberts, B., Bray, D., Lewis, J., Raff, M., Roberts, K., and Watson, J.D.: *Molecular Biology of the Cell.* Garland Publishing, Inc., New York and London 1983.

on glycolipid storage disorders: Brady, R.O., "Inherited metabolic storage disorders", *Ann. Rev. Neurosci.* **5**, 33–56 (1982).
on phosphatidyl inositol and Ca^{2+}-channel gating: Fisher, S.K., Van Rooyen, L.A., and Agranoff, B.W., "Renewed interest in the polyphosphoinositides", *TIBS* **9**, 53–56 (1984).
Berridge, M.J., and Irvine, R.F., "Inositol trisphosphate, a novel second messenger in cellular signal transduction", *Nature* **312**, 315–321 (1984).

Chapter 3

Membranes

We have defined the nervous system as the organ of communication in a complex organism. This function is carried out specifically by the nerve membrane. For example the conduction of a nerve impulse along the axon, which can be as much as a meter long, is signaled by an action potential resulting from a flow of ions across the axonal membrane (Chapters 5 and 6). Similarly, essential steps in the transmission of the impulse from one cell to another involve chemical and electrical phenomena at the synaptic membrane (Chapters 8 and 9). Nerve membranes also play a vital role during the development of the nervous system, and in its interaction with the environment. So it is clear that a significant part of neurochemistry today is the biochemistry of the nerve membrane. This chapter is largely a summary from the special point of view of the neurochemist of the facts of modern membranology. For a more complete presentation two recent monographs [1,2] may be consulted.

The neuronal membrane as a plasma membrane

We know little about the structure of the neuronal membrane, but when it is understood that it is in principle a plasma membrane, and has much in common with the plasma membranes of other cell types, then we can transfer to it many of the results of membrane research in general, whether these are obtained from bacteria, fungi, erythrocytes or myelin.

We will now consider some of the questions of "common membranology" essential for neurochemists. Some of the basic questions about the relation between structure and function that any satisfactory membrane model has to answer are:
1. Can it integrate all the facts concerning membrane composition with the information derived from electron microscopy, X-ray crystallography and other physical methods?
2. How does the structure so arrived at account for the basic biological properties of the membrane:
a) as a barrier between the outside and inside of the cell, not only for organic molecules but also – of particular importance in the case of the neuron – for ions;

b) as an active, co-operative system concerned with the maintenance of the cell's internal environment by the active transport of metabolites and ions;
c) as an aspect of the individuality of the cell expressed for example by its specific antigenicity. This is significant in intercellular communication and recognition and may involve temporary contact, adhesion or fusion. In addition it is particularly important in the development of the nervous system for the establishment of the correct synaptic contacts within the neural network and also at an intracellular level, in exocytosis of synaptic vesicles and transmitter release;
d) as a receptive surface which by specific receptors embedded in it recognizes signals from the external milieu (hormones, trophic factors, transmitters).

From Gorter and Grendel to Singer and Nicolson: membrane models

More than 50 years ago Gorter and Grendel extracted the lipids from erythrocyte membranes and prepared from it a monomolecular film on a water surface. They showed that the surface of this film was twice as great as that of the intact erythrocytes and concluded that erythrocytes are enclosed not by a monomolecular but by a bimolecular lipid layer (Fig. 3.1A). This concept of the *lipid bilayer* as the basis of biological membranes can be compared in its significance to the discovery of the DNA double helix. The basis of the lipid bilayer is the construction of the lipid molecule from a polar, hydrophilic head and nonpolar hydrophobic fatty acid chains (Fig. 3.2). In an aqueous milieu these molecules arrange themselves into a thermodynamically favourable structure in which the polar heads interact with the water molecules but the hydrophobic residues are repulsed by the water and interact with each other as in a fat droplet. Possible structures – *micelles, monolayers, liposomes (lipid vesicles)* – are shown in Fig. 3.3. The liposomes already possess the characteristic bilayer structure of biological membranes.

We will not state the arguments for and against the bilayer model here. This stimulating chapter of scientific history can be read in numerous specialist monographs on membranes. The basic concept was soon accepted, although some controversy continued particularly over the position of the proteins. Danielli and Davson's model placed them mainly on the surface of the bilayer where they were held by electrostatic forces. Since according to this model the lipid bilayer is coated on both sides with protein like butter between two slices of bread it was known as the "sandwich" or scientifically as the "unit membrane" model. In order that the protein coverage might be as complete as possible Danielli had to assume that the molecules were not globular but extensively unfolded. In Chapter 4 we shall see that myelin comes closest to this model. Specifically there is, in myelin, an extended basic protein which is adsorbed on to the surface of the bilayer as a result of interaction with the acid phospholipid heads. But myelin also contains other proteins which are arranged differently from the predictions of the Danielli-Davson model.

Theoretical considerations and experimental observations do not support the sandwich model. It does not seem likely thermodynamically since the water would effectively compete with protein for the polar heads of the lipid molecules and the aqueous environment, from which they are screened by the protein layer. Experimentally there is much counter evidence of which the clearest is the electronmicroscopic picture of the freeze-fractured membrane (Fig. 3.4). This shows particles, probably proteins, appearing on the inner surface of the bilayer. Biochemical methods also support the conclusion that proteins are partially or totally inserted into the membrane. We will return to this later. In 1972 Singer and Nicolson summarized this and various other evidence to produce a model called the "fluid mosaic model" [3] and this is now generally accepted as the best description to date of the plasma membrane.

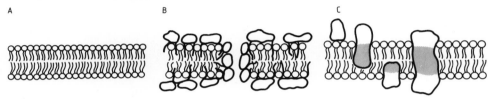

Fig. 3.1. Membrane models. (A) Bilayer model of lipid membranes. (B) Danielli-Davson model. In this and in a later modified version, the proteins were placed mainly on the surfaces of the bilayer. The hydrophobic regions of the protein penetrate into the lipid phase. Pores are formed by proteins in the membrane. (C) Singer-Nicolson model, distinguishing integral and peripheral proteins (see text). The hydrophobic regions of the integral proteins are embedded in, or penetrate through the lipid phase (see also Fig. 3.4).

Models are not copies of reality

The fluid mosaic model is however no more than a model. Models are schematic representations and essentially not much more than working hypotheses. This book like every scientific textbook, is full of models, with which the author is not entirely happy. Their clarity and plausibility can easily distract from the fact that they are only temporarily valid. Any good experiment may modify them or challenge their validity. The use of a model in a textbook can easily create a false impression of certainty in what is really a far from settled theory, an "idée fixe" which can inhibit the discussion of experimental tests of the theory. A further danger is their overgeneralization. Models tend to put undue stress on the common properties of, in this case, diverse membranes and thus obscure differences which may be important in specific functions.

Membrane properties: the lipid phase

From Singer's model we will pass on to the subject of the lipid bilayer as a matrix or structural phase into which the proteins are inserted. The lipid bilayer will be seen to be in no way a passive or rigid structure.

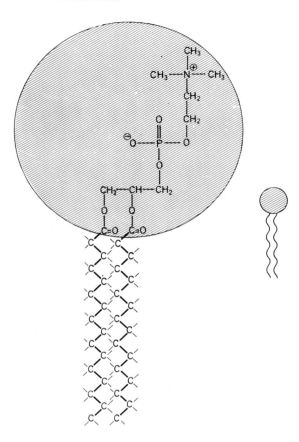

Fig. 3.2. Amphipathic phospholipid molecule; symbol: ⚇. The hatched part of the molecule is hydrophilic, the remainder hydrophobic; not in scale, the hydrophilic part is exaggerated.

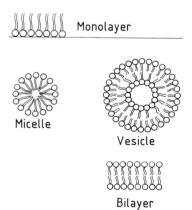

Fig. 3.3. Lipid structures in a hydrophilic environment.

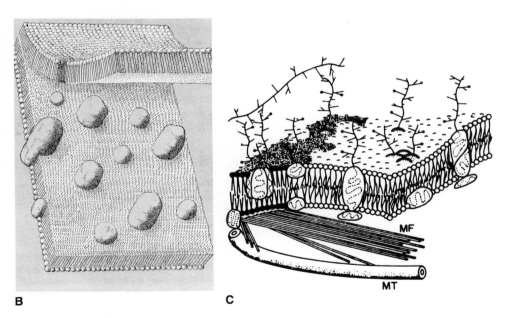

Fig. 3.4. Fluid mosaic model. (A) Electronmicrograph of a freeze-fractured cell membrane. In several places this has split the bilayer and gives a view of the interior of the membrane. The particles visible there are probably proteins or protein-aggregates. (B) Schematic representation of the freeze-fracture with embedded proteins. (C) Diagram showing proteins embedded in the lipid phase; oligosaccharides extend to the extracellular and the cytoskeleton to the intracellular space. MF, microfilaments; MT, microtubules. (A) from Silva and Branton, (B) from Tagawa, from [1]. (C) from Nicolson, G.L., *Biochim. Biophys. Acta* **457**, 57-108 (1976).

Bio-membranes are "fluid-crystalline"

If an artificial membrane is made from a lipid, it exists depending on the temperature, in two forms [4], a *"more fluid"* and a *"more rigid"* form.

It is only partially correct to describe the transition from one to the other in terms of melting and solidifying; it is more accurate to speak of a *fluid crystalline* phase above the phase transition temperature and a *crystalline* phase below. In both, the lipid molecules exist to a considerable extent in an ordered state with the polar head towards the outside and the hydrophobic fatty acid side-chain towards the inside. In the fluid crystalline phase the hydrophobic chains – particularly their ends – are relatively mobile. In the crystalline phase they are stationary.

The phase transition from crystalline to fluid crystalline is an endothermic process; the heat needed for the "melting" of the fatty acid chains can be determined in a calorimeter (Fig. 3.5). If the lipid bilayer consists of only one lipid, the phase change takes place over a narrow temperature range. Since biological membranes normally consist of many different lipids they do not show a sharp phase transition and are fluid crystalline at physiological temperatures. However it is certain that the fluidity of biological membranes can be very different, both from organ to organ and also in different regions of a single cell membrane. The differing lipid composition indicates this. Although a general relationship between membrane fluidity and biological function has not yet been established, some of the factors influencing fluidity have been identified utilizing artificial lipid membranes; there is increasing evidence that the same factors operate in biological membranes. The phase transition temperature depends on the nature of the fatty acid side chains. The longer the chain and the fewer unsaturated double bonds, the higher the temperature. This implies that the longer the chain, the greater the Van der Waals interaction between neighbouring chains. The longer chains are more strongly anchored, and the fluidity of the membrane is thereby decreased. By contrast, a double bond, which occurs *in vivo* predominantly in its *cis*-configuration, produces a disturbance of the interaction between the chains and an increase in the fluidity of the membrane.

This applies equally to biological membranes. Bacteria which grow at low temperatures maintain the fluidity of their membranes through increased incorporation of short chain

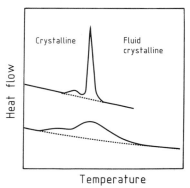

Fig 3.5. Phase transition of a phospholipid membrane. At a definite temperature the membrane changes from crystalline to fluid crystalline state. The transition is an endothermic process, which can be measured calorimetrically as an increase in heat absorption. The addition of cholesterol broadens the transition point (lower curve) and abolishes it when the cholesterol content exceeds 50%. Under physiological conditions the biological membrane is always above the phase transition temperature and therefore fluid crystalline.

and unsaturated fatty acids. In the same way fish adapt their lipid composition to the needs of their environmental temperature. The lipid-dependent fluidity of the membrane has effects on its function. For example in experiments involving mutants of *E. coli* which are unable to synthesize fatty acids and are dependent on the surrounding medium for them, galactoside transport across the membrane showed a temperature dependence which was determined by the lipid of the growth medium [5].

It is not so easy to describe the effect of single lipids on the properties of a membrane. In general one can only say that the fluidity of biological membranes is determined by the phase transition temperature of single lipids. Factors which increase the fluidity (see above) produce the following changes:
a) a lateral expansion of the membrane which decreases its thickness;
b) an increase of the rotational isomerization of the CH_2- and terminal CH_3-groups around the C-C bonds;
c) an increase in the mobility of the $-N^+(CH_3)_3$-groups;
d) an increase in the lateral diffusion of the lipids.

Lysolecithin damages membranes

Lysolecithin originates from lecithin through the splitting off of a fatty acid chain by the action of phospholipases A_1 or A_2. We have already mentioned in Chapter 2 that lysolecithin is an intermediate in the formation and breakdown of lipids, that it is very rapidly reacylated and is probably important for the maintenance of the specific lipid composition of the membrane. Lysolecithin cannot be allowed to accumulate as it markedly disrupts the bilayer structure of the cell membrane. A schematic representation of this is shown in Fig. 3.6.

It is easy to demonstrate this process in erythrocytes: the haemolytic effect of snake venom is due primarily to the production of lysolecithin by the lipases it contains. Lysolecithin can play another interesting role. If it is added in low concentration to liposomes or to chicken erythrocytes fusion is produced instead of complete lysis. Perhaps lysolecithin also plays a part in membrane fusion *in vivo* as has been proposed for the exocytotic release of transmitter molecules from chromaffin granules. Lysolecithin may soften the dividing membrane so that they can more easily penetrate to the other side.

Lecithin

Lyso-Lecithin

Fig. 3.6. Lysolecithin damages lipid membranes. Schematic diagram of the lytic effect on a bilayer (right).

Cholesterol-containing membranes are neither crystalline nor fluid crystalline

The insertion of cholesterol among the fatty acid chains of the membrane disturbs their arrangement and disperses them [6]. In the crystalline phase this leads to an increase in fluidity whereas in the fluid crystalline phase the membrane becomes stiffer i.e. cholesterol reduces the mobility of the fatty acid chains. Due to this inhibition of crystallization and increase of fluidity, membranes containing more than 20% cholesterol, like myelin, have practically no phase transition. They exist in an intermediate semicrystalline state. The biological significance of this might be that cholesterol, by suppressing the interaction between other lipid molecules, blocks the transfer of "information" in the membrane.

It is not only the fatty acid chains but also the polar head groups that characterize a membrane; the latter can be very different in individual biological membranes. They may be more important for lipid-protein than for lipid-lipid interactions, i.e. certain proteins may need specific lipid head groups in their environment for their activity. This has been shown for a variety of enzymes [7] (see below).

Drugs affect membrane fluidity

There are other factors as well as the endogenous ones, which affect the fluidity of the lipid matrix. Numerous neurotoxins and neuroactive drugs act at the nerve membrane (see Chapters 6, 8, 9), some at specific binding or receptor sites, others through a nonspecific effect on general membrane properties. Local anaesthetics will be taken as an example of the latter.

The mode of action of general or local anaesthetics is complex and as yet little understood [8]; it will be more thoroughly discussed in Chapter 6. One theory suggested by Meyer and Overton at the beginning of the century, is with certain qualifications still valid today. It stated that the anaesthetic potency of a substance depends chiefly on its lipophilicity, i.e. its partition coefficient between membrane and water. In model experiments it was shown that anaesthetics lower the phase transition temperature of some lipids, and so increase the fluidity of the membrane [9,10]. The fluidity is linked with the permeability of the membrane for ions and other low molecular mass substances. In a classical experiment Bangham showed that liposomes containing a radioactive substance are rendered permeable by chloroform or diethyl ether and release their radioactive marker to the surrounding medium. The chloroform concentration needed for this was enough to anaesthetize an animal such as a tadpole. Bangham considered that a common molecular mechanism was responsible for both the permeability of the membrane and the anaesthetizing effect, and he supported this conclusion with the following experiment.

He reduced the fluidity of the liposomes by increasing the pressure ninetyfold: the liposomes were now impermeable and the anaesthetic no longer released radioactivity. The high pressure had a dramatic effect on the anaesthetized tadpoles: they awoke from their anaesthesia. Evidently the pressure had restored the structure of the membrane and therefore its function.

The mode of action of local anaesthetics is however more complex than these experiments appear to show, since for example in axonal membranes sodium permeability is blocked selectively. Various mechanisms for this are discussed (see Chapter 6). In summary there is good evidence for a connection between the efficiency of these drugs and their effect on the fluidity of the membrane. For example, the agglutination of mammalian cells by plant lectins is increased by local anaesthetics [10] again suggesting a relationship with the fluidity of the cell membrane.

Effect of ions on lipid membranes

To consider the relationship between the mobility of the molecules making up a bilayer and its function we must discuss in more detail the forms this mobility may take. This includes the vibration and rotation of single groups or side chains of the lipid molecules, and also the lateral diffusion of the whole lipid molecule in its monolayer. The lipid molecule in a liposome changes places with its neighbour 10^6 times per second. By contrast it transfers from one side to the other of the bilayer only once in 14 days. This so-called *"flip flop"* or third mode of mobility makes virtually no contribution to the mobility of the membrane.

An important factor for all three types of mobility is the ionic composition of the aqueous medium surrounding the membrane [11]. Cations raise the phase transition temperature. The strength of the binding of cations to the negatively charged phospholipid heads depends on the size of their positive charge: alkali-metal ions are more weakly bound than alkaline earth ions, Na^+ about the same as K^+, but Ca^{2+} more strongly than the other ions of its group; the binding strengths are as follows:

$(CH_3)_4N^+ \approx (C_2H_5)_4N^+ \approx$ acetylcholine $< Na^+ \approx K^+ < NH_4^+ < Li^+$
$< Ba^{2+} < Sr^{2+} < Mg^{2+} < Ca^{2+} <$ (transition elements)$^{2+}$

The effect of Ca^{2+} is particularly interesting for neurobiology. It raises the electrical resistance of artificial lipid membranes if it is present in the same concentration on both sides of the membrane. This stabilization is in contrast to the destabilization occurring when it is only on one side: resistance is lowered and the membrane breaks down when the concentration exceeds 1 mmol/l. Electrophysiologists have observed something similar in the nerve membrane. They have shown that the "threshold" for generating an action potential and thus for a temporary rise of axonal membrane ionic permeability decreases when the calcium concentration in the external medium is reduced (see Chapter 6). Ca^{2+} affects the packing and mobility of lipid molecules in the bilayer. It raises the phase transition temperature, thereby stabilizing the crystalline state. But to apply the results from artificial membranes to biological membranes is to transfer them from a simple biophysical system to a biological one of far greater complexity. For instance, the ionic effects already described will be profoundly modified by the anions, proteins and lipid heterogeneity of the biological membrane.

Asymmetry of biological membranes

Although the bilayer model suggests a symmetrical structure with the area of contact of the monolayers as the axis of symmetry, we now know that this concept is not correct. Proteins, carbohydrates and the lipids of the bilayer are asymmetrically distributed. This information was derived from erythrocytes, like many basic ideas of biological membrane structures [12]. There is much evidence to indicate that the neuronal membrane is essentially similar.

Bretscher, Gordesky and van Deenen and their groups devised methods for detecting lipid asymmetry in the bilayer structure [13]. They investigated the lipids which are attacked from the outside of the cell by phospholipases, those that exchange with lipid vesicles and the erythrocyte membrane lipids which react with reagents like FDNB (1-fluoro-2,4-dinitrobenzene) or TNBS (2,4,6-trinitrobenzyl sulphonic acid) (FDNB diffuses easily through the membrane and thus can react with the lipids of both monolayers, while TNBS fails to penetrate and so reacts only with the outer monolayer). These experiments showed sphingomyelin and phosphatidylcholine to be localized primarily on the outside, and phosphatidylserine and phosphatidylethanolamine on the inner (cytoplasmic) side (Fig. 3.7.). Part of the lipid is perhaps not restricted to either side, but is bound to proteins, possibly surrounding them like rigid "collars".

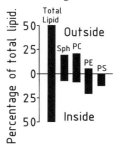

Fig. 3.7. Asymmetry of biological lipid membranes. The individual phospholipids are unequally distributed over the two monolayers of the bilayer. (Analysis of an erythrocyte membrane [30]). Sph, sphingomyelin; other abbreviations PC, PE, PS, see Table 2.2.

Various theories have been put forward about the biological significance of this asymmetry: for example, it may have something to do with the maintenance of the membrane potential, or it may reflect the protein asymmetry of the membrane. Thus phosphatidylserine which is needed for the activity of ATPase could be concentrated in the inner monolayer by its association with this protein.

Carbohydrates and proteins are also distributed asymmetrically in the bilayer

The carbohydrates of gangliosides and cerebrosides occur exclusively on the outer side of the cell membrane. Although this membrane asymmetry is so striking, there is no clear concept of its biological significance. We have already discussed this in Chapter 2. More is known about membrane proteins. During the last decade the same conclusion has been reached by many different methods: the membrane proteins are asymmetrically arranged in the membrane. Danielli's model of 1934 showed the proteins in a layer of spheres on the polar heads of the lipids on either side of the membrane (sandwich model). Calculations proved that there was insufficient protein to cover the bilayer with protein in globular form. His revised model of 1937 showed the protein molecules partly in extended, i.e. β-configuration with hydrophobic regions penetrating the membrane. We know now that even this model is too simple and is not in accord with many observations.

The fluid mosaic model of Singer and Nicolson [3] distinguishes two kinds of membrane proteins, the peripheral and the integral. Peripheral proteins are retained predominantly on the surface of the membrane by ionic interactions and are relatively easily solubilized e.g. by raising the ionic strength. Integral proteins are embedded in the lipid phase and cannot be released from the membrane unless this is at least partially destroyed. They are water insoluble, hydrophobic and lipophilic. This characterization of the two classes of membrane proteins implies that they are asymmetrically distributed in the cell membrane: peripheral proteins are only on one side of the bilayer. Integral proteins frequently penetrate only one monolayer; if they span the whole bilayer they are functionally asymmetric. Transport systems like the Na^+,K^+-ATPase (see Chapter 7) are an example of this.

Various methods are available for investigating the localization of integral proteins in the membrane [14]. Enzymatic methods are preferred for their selectivity. For example one can test whether a protein is altered in structure and function by proteases when the protease is applied first to the outer side and then to the inner side of the membrane. It has been shown, for example, that when the axon is perfused with pronase the proteins concerned with the inactivation of the axonal sodium channel are attacked and therefore must be located on the inside of the membrane. Applied from the outside pronase has little effect on sodium channel inactivation (see Chapter 6).

The lactoperoxidase (LPO) method has proved particularly successful for this type of localization [15] (Fig. 3.8).

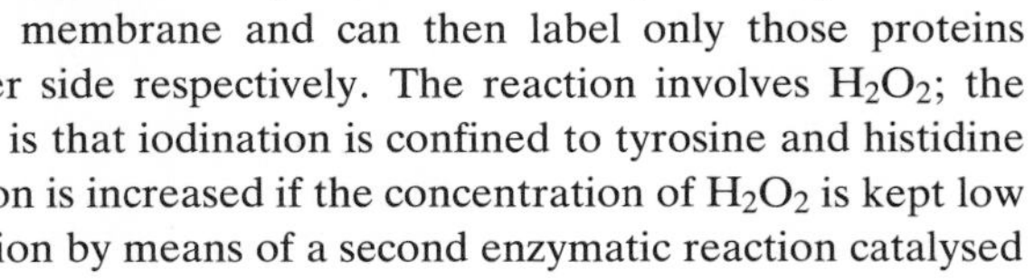

Fig. 3.8. Lactoperoxidase method for iodinating the tyrosine residues of proteins. Only surface proteins can be affected by enzymatic iodination. Therefore the method can be used to localize proteins in the cell membrane.

When protein is iodinated with radioactive ^{125}I by means of the enzyme lactoperoxidase the enzyme does not penetrate the membrane and can then label only those proteins accessible to it on the outer or inner side respectively. The reaction involves H_2O_2; the mechanism is not clear but the result is that iodination is confined to tyrosine and histidine residues. The selectivity of the reaction is increased if the concentration of H_2O_2 is kept low by generating it during the halogenation by means of a second enzymatic reaction catalysed by glucose oxidase.

The only non-enzymatic method to be mentioned here is that of the labelling of SH-groups. The ionic SH-reagent p-chloromercuribenzoate does not penetrate the membrane, and therefore only reacts with the protein on the side of the bilayer to which it has been applied. Other less polar reagents like N-ethylmaleimide (NEM) react with SH-groups throughout the membrane.

Integral membrane proteins are probably always asymmetrical in themselves. There is no known example of a whole protein molecule embedded between the lipid monolayers of the membrane which is not accessible from either side. Invariably one part of the molecular surface seems to be hydrophobic and inserted into the membrane, and the remainder is hydrophilic and is in contact with the surrounding aqueous medium. Integral proteins are thus *amphiphilic*. Sometimes integral proteins have a particularly high content of hydrophobic amino acids. The amphiphilic structure can also arise as a result of the accumulation of hydrophobic residues in one part of the amino acid sequence. There is evidence that this hydrophobic region spans the membrane in helical form. The α-helix configuration provides a particularly flexible solution to the structural problem as in it the polar groups of the peptide bonds are directed towards the inner side but the non-polar side chains of amino acids towards the outside. This amphiphilic construction was first shown in the case of glycophorin A of the erythrocyte membrane [12] (Fig. 3.9). The amino acid sequence shows all the fundamental points of the model in Fig. 3.9: a hydrophobic region, comprising the sequence between amino acid residues 73 to 92, is inserted between two hydrophilic sequences at the N and C terminals. The hydrophilicity of the N-terminal sequence (1-73) is greatly enhanced by 16 oligosaccharide side chains which *in toto* comprise 60% of the molecular mass. This part of the molecule is directed in an extended configuration to the outside of the membrane. The central hydrophobic sequence presumably forms a cylindrical α-helical region spanning the membrane; and the C-terminal sequence with its remarkably high number of acid amino acid side chains (93-131) is another hydrophilic region which forms a globular configuration on the inside surface of the membrane. In Chapter 7 we shall encounter bacteriorhodopsin which penetrates the membrane of halophilic bacteria and constitutes a light-driven proton pump. In this case the α-helical structure is made strikingly visible by means of high-resolution electron microscopy (see Fig. 7.12). Cytochrome b_5 of the endoplasmic reticulum also shows some structural analogies [17]. It is released from the membrane by proteases leaving behind a hydrophobic "anchor". This was recently sequenced and shows the expected hydrophobic structure. A further example for this structural principle – hydrophobic helices spanning the membrane and anchoring hydrophilic protein structures – is shown by the nicotinic acetylcholine receptor (Chapter 9).

Protein-lipid interactions cause phase separation and asymmetry of the membrane

The hydrophobic polypeptide chain can bind specifically a single type of lipid, and in this way produces phase separation and asymmetry in the lipid matrix.

Membrane proteins, with few exceptions, are non-covalently bound to the lipids of their environment. It is assumed that they accumulate around themselves specific lipids in the form of a collar or halo. Evidence for this has been obtained using ESR and spin-labelled lipids. In addition model studies with artificial liposomes made from phosphatidylserine and phosphatidylcholine have shown that the basic protein of human myelin (see Chapter 4) binds to acidic and neutral lipid molecules thereby bringing about phase separation [18]. The lipoprotein of the myelin membrane in model experiments with artificial liposomes shows a similar effect [19]. The nicotinic acetylcholine receptor from the *Torpedo* electric organ preferentially binds cholesterol rather than phospholipids. This has been established from model experiments with artificial lipid monolayers (see below) [20]. Only in a few cases has it been clearly demonstrated that a specific lipid is essential for the activity of a membrane protein [21]. One well established example is the mitochondrial β-hydroxybutyrate dehydrogenase, which is activated by the choline group of lecithin and lysolecithin. In general the presence of any lipid is probably sufficient to maintain the structure and function of a membrane protein.

Lipid-exchange proteins

An interesting class of lipoproteins, the lipid-exchange proteins discovered by van Deenen's group, are able to remove lipid from membranes or insert it into them. In the liver, for example, an exchange protein was found that preferentially transfers PC between liposomes and cell membranes. In the brain two proteins are found with a specificity for PI [22]. Although no net transport has been observed, they are certainly not concerned with the synthesis of membranes; apparently they have an important function in maintaining the correct lipid composition. More of these proteins with a relative molecular mass of between 12 000 and 30 000 have been purified to homogeneity [22,23]. But here we have digressed rather far from the theme of protein-lipid interactions of integral membrane proteins. We shall now return to this subject.

Mild detergents can replace lipids

Membrane proteins have for a long time been discarded by biochemists as an insoluble fraction of proteins. Interest in them suddenly increased when it was discovered that they could be rendered soluble without denaturing them. This was done by means of so-called mild detergents [24]. Detergents can be viewed as a special class of lipids. They are amphipathic like membrane lipids with a hydrophilic head and a hydrophobic residue. Above a certain concentration – the critical micellar concentration (CMC) – they form stable spherical aggregates (micelles) (see Fig. 3.3). They are classified according to the nature of their hydrophilic heads into non-ionic, ionic, and zwitterionic detergents (Fig. 3.10). In ionic detergents the CMC is in the millimolar region, for the non-ionic one order of magnitude less.

Detergents dissolve natural membranes by forming a mixed micelle with their lipids. With membrane proteins they replace the endogenous lipids. They form micelles around the hydrophobic region of proteins (Fig. 3.11). If the detergents are then removed, e.g. by dialysis or simply by dilution, the protein becomes insoluble. Probably it aggregates

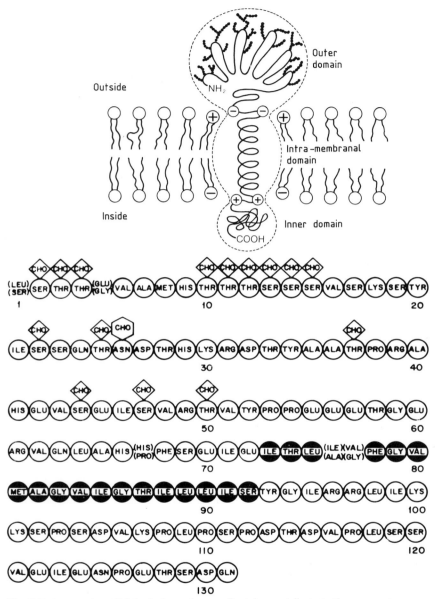

Fig. 3.9. Asymmetry of biological membranes. Proteins contribute to the asymmetry, exemplified here by glycophorin of the erythrocyte membrane. The hydrophilic part of the protein's amino acid sequence interacts with the aqueous environment and a hydrophobic presumably helical region spans the membrane. This region has been localized (below – marked in black); the residues containing carbohydrate side chains (CHO) are also marked. Reproduced, with kind permission, from the Annual Reviews Inc. [12].

because its molecules come in contact with their own hydrophobic areas and themselves form a kind of micelle.

Some detergents e.g. Triton X-100, Lubrol, Brij or cholate and deoxycholate simply dissolve the membrane as already described and solubilize the integral proteins but others severely damage them. They considerably affect the structure of the proteins to the extent of irreversible denaturation. Therefore the action of detergents is unpredictable, and must be tested experimentally in each case.

A) Ionic detergents

Structure	Short name	CMC (mmol/l)
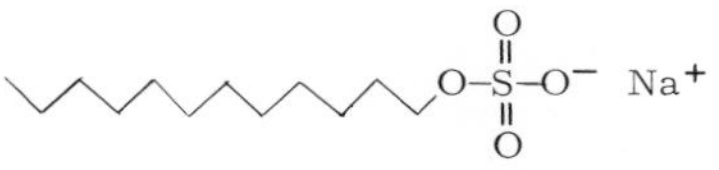	Sodium dodecylsulfate (SDS)	8.2
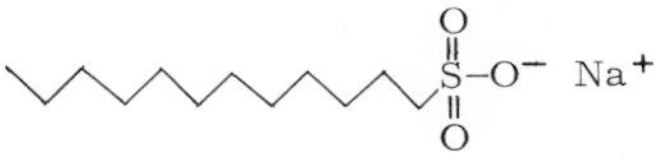	Sodium dodecylsulfonate	9.8
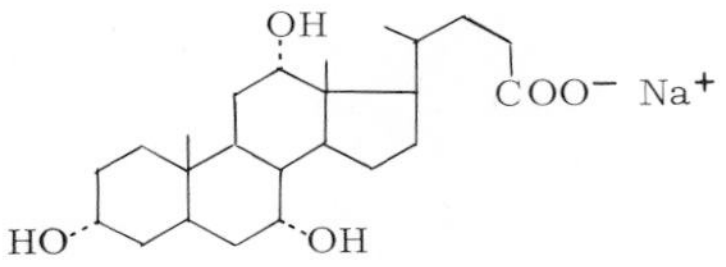	Sodium cholate	10.0
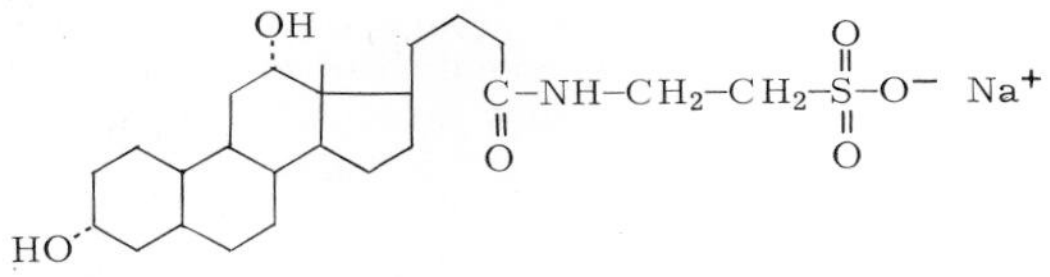	Sodium taurodesoxycholate	2.0

B) Nonionic detergents

Structure	Short name	CMC (mmol/l)
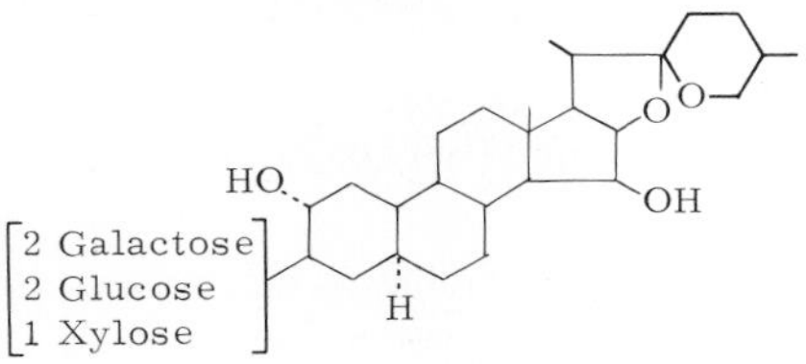	Digitonine	
	Brij, Lubrol W, AL Emulphogen BC, Renex 30, Sterox AJ, AP	
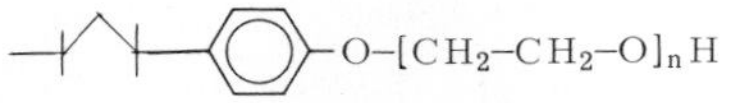	Triton X,	0.24

Fig. 3.10. Detergents. (Continued next page)

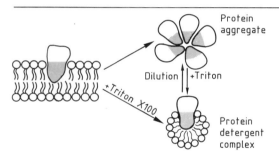

Tween

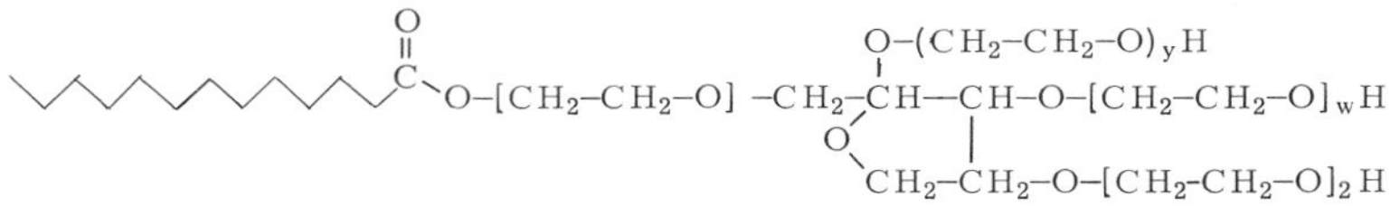

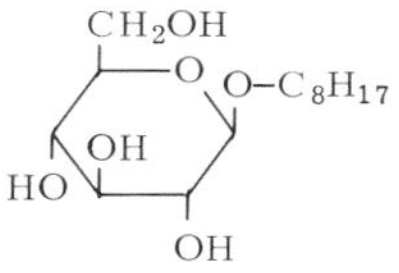

β-Octylglycoside (β-OG)   17.4

C) Zwitterionic detergents

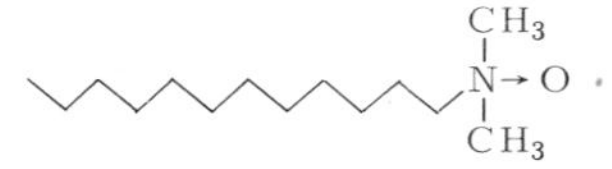

Dodecyldimethylamine oxide (DDAO)   8.4

$R-CO(NH)(CH_2)_3\overset{\oplus}{N}(CH_3)_2-(CH_2)_3SO_3^-$ CHAPS

R: cholate $-CH_2CHOHCH_2SO_3^-$ CHAPSO   4-6

Fig. 3.10. (Continued)

Fig. 3.11. Solution of a membrane by the non-ionic detergent Triton X-100. The detergent arranges itself on the hydrophobic area of the protein and substitutes for the membrane lipid. If the detergent is diluted below its CMC, the protein itself forms micelles or aggregates which often precipitate.

Artificial lipid membranes are models of biological membranes

A recurring dilemma for neurochemists, and for that matter membrane biochemists in general, is that in order to characterize a functionally important membrane molecule it must be isolated, i.e. removed from its functional environment. Important properties can thus be lost. This is especially true for transport proteins whose function by definition cannot be separated from those of the membrane. The biochemist can only investigate structural properties of the purified membrane protein and not the mechanism of its function. One solution to this dilemma is by reconstitution i.e. the insertion of a purified membrane protein into an artificial lipid membrane so restoring its function. The advantage of this reconstituted system compared with a biological one is that one can work under well defined conditions: the purified molecule, such as the transport protein is the only protein in the artificial membrane and the lipid environment can be defined by the experimenter. In this way it is possible to test which component of the biological membrane is necessary for the function under investigation, and under what other conditions such as ionic strength, pH and membrane potential the system functions. As this method is increasingly important for neurochemistry we will give a short review of the advantages and problems of artificial lipid systems and methods for reconstitution.

Artificial lipid systems are divided into *vesicular* and *planar* (see Fig. 3.3). Artificial lipid vesicles provide an easier experimental method involving less expensive apparatus but they usually yield less information than the planar. In many cases they are a compromise, a preliminary step on the way to the reconstitution of the planar membrane system. Their evidence is more qualitative than quantitative. They are particularly suitable for measurements of the flux of radioactively labelled molecules; spectroscopic and resonance methods (NMR, ESR) also provide valuable data. Protein-lipid interactions and particularly ionic permeability can be measured quantitatively most effectively with planar lipid systems. The number of genuinely reconstituted systems is still small because of the experimental difficulties. The basis of this technique is an unproved assumption of molecular biology: biomolecules contain within their structure all the necessary information for the formation of higher structures. Biomembranes are *self assembling systems*.

Three most commonly used methods for reconstitution of vesicular protein-lipid systems are (Fig. 3.12):
1. Sonication of lipids suspended in a protein-containing medium [25].
2. Dialysis of a solution of lipid and protein in detergent [26].
3. Gel filtration of a solution of lipid and protein in detergent [27].

Common to the three methods is the following hypothesis: the lipid vesicle (liposome) in aqueous solution is thermodynamically more stable than either the single lipid molecule or the planar film. During its formation by sonication or by removal of the detergent it captures protein molecules if these are hydrophobic and thereby in a thermodynamically unfavourable state. There are two ways in which hydrophobic proteins may become stabilized – by aggregation with each other or by incorporation in the lipid phase. If the experimenter is lucky it is the latter. A recent development is the preparation of large (50 μm diameter) vesicles which may be accessible to the electrophysiological method of impalement by microelectrodes or even to the patch clamp method [31].

Planar lipid membranes can be used either as a mono- or as a bilayer in reconstitution experiments. The monolayer is formed at the air-water interface. Each lipid has a characteristic structure-dependent surface area; i.e. instead of measuring directly the monolayer surface, a torsion balance can be used to define the tendency of the lipid to spread, the so-called *surface pressure*.

The apparatus for such experiments is called a *Langmuir trough*. With it the reconstitution of a protein-lipid system can be followed quantitatively, for when a protein is built into a monolayer a spreading pressure occurs corresponding to the spatial requirement of the molecule. Thus, if a protein is introduced into the subphase under the monolayer a change in the surface pressure in the Langmuir trough will record its incorporation into the membrane. By varying the composition of the lipid monolayer one can identify which lipid interacts with a specific protein. For example the acetylcholine receptor of the postsynaptic membrane has been found by this method to interact better with cholesterol than with phospholipids [20]. This kind of information is not only of theoretical interest; it can be vital for successful reconstitution.

Artificial bilayer membranes can be obtained in two ways:

1. A drop of a solution of lipid in decane is put into a hole in a partition wall separating two compartments of a plastic cell (Fig. 3.13). As the solvent diffuses away the lipid droplet spontaneously thins to form a bilayer known as a *black lipid membrane* (BLM) from the change in colour of the lipid from brown to black (Mueller-Rudin-method [28]).

2. Lipid is spread as a monolayer on a water surface and two of these monolayers are pulled over an opening in a hydrophobic carrier (Montal-Mueller-method [29]).

Proteins can be incorporated by either adding them to the lipid solution before the formation of the membrane or introducing them by diffusion into the completed bilayer. The black membrane has been particularly successful for small peptide ionophores like the antibiotics gramicidin and valinomycin. Their ion transport kinetics can be analysed in detail; these show that valinomycin is an ion carrier; by contrast gramicidin dimerizes thus forming pores in the membrane. The method is so sensitive that one can study quantitatively the properties of single ion channels, their ionic selectivity, their maximum conductance, and their life-time.

By using one of these methods reconstitution has been achieved of the sodium/potassium-pump (Na^+,K^+-ATPase), calcium-ATPase (Chapter 7), rhodopsin and bacteriorhodopsin, also of nerve and muscle proteins like the nicotinic acetylcholine receptor and the voltage-dependent sodium channel of the axonal membrane. Many published reports of

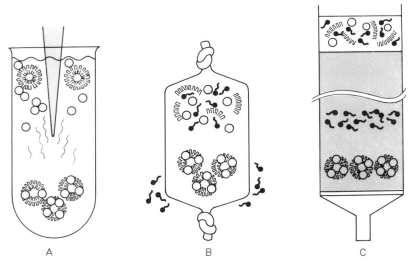

Fig. 3.12. Reconstitution of vesicular protein-lipid systems. ○ protein; ⅡⅡ lipid. (A) Sonication of an aqueous suspension of membrane protein and lipid produces lipid vesicles, containing protein in the bilayer. (B) Reconstitution of the protein-lipid vesicle in an aqueous solution of lipid and protein in detergent (𝟋). The detergent is removed by dialysis. (C) As (B) but detergent removed by gel chromatography.

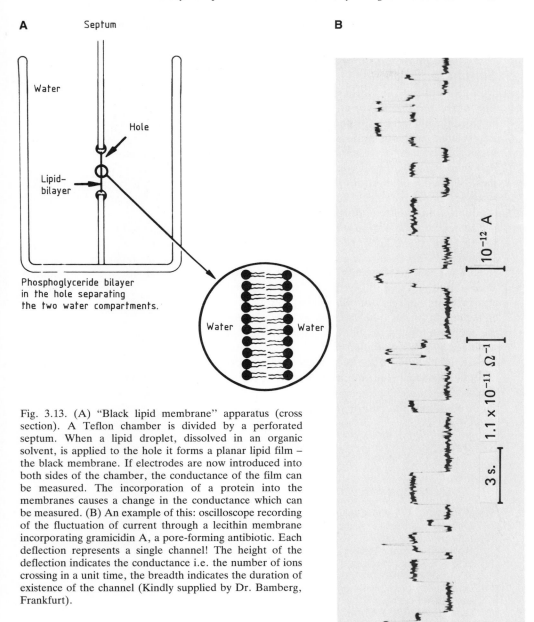

Fig. 3.13. (A) "Black lipid membrane" apparatus (cross section). A Teflon chamber is divided by a perforated septum. When a lipid droplet, dissolved in an organic solvent, is applied to the hole it forms a planar lipid film – the black membrane. If electrodes are now introduced into both sides of the chamber, the conductance of the film can be measured. The incorporation of a protein into the membranes causes a change in the conductance which can be measured. (B) An example of this: oscilloscope recording of the fluctuation of current through a lecithin membrane incorporating gramicidin A, a pore-forming antibiotic. Each deflection represents a single channel! The height of the deflection indicates the conductance i.e. the number of ions crossing in a unit time, the breadth indicates the duration of existence of the channel (Kindly supplied by Dr. Bamberg, Frankfurt).

successful reconstitution must be treated with reservation since the properties they show differ too much from those of the biological system.

Summary

The nerve membrane considered as a cytoplasmic membrane is not simply a passive supporting membrane for the cell. It serves as a barrier for the maintenance of the intracellular environment (ions, electric potential, metabolites) and compartmentation (cell organelles, transmitter vesicles). It takes an active part (ion pumps, enzymes) and a passive part (ion channels, transmitter release) in the transmission of the nerve impulse. It carries the individual characteristics necessary for the development of the nervous system and its synaptic connectivity (cell adhesion and recognition). It transmits extracellular signals (hormones, transmitters, drugs).

The neuronal membrane is so far best described by S.J. Singer's fluid-mosaic model. It is founded on the bilayer concept and has the following characteristics: fluidity and asymmetry. The fluidity is influenced by the composition of the fatty acid chains, the cholesterol content, ions (e.g. Ca^{2+}) and drugs (e.g. local anaesthetics). The asymmetry affects lipids (PS and PE inside, PC and sphingomyelin outside), carbohydrates (outside) and proteins (integral or peripheral). Proteins can span the membrane or be partly embedded in it. An α-helical hydrophobic sequence is presumably the form favoured for the spanning of a lipid bilayer. The structure and activity of membrane proteins are stabilized by lipids. So-called mild detergents can take over the role of membrane lipids and cause solubilization of the membrane; under these conditions however many membrane molecules lose their function by definition (example: transport proteins). For investigating "isolated" membrane proteins attempts have been made to reconstitute them in an artificial planar (black membrane) or vesicular lipid system.

References

Cited:

[1] Weissmann, G., and Claiborne, R. (eds.): *Cell Membranes. Biochemistry, Cell Biology and Pathology.* Hospital Practice Publishing Co, New York 1975.
[2] Finean, J.B., Coleman, R., and Mitchell, R.H.: *Membranes and Their Cellular Function.* Blackwell Scientific Publications, Oxford 1974.
[3] Singer, S.J., and Nicolson, G.L., "The fluid mosaic model of the structure of cell membranes", *Science* **175**, 720-731. (1972).
[4] Lee, A.G., "Lipid phase transitions and phase diagrams", *Biochim. Biophys. Acta* **472**, 237-282 and 285-344 (1977).
[5] Overath, P., Hill, F.F., and Lamnek-Hirsch, I., "Biogenesis of E. coli membrane: evidence for randomization of lipid phase", *Nature (London) New Biol.* **234**, 264-267 (1971).
[6] Demle, R.A., and De Kruyff, B., "The function of sterols in membranes", *Biochim. Biophys. Acta* **457**, 109-132 (1976).
[7] Coleman, R., "Membrane-bound enzymes and membrane ultrastructure", *Biochim. Biophys. Acta* **300**, 130 (1973).

[8] Lee, A.G., "Model for action of local anaesthetics", *Nature* **262**, 545-548 (1976).
[9] Papahadjopoulos, D., Jacobson, K., Poste, G., and Shepard, G., "Effects of local anaesthetics on membrane properties. I. Changes in the fluidity of phospholipid bilayers", *Biochim. Biophys. Acta* **394**, 509-519 (1975).
[10] Poste, G., Papahadjopoulos, D., Jacobson, K., and Vail, W.J., "Effects of local anaesthetics on membrane properties. II. Enhancement of the susceptibility of mammalian cells to agglutination by plant lectins", *Biochim. Biophys. Acta* **94**, 520-539 (1975).
[11] Hauser, H., Levine, B.A., and Williams, R.J.P., "Interactions of ions with membranes", *TIBS* **1**, 278-281 (1976).
[12] Marchesi, V.P., Furthmayr, H., and Tomita, M., "The red cell membrane", *Ann. Rev. Biochem.* **45**, 667-698 (1976).
[13] Bergelson, L.D., and Barsukow, L.I., "Topological asymmetry of phospholipids in membranes", *Science* **197**, 224-230 (1977).
[14] Maddy, A.H. (ed.): *Biochemical Analysis of Membranes.* John Wiley and Sons, New York 1976.
[15] Morrison, M., and Schonbaum, G.R., "Peroxidase-catalyzed halogenation", *Ann. Rev. Biochem.* **45**, 861-888 (1976).
[16] Hubbard, A.L., and Cohn, Z.A., "The enzymatic iodination of the red cell membrane", *J. Cell Biol.* **55**, 390-405 (1977).
[17] Ozols, J., and Gerard, C., "Primary structure of the membranous segment of cytochrome b_5", *Proc. Natl. Acad. Sci. USA* **74**, 3725-3729 (1977).
[18] Boggs, J.M., Moscarello, M.A., and Papahadjopoulos, D., "Phase separation of acidic and neutral phospholipids induced by human myelin basic protein", *Biochemistry* **16**, 5420-5426 (1977).
[19] Boggs, J.M., Wood, D.D., Moscarello, M.A., and Papahadjopoulos, D., "Lipid phase separation induced by a hydrophobic protein in phosphatidylserine-phosphatidylcholine vesicles", *Biochemistry* **16**, 2325-2329 (1977).
[20] Gennis, R.B., "Protein-lipid interactions", *Ann. Rev. Biophys. Bioeng.* **6**, 195-238 (1977).
[21] Sandermann, H., "Regulation of membrane enzymes by lipids", *Biochim. Biophys. Acta* **515**, 209-237 (1978).
[22] Helmkamp, G.M., Harvey, M.S., Wirtz, K.W.A., and van Deenen, L.L.M., "Phospholipid exchange between membranes. Purification of bovine brain proteins that preferentially catalyze the transfer of phosphatidylinositol", *J. Biol. Chem.* **249**, 6382-6389 (1974).
[23] Bloj, B., and Zilversmit, D.B., "Rat liver proteins capable of transfering phosphatidylethanolamine", *J. Biol. Chem.* **252**, 1613-1619 (1977).
[24] Helenius, A., and Simons, K., "Solubilization of membranes by detergents", *Biochim. Biophys. Acta* **145**, 29-79 (1975).
[25] Kagawa, Y., and Racker, E., "Partial reconstitution of the enzyme catalyzing oxidative phosphorylation", *J. Biol. Chem.* **246**, 5477-5487 (1971).
[26] Racker, E., "A new procedure for the reconstitution of biological active phospholipid vesicles", *Biochem. Biophys. Res. Commun.* **55**, 224-230 (1973).
[27] Brunner, J., Skrabal, P., and Hauser, H., "Single bilayer vesicles prepared without sonication. Physicochemical properties", *Biochim. Biophys. Acta* **455**, 322-331 (1976).
[28] Mueller, P., Rudin, D.O., Tien, H.T., and Wescott, W.C., "Methods for the formation of single bimolecular lipid membranes in aqueous solution", *J. Phys. Chem.* **67**, 534-535 (1963).
[29] Montal, M., and Mueller, P., "Formation of bimolecular membranes from lipid monolayers and a study of their electrical properties", *Proc. Natl. Acad. Sci. USA* **69**, 3561-3566 (1972).
[30] Verklej, A.J., Zwall, R.F.A., Roelofsen, B., Kastelijn, D., and van Deenen, L.L.M., "The asymmetric distribution of phospholipids in the human red cell membrane", *Biochim. Biophys. Acta* **323**, 178-193 (1973).
[31] Hub, H.H., Zimmermann, U., and Ringsdorf, H., "Preparation of large and unilamellar vesicles", *FEBS Lett.* **140**, 254-256 (1982).

Further reading:

Houslay, M.D., and Stanley, K.K.: *Dynamics of Biological Membranes.* J. Wiley & Sons, New York 1982.

Azzi, A., Brodbeck, U., and Zahler, P. (eds.): *Membrane Proteins, a laboratory manual,* Springer Verlag, Berlin, Heidelberg, New York 1981.

Coronado, R., and Labarca, P.P., "Reconstitution of single ion channel molecules", *TINS* **7**, 155–160 (1984).

Weissmann, G., and Clairborne, R. (eds.): *Cell Membranes. Biochemistry, Cell Biology, and Pathology.* Hospital Practice Publishing Co., New York 1975.

See also ref. 2, 3, 4, 8, 14, 20, 24.

Chapter 4

Myelin

Much of what we now know about the general structure of biomembranes is derived from investigations of a special membrane of the nervous system – myelin. Because of its relatively simple structure it has been used for the development of experimental methods and theoretical models in membrane research. Myelin is a multi-lamellar system which insulates central and peripheral nerve fibres. The white matter of the brain of higher organisms consists of more than 50% myelin and disturbances in myelin formation during ontogeny as well as changes in the myelin of the fully developed nervous system give rise to severe neuropathies. The structure, function and origin of myelin is therefore of considerable relevance both to membranology and to neurology.

Functions of myelin: (I) insulation; acceleration of conduction velocity

Myelin ensheaths the axon in a thick layer and so prevents reciprocal electrical interference between the single nerve fibres in a thickly packed nerve bundle, (Fig. 4.1). Its insulating function is well served by its unusually high content of lipid, compared with other membranes. But it has other functions as well. Myelin increases the efficiency with which the impulse is conducted, in two different ways: it accelerates it, and saves energy. The former depends on the fact that the sheath is not continuous, but is interrupted at regular intervals (usually 1-2 mm) (Fig. 4.1B). These interruptions in the insulation are called *nodes of Ranvier* (from their discoverer), the region between the nodes, the *internode*. Details of the mechanism of transmission along the nerve fibre will be covered in Chapter 5. It has been discovered by electrophysiological techniques that the course of the impulse along the myelinated fibre is by saltatory conduction (from *saltare* – to jump) i.e. instead of a slow progression of the impulse by the excitation of one ion channel after another, it is forced, by the insulation of the internodal region to jump 1-2 mm at a time from node to node (Fig. 4.2B).

A considerable acceleration of transmission is achieved varying from 5 to 10-fold in fibres of comparable diameter. The speed of conduction increases with the diameter of the fibre – in unmyelinated fibres it increases in proportion to the square of the diameter (because the electrical resistance decreases in proportion to the square of the radius), in

74   4 Myelin

myelinated fibres in direct proportion. The advantage of myelination is so great that if our spinal cord were composed of unmyelinated fibres instead of myelinated fibres, it would have to have the diameter of a medium sized tree trunk. In vertebrates all nerve fibres are myelinated and the impulse is conducted with a speed of over 3 m/s.

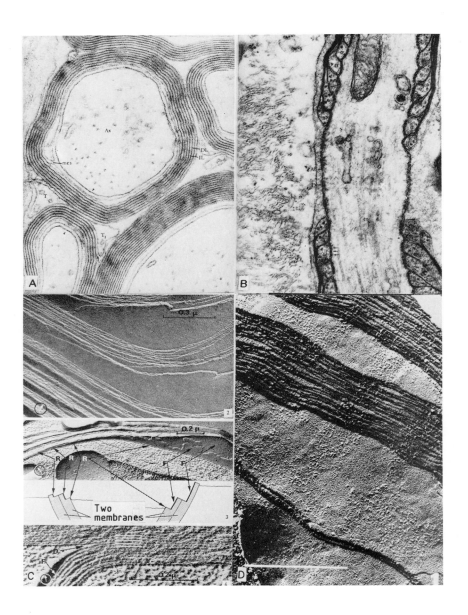

Fig. 4.2. The conduction of a nerve impulse in (A) unmyelinated (B) myelinated nerves (saltatory conduction). The internode is the region between two nodes of the nerve fibre.

Functions of myelin: (II) space and energy saving

Invertebrates (except crayfish) have no myelinated fibres. However early stages do occur in which axons have many layers of loosely wrapped Schwann cell processes around them – a kind of protomyelin. As evolution progresses myelinated fibres become dominant. Their advantage consists in their improved performance achieved without undue extension of their space requirements. This space saving is even more important for the brain than for the spinal cord.

It is not only space which is saved by myelin. A nerve impulse involves ion fluxes through the axolemma. They flow through ion channels which are opened by depolarization of the membrane involving the displacement of charges. This process is stabilized and slowed down by the membrane itself, which, in an electrical sense, is a capacitance. If the capacitance is large it absorbs many of the displaced charges until the net depolarization necessary for the release of a nerve impulse can be achieved. This will be further explained in Chapters 5 and 6; here the point will only be made that the capacitance with myelinated fibres is smaller than that with unmyelinated; fewer charge displacements are needed for a depolarization, and also fewer charges need to be transported back for the restoration of the resting state. The latter involves ion pumps (Chapter 7) and uses metabolic energy in the form of ATP. The result is that due to their smaller capacitance myelinated fibres use less energy per impulse than unmyelinated.

But why is the capacitance of the myelin sheath less than that of a simple axon membrane? The explanation is derived from physics: the many concentric layers of myelin around the axon represent many capacitances connected in series (Fig. 4.3) and the total value of a system of capacitances in series is less than that of any single capacitance.

Fig. 4.1. Myelin. (A) Cross-section through the rat optic nerve; Ax, axon; mes, mesaxon and beginning of myelin spiral; DL, dense line, formed by the apposition of the cytoplasmic side of the glial cell membrane; IL, intraperiod line, formed by the apposition of the outer side of the cell membrane; it is also the intermediate area between two turns of the spiral; T, tongue process, the outer end of the spiral. Reproduced, with kind permission, from Peters [22]. (B) Node of Ranvier in rat optic nerve. Al, axolemma (axon membrane); D, optically dense zone parallel to the node membrane; P, pockets which represent the ends of the myelin lamellae; m, microtubules; mit, mitochondria. (C) Freeze fracture through a myelin sheath. (D) Freeze fracture showing particles, probably proteins, embedded in the myelin bilayer. These seem to be concentrated in positions corresponding to the myelin lamellae. Reproduced, with kind permission, from Pinto da Silva [3]. Electron micrographs; A,B: negatively stained; C,D: freeze etched sample.

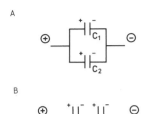

Fig. 4.3. Condensers connected (A) in parallel (B) in series. (B) represents the arrangement in a myelinated axon.

Myelin is a compact spiral made from plasma membranes

In 1954 Geren observed the formation of myelin around the ischial nerve of a chick embryo [1]. She established that the number of layers depended on the age of the embryo and that in the early stages there was a clear spiral structure. The following picture (Fig. 4.4) shows Geren's original observations of a peripheral nerve. A similar situation is presumed to exist with central myelinated nerves: the axon generates a depression in the surface of a Schwann cell and the latter grows to encircle it, appearing to roll round it. In tissue cultures it takes about 44 hours to complete one turn. As the number of turns increases the cytoplasm is squeezed out so that the Schwann cell plasma membrane becomes more and more tightly packed. The completed myelin sheath consists therefore of many layers of Schwann cell membrane wrapped round the axon. Each internode is formed from a single Schwann cell (or in the central nervous system from one process of an oligodendroglial cell). Although the ontogeny of myelin has not yet been described biochemically nevertheless, parallel to the visible morphological development, molecular changes occur which lead to the characteristically different composition of myelin membrane compared with other plasma membranes including that of the Schwann cell (see below).

The bilayer structure of myelin

Developmentally myelin is a scientific fossil. Although Danielli and Davson's membrane model of 1934 (see Chapter 3) is no longer appropriate for the majority of cell membranes, in the case of the formation of myelin it seems to apply. In particular the freeze – etch electron micrographs of Branton [2] seem to confirm the myelin membrane structure as a sandwich – a lipid bilayer inside, and on each outer side a layer of protein (Fig. 4.1C). He found on the fracture face corresponding to the inside of the bilayer no particles resembling those identified as integral proteins in other plasma membranes. (However, more recent electron microscopic investigations of Pinto da Silva and Miller directly dispute this; these authors do find particles on the inner side of the fracture face (Fig. 4.1D [3]). Other authors show that the apparently lamellar structure of myelin remains even when 98% of the lipid has been extracted [4].

The bilayer structure of myelin 77

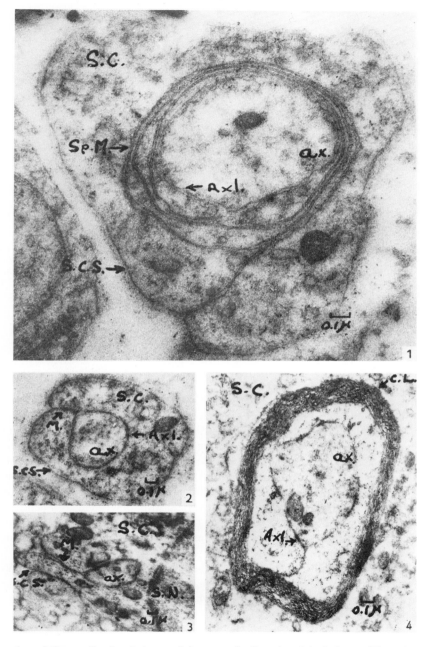

Fig. 4.4. Formation of the myelin sheath from a Schwann cell. Geren's original picture [1], and a schematic diagram. (Continued next page)

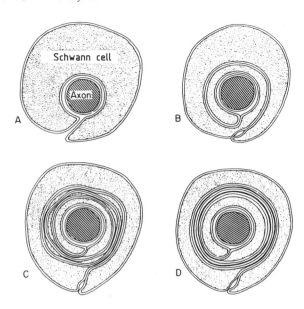

Fig. 4.4. (Continued)

We will next summarize the available data before constructing models: from the electron microscopic picture a "repeat unit" of 17 nm can be clearly recognized. It is formed from two cell membrane bilayers of one turn of a Schwann cell spiral. X-ray diffraction studies with a resolution of 1 nm (Fig. 4.5) have given more detailed information [5]: the lipid bilayer is about 5 nm thick, on either side proteins occupy about 1.5 nm so the whole is 8 nm thick. The electron density is least in the interior of the membrane; it increases on each side to a maximum and then declines again towards the outside. From this it can be concluded that in a 1.5 nm thick central layer there are mainly hydrocarbon chains then hydrophilic groups of lipids and finally proteins. An important conclusion from the X-ray diffraction pattern analysis is the asymmetric distribution of electron density which shows that cholesterol is concentrated more on the outside of the membrane; other lipids may be also asymmetrically distributed. This asymmetry of the myelin membrane was later confirmed by biochemical experiments which showed that in guinea pig brain myelin phosphatidylcholine is concentrated on the outer side [6]. The uneven distribution of lipids between the two layers of the myelin bilayer is perhaps a consequence of their different affinities for membrane proteins [7,8] which are also asymmetrically distributed (see also Chapter 3).

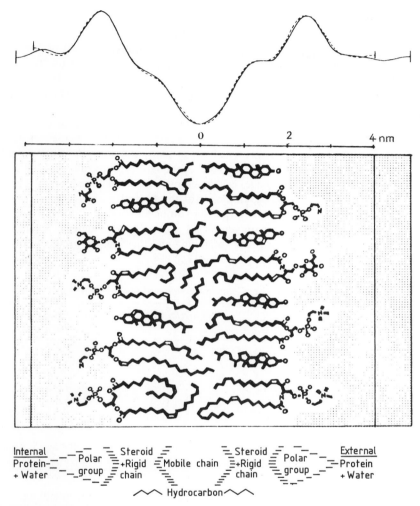

Fig. 4.5. Top: Distribution of the intensity of low angle X-ray scattering by the myelin membrane at 1 nm resolution. The peaks of maximum electron density represent the polar lipid heads; the minimum in the middle the hydrophobic chains of the bilayer. Note the higher shoulder on the right (outer) side, which is probably caused by the higher cholesterol content of the outer monolayer. Below: Interpretation of the model. Reproduced, with kind permission, from Caspar and from Macmillan Journals Ltd. [5].

Chemical composition of myelin

The model shown in Fig. 4.5 is obtained through the application of X-ray scattering and includes the biochemical analysis of myelin. The lipids are most conspicuous; they comprise 70-85% of the dry mass which is double that in other plasma membranes [9,10]. There are no myelin-specific lipids and, if one excludes cardiolipin, the specific lipid of mitochondrial membranes, there are no lipids in other plasma membranes which are not present in myelin. Myelin membrane is distinguished from other membranes by the percentage composition of its lipids (Tabs. 4.1, 4.2). The high proportion of cerebroside is noteworthy. There is a somewhat lower content both of phospholipids and of ganglioside. The latter consists almost completely of G_{M1}. Although cerebroside can be found in other membranes, its high percentage is characteristic for myelin membrane. The increase in its amount during embryonic development of the central nervous system is correlated with progressive myelination. In contrast the proportion of cerebroside is characteristically decreased in pathologically or genetically caused myelin defects. Mouse mutants with genetic myelin disorders (e.g. the "quaking" and the "jimpy" mutants) have only 10% of the brain cerebrosides of normal strains [11].

Table 4.1. Composition of myelin. Comparison with erythrocyte membrane [10].

Composition	CNS myelin (Ox)	Erythrocyte membrane
Protein (% of total mass)	25	63
Lipid (% of total mass)	75	37
Enzymes	few	many
Lipid (% of lipid-fraction):		
Cholesterol	28	28
Phospholipids	42	61
Cerebrosides	25	11
Cerebroside sulphates	5	0
Gangliosides	Trace	+

Myelin is prepared by homogenization of nervous tissue and fractionation by means of density gradient centrifugation. Its density of only 1.08 g·ml^{-1} favours the use of isopycnic methods. The analyses of different authors vary; the reason may be varying amounts of contamination by axonal membranes.

The high lipid content implies a low percentage of protein. The protein composition is relatively simple. The main component is a proteolipid (35-50% of myelin protein), a basic protein (30%) and a group of acidic proteins (called Wolfgram proteins) whose function is unknown (see below).

Only a few enzyme activities are present: a cholesterolesterase, a 2',3'-cAMP hydrolysing phosphodiesterase and a cAMP-dependent protein kinase which phosphorylates the basic protein. In addition, as a counterpart to the kinase, a phosphatase occurs whose activity is mainly in the cytosol but which is localized in the myelin, and as a counterpart to the

Table 4.2. Protein and lipid analysis of myelin in different organisms; comparison of white matter containing predominantly myelinated nerve fibres with grey matter containing the neuron cell bodies [10] (all single lipid numbers mean % of total weight of lipid). Differences from other tables in this book are due to the variation in analytical values obtained in different laboratories.

	Myelin			White Matter		Grey Matter	Total brain
	Human	Bovine	Rat	Human	Bovine	Human	Rat
Protein (% of total dry weight)	30.0	24.7	29.5	39.0	39.6	55.3	56.9
Lipid (% of total dry weight)	70.0	75.3	54.9	54.9	55.0	32.7	37.0
Cholesterol	27.7	28.1	27.3	27.5	23.6	22.0	23.0
Cerebrosides	22.7	24.0	23.7	19.8	22.5	5.4	14.6
Sulphatides	3.8	3.6	7.1	5.4	5.0	1.7	4.8
Total galactolipids	27.5	29.3	31.5	26.4	28.6	7.3	21.3
Total phospholipids	43.1	43.0	44.0	45.9	46.3	69.5	57.6
Lecithin	11.2	10.9	11.3	12.8	12.9	26.7	22.0
Sphingomyelin	7.9	7.1	3.2	7.7	6.7	6.9	3.8
Phosphatidylserine (PS)	4.8	6.5	7.0	7.9	11.4	8.7	7.2
Phosphatidylinositol (PI)	0.6	0.8	1.2	0.9	0.9	2.7	2.4
Plasmalogens	12.3	14.1	14.1	11.2	12.2	8.8	11.6

Table 4.3. Some of the enzymatic activities present in the myelin membrane.

Enzymatic Acitivity	Animal	Ref.
1. 2'3'-Cyclonucleotide- 2'-phosphodiesterase (EC 3.1.4.16)	Ox (White matter)	[14]
2. Kinase (A1-phosphorylating)	Rat (White matter)	[25]
3. Phosphatase (A1-dephosphorylating)	Rat (CNS)	[25]
4. Sphingomyelinase (EC 3.1.4.12)	Rat (CNS)	[13]
5. Cholesterolesterase (EC 3.1.1.13)	Rat (CNS)	[26]
6. Cholesterol acetyltransferase (EC 2.3.1.26)	Rat (CNS)	[12a]
7. Carbonic anhydrase (EC 4.2.1.1)	Rat (CNS)	[27]

cholesterolesterase a cholesterol acetyltransferase is found in rat central nervous system myelin [12a]. A sphingomyelinase [13] and a carbonic anhydrase are also found. The phosphodiesterase has been purified to homogeneity and some of its enzymatic properties characterized [14]. Its designation as a myelin enzyme is confirmed by the fact that its activity is clearly reduced in the brain of mouse mutants with disordered myelination [15]. The lipid dependency of the cholesterolesterase has been investigated: galactosylcerebroside and PS activate, ganglioside and lysolecithin inhibit it. No clear physiological function for any of the myelin enzymes has yet been discovered.

Myelin is not metabolized as a whole

In view of the fact that myelin contains little cytoplasm and there are often fifty or more densely packed layers of membrane it is not surprising that myelin is associated with a low turnover rate, substantially lower than that of other plasma membranes. The half-life of lipids vary between 5 weeks (PI), 2 to 4 months (PC and PS) and about one year (PE, cholesterol, cerebroside, sulphatide, sphingomyelin). For proteins the half-life ranges from 2 to 6 weeks (myelin basic protein and proteolipid). More interesting than the relative stability of myelin is the fact that the half-lives are different for individual components so that they are renewed independently of each other. Thus myelin is not metabolized as a unit. In addition it is not simply the outer layer or those nearest to the axoplasm that are replaced. Experiments with radioactive lipid precursors show that exchange occurs between all layers.

Proteins of unknown function

In 1951 Folch-Pi isolated a protein fraction by extracting brain myelin with chloroform/methanol in the ratio of 2:1 [16]. Since then proteins extracted in this way have been classified operationally as *proteolipid* (PLP). Contrary to the original belief the Folch-Pi protein which contains about 2 to 3% covalently bound lipid is heterogeneous. Besides the main fraction, a PLP with a relative molecular mass of 23 500, two other lipophilic proteins were characterized, one with a relative molecular mass of 20 500 (sometimes called DM-20) and the other with a relative molecular mass of 12 000 (P-12). The proteolipid contains more than 50% of hydrophobic amino acid, is α-helical and monomeric in chloroform/methanol and aggregates in water with the loss of its helical structure. Since it produces lipid phase separation in model experiments and binds acid lipid preferentially [7], it contributes to the above mentioned asymmetry of the myelin membrane. Pictures of freeze-etched myelin [3] as well as artificial liposomes into which proteolipid has been incorporated [17], show that it is an integral protein embedded in the lipid phase of the membrane.

A second protein fraction which was named after its discoverer, the Wolfgram protein, is differentiated by the fact that it is soluble in chloroform/methanol acidified with HCl. It has a high molecular mass and makes up 20% of myelin protein. In contrast to the Folch-Pi protein, Wolfgram protein is hydrolysed by trypsin and consists of at least three components.

The myelin proteins on which most attention has been focused are the basic proteins for it has been found that after their injection symptoms are developed similar to those of multiple sclerosis [18]. This was called experimental allergic encephalomyelitis (EAE) and will be dealt with later. The myelin basic protein, also called A1-protein and extracted with buffer of high ionic strength and weakly acidic pH, forms 30% of myelin protein and has a relative molecular mass of 18 000. Its primary structure is shown in Fig. 4.6 and it probably has little clearly defined secondary and tertiary structure. Its isoelectric point is at pH 10, indicating a high content of arginine and lysine. The latter amino acids are distributed uniformly over the whole sequence which allows a uniform interaction of the whole molecule with acidic groups of membrane lipids.

Viscosity measurements give an axial ratio of 10:1; this means that the protein backbone is highly extended and ensures that the antigenic properties of the protein survive heating to 95°C for 1 h. Optical rotatory dispersion (ORD) measurements give no evidence of a helical structure.

```
                 ( )                  (His-Gly)                    Thr
N-Ac-Ala-Ser-Ala-Gln-Lys-Arg-Pro-Ser-Gln-Arg-Ser-Lys-Tyr-Leu-Ala-Ser-Ala-Ser-Thr-Met-
      ( )        5                   (His-Gly)       15            Thr            20
                                                                          Ile
Asp-His-Ala-Arg-His-Gly-Phe-Leu-Pro-Arg-His-Arg-Asp-Thr-Gly-Ile-Leu-Asp-Ser-Leu-Gly-Arg-
                 25                 30              35                         Ile
              Gly                                      Ser
Phe-Phe-Gly-Ser-Asp-Arg-Gly-Ala-Pro-Lys-Arg-Gly-Ser-Gly-Lys-Asp-Gly-His-His-Ala-Ala-Arg-
            45            50              52              60
    Ala                         Ser ( )
Thr-Thr-His-Tyr-Gly-Ser-Leu-Pro-Gln-Lys-Ala-Gln-Gly-His-Arg-Pro-Gln-Asp-Glu-Asn-Pro-
65              70              Ser ( )        80                          85

Val-Val-His-Phe-Phe-Lys-Asn-Ile-Val-Thr-Pro-Arg-Thr-Pro-Pro-Ser-Gln-Gly-Lys-Gly
              90              95              100             105
                                                      Arg
Arg-Gly-Leu-Ser-Leu-Ser-Arg-Phe-Ser-Trp-Gly-Ala-Glu-Gly-Gln-Lys-Pro-Gly-Phe-Gly-Tyr-
    Thr-Val         115             120                 125
                                          Phe         Val
Gly-Gly-Arg-Ala-Ser-Asp-Tyr-Lys-Ser-Ala-His-Lys-Gly-Leu-Lys-Gly-His-Asp-Ala-Gln-Gly-Thr-
    130       Ala       135             140             Ala 145

Leu-Ser-Lys-Ile-Phe-Lys-Leu-Gly-Gly-Arg-Asp-Ser-Arg-Ser-Gly-Ser-Pro-Met-Ala-Arg-Arg-COOH
150       Leu    155             160             165                 170
```

Fig. 4.6. Primary structure of myelin A1 protein (myelin basic protein). Ox protein is shown with substitutions for human (above) and rabbit (below).

The A1-protein is not however completely lacking in tertiary structure [9]. Four proline residues (positions 96 and 99-101) fold the protein backbone into parallel chains; this is not stabilized as usual by a disulphide bridge (A1 contains no Cys) but is perhaps by specific chemical protein interactions. In position 107 there is an arginine residue which has become hydrophobic by the methylation of the side chain. The model (Fig. 4.7) shows that the kink created by the four proline residues places the methyl groups opposite two phenylalanine residues (Phe 89 and 90). Eylar [9] believes that at this point a hydrophobic interaction stabilizes the structure.

84   4 Myelin

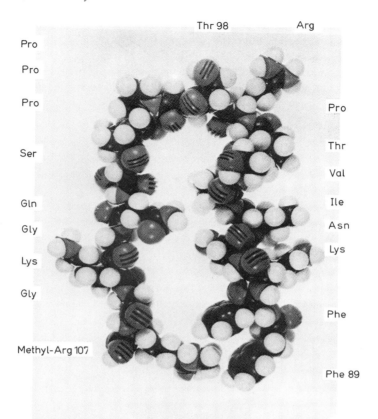

Fig. 4.7. Space filling atomic model of the region of the basic (A1) protein which carries encephalitogenic determinants. The sequence Phe 89 (right) to methyl-Arg 107 (left) is shown (see text).

Hydrophobic sequences, for example in positions 86-90, may interact with hydrophobic regions of the membrane (Fig. 4.8). One encephalitogenic determinant is a peptide around the amino acid tryptophan in position 116. By synthesizing various peptides it has been found that the sequence – Trp----Gln-Lys(Arg) is essential for the genesis of symptoms [19,20].

The protein composition of myelin from the peripheral nervous tissue differs from that of the central nervous system that has been considered so far. Proteolipid is absent, A1 is present, but the majority consists of another basic protein called P0. It is insoluble in water and organic solvents and has a relative molecular mass of 30 000. Another basic protein called P2 (relative molecular mass 12 000) appears to carry the encephalitogenic determinant (see below). The relative quantity of these proteins varies in the peripheral nerves of different organisms.

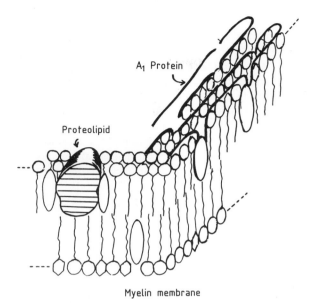

Fig. 4.8. Eylar's model of myelin membrane. The basic (A1) protein interacts with the polar lipid heads and stabilizes the membrane horizontally by its unfolded structure. The hydrophobic region of A1 anchors it additionally in the lipid phase [9]. Reproduced, with kind permission, of Raven Press, New York, 1980.

Experimental autoimmune disease: a model for multiple sclerosis?

Autoimmune diseases are disturbances of the mechanism by which the immune system distinguishes "self" from "non-self". Two classical examples are *myasthenia gravis* (Chapter 9) and *multiple sclerosis*. For neither is the triggering factor or the exact mechanism known. However there is now hope of progress through the discovery of models, i.e. the experimental production of essentially similar symptoms in animals. The primary symptom of multiple sclerosis is a progressive demyelination with resultant disturbance of nerve function. The same symptoms occur when brain tissue suspended in Freund's adjuvant is injected into an animal's CNS. Research in the field of myelin components has established that basic myelin protein A1 is responsible for this effect. Injections of less than 5 μg of A1 cause paralysis and death in guinea pigs. It evidently sensitizes the immune system and leads to the production of the condition known as *experimental allergic encephalomyelitis* (EAE) when it affects the central nervous system. This corresponds to *experimental allergic neuritis* (EAN) of the peripheral nerves. Here P2 protein is the triggering factor. A1 of peripheral myelin injected into the central nervous system gives rise to EAE, while an injection of whole peripheral nerve myelin has no effect on the brain. In this intact peripheral myelin the antigenic determinant is evidently masked. There is evidence that the cause of the disease is not antibody directed against the protein but sensitization of the T-system.

It is not clear why A1 (or P2) triggers the disease nor, moreover, why the symptoms are restricted to a relatively small region of the affected organism. It has been suggested [20] that it may be associated with the structural analogy between tryptophan in position 116 on A1 (see above) and the transmitter serotonin (5-hydroxytryptamine); EAE would thus be an immune response to the serotonin receptor preventing serotonin from acting as a transmitter in specific areas of the central nervous system. If this theory is substantiated, it is perhaps a starting-off point for chemotherapy.

Diseases caused by myelin defects

Numerous neuropathies cause a defect of the myelin sheath. It can also be affected by infections, genetic metabolic disorders or toxic and dietary factors. Although it has not yet been proved, a viral origin has been suggested for multiple sclerosis and its counterpart in the peripheral nervous system, the *Landry-Guillain-Barré syndrome*. In connection with viruses one should perhaps mention the series of other degenerative diseases of the central nervous system, the so-called "slow virus infections". These are the established causes of *Creutzfeldt-Jacob disease* and *Kuru*, the fatal disease of headhunters in New Guinea. In Kuru the Papuan mothers and children are infected during the preparation of a ritual meal of human brains [23]. Kuru viruses behave so unusually that it has been doubted whether they are true viruses. They are smaller than all other known viruses, resistant to proteases, nucleases, heat (80°C), formaldehyde, UV-irradiation (254 nm) and ultrasound. They are not antigenic. Kuru means "trembling" in the native language which describes the disease symptoms. The counterpart of this human brain disease in animals is "scrapie" (scratching) which occurs in sheep and has been recently used intensively as a model for this group of diseases [24].

We have already met genetically inherited diseases of myelin under the sphingolipidoses in Chapter 2. *Metachromatic leucodystrophy* and *Krabbe's leucodystrophy* are examples of this group in which there is an accumulation of certain glycolipids due to an inherited defect in their formation. The name leucodystrophy means the degeneration of the white matter. In addition to these demyelination diseases there are others in which damage to the myelin sheath is only one of many symptoms, perhaps not even the decisive one. *Phenylketonuria*, for example, in which damage to myelination occurs, is primarily the result of an accumulation of phenylpyruvic acid. This is caused by a defect in phenylalanine hydroxylase which blocks the normal metabolism of phenylalanine.

An interesting model for the investigation of genetically caused disorders of myelin metabolism are the "quaking" and "jimpy" mouse mutants referred to on page 80.

The toxic and dietary factors which can give rise to demyelination diseases include such varying things as diphtheria toxin, hexachlorophene or lead. Starvation damages orderly myelination particularly during certain critical phases in the development of the central nervous system. In man this is in the first year of life, as at birth myelination of the central nervous system is not complete.

Summary

Myelin is a multilamellar structure derived from membrane processes of oligodendroglial cells (CNS) or Schwann cells (PNS). Its function includes insulation of the axon, elevation of the speed of conduction of the nerve impulse (saltatory conduction) and conservation of ionic currents by reduction of membrane capacitance thereby saving energy, since fewer ions need to be pumped out of the axon after depolarization of the membrane. Myelin also saves space since myelinated fibres can be appreciably thinner than unmyelinated fibres for a given conduction velocity. Myelin appears relatively late both in phylogeny and ontogeny.

Myelin was thought to support Danielli's and Branton's membrane model; but it is now known that protein is also embedded in the lipid phase of the membrane (Pinto da Silva and Miller). Myelin membrane is asymmetrical (PC and cholesterol mainly on the outside). It consists of ca 75% lipid (galactosylcerebroside is characteristic) of which 28% is cholesterol. Myelin lipids have a comparatively slow turnover.

Myelin proteins: among others three enzymes of unknown function are recognized (2',3'-cAMP phosphodiesterase, cholesterolesterase and a protein kinase). CNS myelin contains three proteins distinguished by their solubility (Tab. 4.4):

Table 4.4. Three proteins contained in CNS myelin distinguished by their solubility.

	Proteolipid	Basic (A1) protein	Wolfgram protein
soluble in	org. solvents	weak acids	org. solvents/HCl
molecular mass	25 000 heterogeneous	18 000	high, heterogeneous
% of total protein mass	30-50	30	20
structure	50% nonpolar amino acids 2% fatty acids, covalently linked	no secondary or tertiary structure sequence know, encephalitogenic (EAE)	

PNS contains A1, besides a further basic protein (P2). The main constituent is a protein (P0) insoluble in water and in organic solvents (M_r 30 000). A1 evokes EAE (model for multiple sclerosis); P2 evokes EAN.

References

Cited:

[1] Geren, B.B., "The formation from the Schwann cell surface of myelin in the peripheral nerves of chick embryos", *Exp. Cell Res.* **7**, 558-562 (1954).
[2] Branton, D., "Fracture faces of frozen myelin", *Exp. Cell Res.* **45**, 703-707 (1967).
[3] Pinto da Silva, P., and Miller, R.G., "Membrane particles on fracture faces of frozen myelin", *Proc. Natl. Acad. Sci. USA* **72**, 4046-4050 (1975).
[4] Napolitano, L., LeBaron, F., and Scaletti, J., "Preservation of myelin lamellar structure in the absence of lipid. A correlated chemical and morphological study", *J. Cell Biol.* **34**, 817-826 (1967).
[5] Caspar, D.L., and Kirschner, D.A., "Myelin membrane structure at 1 nm resolution", *Nature New York Biol.* **231**, 46-52 (1971).
[6] Brammer, M.J., and Sheltawy, A., "The role of lipids in the observed lack of phosphatidylcholine exchange in myelin", *J. Neurochem.* **27**, 937-942 (1976).
[7] Boggs, J.M., Wood, D.D., Moscarello, M.A., and Papahadjopoulos, D., "Lipid phase separation induced by hydrophobic protein in phosphatidylserine-phosphatidylcholine vesicles", *Biochemistry* **16**, 2325-2329 (1977).
[8] Boggs, J.M., Moscarello, M.A., and Papahadjopoulos, D., "Phase separation of acidic and neutral phospholipids induced by human myelin basic protein", *Biochemistry* **16**, 5420-5426 (1977).
[9] Eylar, E.H.: "Myelin-specific proteins". In: *Proteins of the Nervous System*. Schneider, D.J. (ed.), 3rd edition, Raven Press, New York 1976.
[10] Norton, W.T.: "Formation, structure, and biochemistry of myelin". In: *Basic Neurochemistry*. Siegel, G.J., Albers, R.W., Agranoff, B.W., and Katzman, R. (eds.), 3rd edition, p. 63-92, Little Brown and Co, Boston 1981.
[11] Hogan, E.L., and Joseph, K.C., "Composition of cerebral lipids in murine leucodystrophy: the quaking mutant", *J. Neurochem.* **17**, 1209-1214 (1970).
[12] McNamara, J.O., and Appel, S.H., "Myelin basic protein phosphatase activity in rat brain", *J. Neurochem.* **29**, 27-35 (1977).
[12a] Choi, M.U., and Suzuki, K., "A cholesterol-esterifying enzyme in rat central nervous system myelin", *J. Neurochem.* **31**, 870-885 (1978).
[13] Yamaguchi, S., and Suzuki, K., "A novel magnesium-independent neutral sphingomyelinase associated with rat central nervous system myelin", *J. Biol. Chem.* **253**, 4090-4092 (1978).
[14] Clapshaw, P.A., Müller, H.W., and Seifert, W., "Characterization of 2',3'-CNPase: Rapid isolation, native enzyme analysis, identification of a serum soluble activity, and kinetics", *J. Neurochem.* **36**, 1996-2003 (1981).
[15] Kurihara, T., Nussbaum, J.L., and Mandel, P., "2',3'-cyclic nucleotide 3-phosphohydrolase in brain of mutant mice with deficient myelination", *J. Neurochem.* **17**, 993-997 (1970).
[16] Folch-Pi, J.: "Proteolipids". In: *Proteins of the Nervous System*. Schneider, D.J. (ed.), p. 45-66, Raven Press, New York 1973.
[17] Boggs, J.M., and Moscarello, M.A., "Structural organization of the human myelin membrane", *Biochim. Biophys. Acta* **515**, 1-12 (1978).
[18] Kies, M.: "Immunology of Myelin Basic Proteins". In: *The Nervous System*. D.B. Tower (ed.), Vol. 1, p. 637-646, Raven Press, New York 1975.
[19] Westall, F.C., Robinson, A.B., Cacam, J., Jackson, J., and Eylar, E.H., "Essential chemical requirements for induction of allergic encephalomyelitis", *Nature* **229**, 22-24 (1971).
[20] McKhann, G.M., "Multiple sclerosis", *Ann. Rev. Neurosci.* **5**, 219-240 (1982).
[21] Brostoff, S.W., Karkhanis, Y.D., Carlo, D.J., Reuter, M., and Eylar, E.H., "Isolation and partial characterization of the major proteins of rabbit sciatic nerve myelin", *Brain Res.* **86**, 449-458 (1975).

[22] Peters, A., Palay, S.L., and Webster H.F.: *The Fine Structure of the Nervous System*. Harper and Row, New York 1970.
[23] Gadjusek, D.C., "Unconventional viruses and the origin and disappearance of Kuru", *Science* **197**, 943-960 (1977).
[24] Prusiner, S.B., "Novel proteinaceous infectious particles cause Scrapie", *Science* **216**, 136-144 (1982).
[25] Carnegie, P.R., Dunkley, P.R., Kemp, B.E., and Murray, A.W., "*In vitro* and *in vivo* phosphorylation of myelin basic protein by cerebral protein kinase", *Nature* **249**, 147-149 (1974).
[26] Eto, Y., and Suzuki, K., "Cholesterol ester metabolism in rat brain: A cholesterol ester hydrolase specifically localized in the myelin sheath", *J. Biol. Chem.* **248**, 1986-1991 (1973).
[27] Sapirstein, V.S., and Lees, M.B., "Purification of myelin carbonic anhydrase", *J. Neurochem.* **31**, 505-511 (1978).

Further reading

Norton, W.T.: "Formation, structure, and biochemistry of myelin". In: *Basic Neurochemistry*. Siegel, G.J., Albers, R. W., Agranoff, B.W., and Katzman, R. (eds.), 3rd edition, p. 63–92, Little Brown and Co, Boston 1981.
Lees, M.B., and Sapirstein, V.S.: "Myelin-associated enzymes". In: *Handbook of Neurochemistry*. Vol. 4, Lajtha, A. (ed.), 2nd edition, Plenum Press, New York 1983.
Chang, D.C., Tasaki, J., Adelman, W.J., and Leuchtag, H.R. (eds.): *Structure and Function in Excitable Cells*. Plenum Press, New York and London 1983.
Morell, P. (ed.): *Myelin*. 2nd edition, Plenum Press, New York and London 1984.
See also ref. 20 and 24.

Chapter 5

Electrophysiology

Nerve cells transmit signals in the form of electrical impulses over considerable distances at speeds of several hundreds of metres per second. An account of the neurochemical basis of nerve function must therefore include some understanding of the electrical properties of the neuron. This non-biochemical part of the subject, rather imprecisely referred to as electrophysiology, will be dealt with only in its essentials in this chapter. For a systematic and comprehensive account of electrophysiology specialist reviews may be consulted [1-2].

The nerve cell oscillates between two states: rest and excitation. First the electrical properties in the resting state will be described and then the changes in these properties that occur with excitation.

The resting potential

It has been known since the beginning of the century that an electrical potential exists between the outside and the inside of the cell membrane (Fig. 5.1). Julius Bernstein described it for the first time as a "membrane potential" arising from the unequal distribution of ions between the interior and exterior of the cell. The quantitative description of this phenomenon however had to await two methodological developments: in 1936 J.Z. Young's discovery of the squid giant axon seemed like a gift from heaven for electrophysiology and in 1946 Graham and Gerand developed the microelectrode, a glass capillary (of less than 1 µm diameter) filled with concentrated solution of electrolyte which can be inserted into a cell without damaging it (Fig. 5.1A). The enormous advantage of the squid giant axon is its size. It has a diameter of up to 0.5 mm so, as Hodgkin and Huxley showed in 1939, a fine electrode can be inserted lengthways into the axoplasm (Fig. 5.1B). These techniques have made it possible to record the resting membrane potential of the cell and the changes occurring in it on excitation. The giant axon can be emptied of axoplasm by squeezing and if perfused with an artificial salt solution of defined composition the changes during excitation can be studied independently of metabolic processes taking place in the cytoplasm. We will now describe briefly some of the more important results of these experiments.

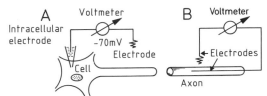

Fig. 5.1. Recording of the membrane potential of a nerve cell. (A) by inserting an intracellular (micro-)electrode. (B) by inserting an electrode lengthways into the interior of the axon. The latter is only possible in axons with a very large diameter e.g. squid giant axon.

A microelectrode, when inserted in a cell, records a membrane potential of ca. -70 mV relative to the external medium (Fig. 5.1). This so called "resting potential" is not a property unique to nerve cells; it is observed in cells as varied as the unicellular organism *Paramecium* and erythrocytes. In muscle cells it reaches -90 mV.

What is the origin of the resting potential? The cell membrane is an effective permeability barrier for ions, but its permeability varies for different ions (Fig. 5.2). Thus although almost impermeable to Na^+ it is permeable to potassium ions, which therefore largely determine the resting potential (70 mV, interior negative). Their concentration in the interior of the cell, for example in the squid giant axon, is twenty times greater than outside (Tab. 5.1). Their positive charges are compensated for by negative counter-ions,

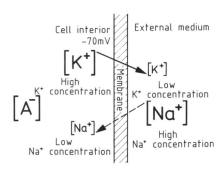

Fig. 5.2. Ion distribution across the nerve membrane. The direction of the slope indicates the respective ion gradients, the steepness of the slope, the permeability of the membrane for the ion concerned. [$\bar{A}$]: anions.

Table 5.1. Ions present in the axoplasm and blood of the squid. The extracellular (blood) ion concentrations are practically the same as those of sea water, which is usually used as the external medium for experiments with the squid giant axon.

| Ion | Concentration (mmol/l) | | Ratio |
	Intracellular	Extracellular	Int. : Ext.
K^+	400	20	20 : 1
Na^+	50	440	1 : 9
Cl^-	50	550	1 : 11
Organic anions	350	–	–
Ca^{2+}	0.4	10	1 : 25

polyanions like proteins, phosphate and isethionate (HO-CH$_2$-CH$_2$-SO$_3^-$) to all of which the membrane is impermeable (Fig. 5.2). Due to the concentration gradient more potassium ions diffuse out of the cell than in, leaving a surplus of negative charge due to the impermeable anions. Naturally this cannot continue until the concentration gradient is abolished since the surplus of negative charges remaining inside holds back the outwardly diffusing positively charged potassium ions. Soon a state is reached in which the same number of potassium ions are travelling in each direction across the membrane. This state is characterized by a balance between two forces: the concentration gradient that strives for equality and as it were drives potassium ions out, and the electrostatic attraction, which draws potassium ions in. For each concentration gradient, i.e. for each concentration ratio [K$^+$ outside] / [K$^+$ inside] there is an electrostatic equilibrium potential that is expressed by the Nernst equation:

$$E_{K^+} = \frac{RT}{nF} \ln \frac{[K^+_o]}{[K^+_i]}$$

or if the numerical value factor $\frac{RT}{nF}$ (at 20°C) is inserted and logarithms to base 10 are substituted for exponential:

$$E_{K^+} = 58 \log \frac{[K^+_o]}{[K^+_i]}$$

where E_{K^+} is the potential in mV between the inside and the outside of the membrane needed to balance the inflow and outflow of potassium ions, so that there is no net ion flow through the membrane. If we substitute the values for the squid giant axon given in Table 5.1 [K^+_o] / [K^+_i] = 1/20 then E_{K^+} = -75 mV.

However it has only been recorded as -70 mV. How does this discrepancy arise? Up till now we have assumed that the membrane only allows K$^+$ through and that it is quite impermeable to Na$^+$. But in fact it is also slightly permeable to Na$^+$ and as the Na$^+$ concentration gradient is opposite to that of K$^+$ the negative potential difference is decreased by the influx of sodium ions (this is called the sodium resting flux to distinguish it from the sodium influx on excitation of the membrane) (Fig. 5.2). The evidence for this is relatively simple: if Na$^+$ is substituted for by another larger positive ion e.g. cholinium [(CH$_3$)$_3$N-CH$_2$-CH$_2$OH]$^+$ there is a membrane potential of -75 mV – the calculated potassium equilibrium potential – as the cholinium ions are too large to permeate.

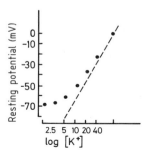

Fig. 5.3. Relationship between the resting potential of a nerve cell and the K$^+$ concentration (mmol/l) outside the cell. The deviation of the points from the straight line at low concentrations indicates that the resting potential is not behaving like a pure potassium potential but that sodium ions are also contributing to the Nernst potential.

Fig. 5.3 shows that the membrane potential of the squid giant axon is almost completely a K^+ equilibrium potential: if the potassium concentration outside the cell is varied, then the resting potential changes according to the Nernst equation. A ten-fold change of $[K^+{}_o]$ produces a change in the value of E_{K^+} of 58 mV (since log 10 = 1) and only at low potassium concentrations is a deviation from the slope 58 observed as then the sodium resting flux relative to the potassium resting flux cannot be ignored.

In addition to K^+, chloride ions Cl^- can also diffuse through the membrane. Their concentration is determined by the membrane potential i.e. $[Cl^-{}_o] / [Cl^-{}_i]$ is the reciprocal of the potassium concentration ratio. A comprehensive description of the membrane potential must take into account the chloride ions as well as the sodium and potassium.

Since the permeabilities of these ions are so different from each other their concentration in the inside and outside of the cell cannot simply be substituted in the Nernst equation but the concentration of each ion must be multiplied by the respective ion permeability coefficients P_{Na}, P_{Cl}, P_K. The membrane potential E_m is thus:

$$E_m = \frac{RT}{F} \ln \frac{P_K[K^+{}_o] + P_{Na}[Na^+{}_o] + P_{Cl}[Cl^-{}_i]}{P_K[K^+{}_i] + P_{Na}[Na^+{}_i] + P_{Cl}[Cl^-{}_o]}$$

This so-called Goldman equation reduces to the Nernst equation when only one ion is considered. The potential across the neuron membrane is practically equivalent to the potassium potential as the sodium permeability in the resting state is small, and the chloride ions, though they are distributed across the membrane according to the potassium potential, have a high permeability and an opposite charge to potassium and thus do not contribute to the net membrane potential.

The Goldman equation expresses the conditions of the cell membrane in terms of an electrical circuit, consisting of several batteries (one each of potassium, sodium and chloride ions). The current in this system is conducted by ions which must overcome a resistance to flow through the membrane. The membrane can be conceptually replaced by an equivalent electrical circuit with the ionic distribution represented as batteries and the ionic permeability as ohmic resistances; to this is added, in the electrical diagram in Fig. 5.4, a capacitance C representing the membrane consisting of a double layer of lipids with embedded protein and polar heads which act like a condenser.

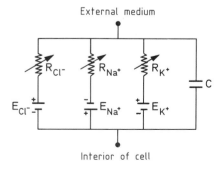

Fig. 5.4. Equivalent circuit. An electrical model of the properties of the nerve membrane: the membrane potential is represented by batteries in parallel (E), the ion conductivities as resistances (R), the capacity of the membrane as a condenser (C). The subscripts Cl^-, Na^+ and K^+ represent the contribution of the respective ions.

The net resting current would lead in the long run to an equal concentration between the inside and the outside of the cell. When the membrane is excited the ion flux speeds up. In order to inhibit it and so maintain a long term constant resting potential, the passive diffusion of cations must be balanced by an active transport mechanism (ion pump). We shall return to this in Chapter 6.

Excitation of the neuron: local potential and action potential

Having discussed the electrical properties of the resting cell we shall now turn to the processes concerned with excitation of the membrane. The state of excitation can be defined as a temporary deviation of the membrane potential from the resting potential caused by an external stimulus. The stimulus, either electrical or chemical, excites the membrane by changing its ionic conductivity i.e. the resistance in our circuit (Fig. 5.4) decreases. The excitation spreads from the site of the stimulus i.e. there is a change in conductivity, and therefore potential, of adjoining areas of the membrane. This propagation of the excitation is called the *impulse*. Two types of impulse can be distinguished: the *action potential* in which the signal is propagated unchanged from the site of excitation to the nerve terminal, and the *local potential* which rapidly diminishes with increasing distance from the site of excitation. Examples of the latter are found at synapses; they are the *excitatory postsynaptic potential* (e.p.s.p.) and the *inhibitory postsynaptic potential* (i.p.s.p.) and at the sensory nerve endings, the *receptor* (or *generator*) *potential*. Local potentials can be summated i.e. they can be augmented by succeeding excitations, whereas action potentials cannot be summated and have an "all-or-nothing" mechanism.

Local potentials, also called *electrotonic potentials* can be made to assume a desired value; if positive charges are injected into the squid giant axon, for example, they give rise to a decrease in the negative charge or a depolarization. This immediately spreads from the site of injection, the distance increasing as the internal resistance of the axon decreases. The resistance of a conductor is in inverse proportion to its diameter; thus very thin axons conduct the local potential less well than those with large diameters. But the conduction of the local potential is not only dependent on the conductivity of the axoplasm and that of the extracellular medium; it also depends on the resistance of the axon membrane. As this is not a perfect insulator some of the injected positive charges (e.g. potassium ions) leak out again and counteract the depolarization of the interior of the cell. With increasing distance from the site of injection the impulse weakens. Losses occur along the axon in the same way as in an electrical cable. We can also symbolize this process in electrical terms, and can imagine the membrane divided into many individual "equivalent circuits", each consisting of a battery, resistance and condenser (Fig. 5.5). If a current flows through one of the circuits (or membrane segments) from outside to inside, it subsequently enters the neighbouring circuit (in the adjoining membrane segment).

During its passage its intensity is, of course, reduced by the resistance of the axoplasm and a further part of the current is used up to charge the condensers (to load the membrane

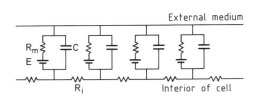

Fig. 5.5. The nerve membrane as circuits connected in parallel. R_m, membrane resistance; R_i, internal resistance.

with ions). As it travels through subsequent circuits it is further reduced in power, and rapidly approaches zero.

To continue briefly with the comparison of an axon with an electric cable: the specific resistance of the axoplasm of the squid giant axon R_i is 30 Ω and that of extracellular space R_o is 20 Ω. A copper wire of the same thickness conducts the current ca. 10^6 times better. The loss is reduced by insulation (the membrane resistance R_m is 1000 Ω; with a membrane thickness of 5 nm it is about 10^9 $\Omega \cdot cm^{-1}$). The quality of a cable can be defined by the exponential decline of a potential along the conductor:

$$V = V_o e^{-x/l}$$

l, the so-called *length constant* is a function of the internal resistance, the membrane resistance and the diameter. It denotes the distance x at which the initial potential V_o has fallen to one e-th of its value; for the cable of the squid giant axon it is only a few millimeters. Thus to conduct signals as graduated electrotonic potentials would be very inefficient. The basic difference between an electrical cable and an axon is the ability of the latter to generate action potentials.

How is an action potential released? Returning to the squid giant axon with its resting potential of -70 mV (interior negative), a displacement of this potential towards a less negative value, is called *depolarization*, a shift to a more negative value *hyperpolarization*. Both can be brought about by applying an appropriately directed current pulse through an electrode inserted in the axon. If the membrane is briefly depolarized to a small extent (e.g. 10 mV) there is little effect; after switching off the depolarizing current pulse the membrane potential rapidly returns to its resting value. The return is not instantaneous; the membrane behaves like a condenser, charging and then discharging, a process which takes time.

If depolarization is more intense, for example, up to -40 mV, a threshold is reached, from which the membrane potential does not simply return to the resting potential; instead the depolarization spontaneously intensifies (Fig. 5.6) and finally the sign of the membrane potential is reversed, i.e. it is now inside positive. When it has reached a value of about +40 mV the process again spontaneously reverses and a repolarization occurs. This overshoots the resting potential and then slowly returns to it.

This account of the action potential is purely phenomenological and needs supplementing with a description of the underlying molecular processes. We will deal in detail with

Fig. 5.6. Action potential, released when the local potential summates to the threshold value (horizontal interrupted line).

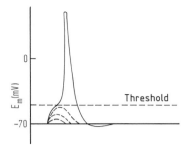

these in Chapter 6 and will only mention the following points here. In the early fifties, the English physiologists Hodgkin and Huxley analysed the action potential and laid the ground for our present day understanding of the phenomenon. They showed that the initial rise of potential (depolarization) was due to the influx of sodium ions (Fig. 5.7). When the threshold is reached, ion channels in the membrane open and allow the passage specifically of sodium ions. The subsequent repolarization was attributed by Hodgkin and Huxley to a flux of potassium ions in the opposite direction, i.e. from inside to outside, due to the opening of specific potassium channels; simultaneously the sodium channels close (inactivation). As can be seen in Fig. 5.7 this repolarization initially exceeds its endpoint, the resting potential, because the equilibrium potential for K^+ has a greater negative charge than the resting potential. The difference, the *afterpotential*, decays slowly as a result of the closing of the potassium channel and the reestablishment of the sodium resting potential. The inactivation of the Na^+ channel implies that the next action potential cannot be immediately released. The period of rest between two action potentials is called the *refractory period*.

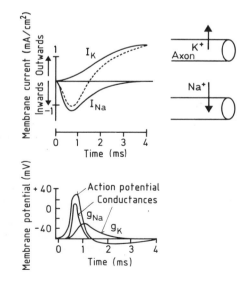

Fig. 5.7. Analysis of an action potential. Above: Ion current. I_{Na} directed inwards, I_K outwards (after a short delay). The interrupted line represents the sum of both ion currents. Below: Change of membrane potential due to a temporary rise of sodium conductance (g_{Na}) followed by one of potassium conductance (g_K).

The duration and direction of the ion flux through the neuronal membrane and its differentiation between Na$^+$ and K$^+$ can be recorded by means of the voltage clamp, a basic piece of electrophysiological equipment (Fig. 5.8). It consists of an electronic device for the maintenance of a predetermined voltage across the membrane. The current from a

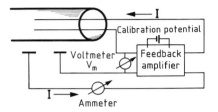

Fig. 5.8. The voltage clamp, a technique for recording the ion currents through a cell membrane at a constant membrane potential; the feedback compensates for potential changes due to in- and out-flowing ions.

microelectrode implanted in a cell can be passed through a circuit regulated by a voltmeter recording differences in membrane potential. Each change in potential is compensated by a corresponding current. The current flow in this compensated circuit can be recorded simultaneously by an oscilloscope and accurately reflects the events in the neuronal membrane during an action potential. If only the potassium component of the action potential is required, the sodium channels of the membrane are blocked with specific toxins (see Chapter 6) like tetrodotoxin (TTX) or saxitoxin (STX), if only the sodium component, the potassium channels are blocked by tetraethylammonium (TEA) or various aminopyridines.

How then is the action potential, i.e. the nerve impulse, propagated from its site of origin? The answer to this is concerned with the voltage dependence of the ion channels: depolarization during the action potential causes the opening of ion channels and thus an inflow of Na$^+$ in neighbouring membrane areas (Fig. 5.9). These cause a further depolarization and an opening of ion channels in the next membrane area and so on until the axon terminal is reached. This propagation proceeds in both directions from the site of origin, orthodromically away from the cell body and antidromically towards it.

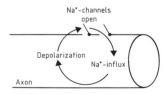

Fig. 5.9. Propagation of an action potential. Sodium ions flowing in through an open sodium channel produce a depolarization of that section of membrane. This causes an opening of the adjoining sodium channel allowing more ions to flow in etc.

The opening of the ion channel is evidently a cooperative process; Hodgkin and Huxley described the observed increases of ion conductivity by exponential equations (see Chapter 6). In the twenty-five years since their pioneering work the kinetics of the sodium and potassium current have been described with a degree of detail for which there is as yet no biochemical counterpart. Recently methods have become available which promise to characterize the relevant molecules of the nerve membrane. The components of the ion channels must have unusual properties which enable them to register potential changes and to react to these in a highly coordinated and cooperative fashion. Up till now the details of these properties can only be formulated rather imprecisely.

Synaptic impulse transmission

Nerve impulses are not only propagated inside a nerve cell, but must be transmitted to neighbouring cells. The site of this transmission is the synapse. We will deal with the biochemistry of synaptic transmission in detail in Chapters 8 and 9, here only the electrical events will be briefly described. Electrical synapses are distinguished from chemical synapses by the nature of the signals transmitted from cell to cell. In the electrical synapse the action potential is propagated directly across a very closely apposed contact site between neighbouring membranes. In the chemical synapse the action potential causes the temporary depolarization of the presynaptic cell membrane, the release of specific transmitter substances and their diffusion across an intercellular space of up to 20 nm wide called the *synaptic cleft* to the postsynaptic membrane. There they are bound by specific receptors. Molecular mechanisms (discussed in Chapter 9) then cause the release of the postsynaptic response, which, if large enough, will trigger off another action potential. The chemical synapse is commoner and more significant, so we shall continue with this.

In 1921 chemical synaptic transmission was first clearly demonstrated by Otto Loewi in a classical experiment: he collected the perfusate of a frog heart before and during the stimulation of the vagus nerve. He transferred the perfusate to a second heart and observed an inhibition of heart rate using the perfusate collected *during* stimulation, and no effect on using the perfusate from *before* stimulation. The stimulated vagus therefore must have produced a heart rate inhibitory substance. Loewi called it *Vagusstoff* ("vagus substance"); it was later identified as acetylcholine. This was shown by Dale in 1936 to be the transmitter in part of the peripheral nervous system and at the neuromuscular junction.

Acetylcholine releases a depolarization in the postsynaptic membrane, the **e**xcitatory **p**ostsynaptic **p**otential (e.p.s.p.). In the special case of the neuromuscular synapse the e.p.s.p. is also called **e**nd**p**late**p**otential (e.p.p.). It behaves like a local potential, which, as we already know, quickly decreases with distance from the site of origin and can be summated; i.e. if the acetylcholine concentration increases, the depolarization becomes stronger. If the concentration of transmitter is plotted against the depolarization, a dose response curve is produced which is S-shaped. As with the corresponding enzyme activity-substrate concentration curve this S-shaped curve indicates that the interaction is cooperative. When the depolarization reaches a predetermined value an action potential can also be released. This acetylcholine-stimulated depolarization indicates the opening of ion channels. It is important to realize that these channels are opened by a chemical transmitter molecule; they are fundamentally different from the action potential channels which are opened electrically. As we shall see later the two channel types are also distinguishable pharmacologically i.e. they react specifically with certain neurotoxins or drugs. Finally, they are distinguished by their ion specificity: the action potential is transmitted by two different voltage-dependent ion channels, separate sodium and potassium channels. The acetylcholine activated channel by contrast is almost equally permeable to sodium and potassium ions. It seems to be of considerable size for in addition it is permeable to organic ions like methylammonium or even guanidinium. It is impermeable to anions (in particular Cl^-). The relationship between the postsynaptic transmitter receptor, in this case the acetylcholine receptor, and the postulated ion channels will be discussed in Chapter 9.

We have mentioned already a second type of local potential, the **i**nhibitory **p**ostsynaptic **p**otential (i.p.s.p.). While the e.p.s.p. represents a depolarization of the postsynaptic membrane, the i.p.s.p. involves a hyperpolarization i.e. a change of membrane potential towards a more negative value. This can be achieved in two ways, either by closing the cation or opening the anion channels. It is called inhibitory because it has the effect of suppressing the excitability of the membrane, and thus inhibiting the depolarizing response of the sodium and potassium channels to the action potential.

One of the most fully investigated inhibitory transmitters is γ-aminobutyric acid, abbreviated GABA, which is utilized in the central nervous system and also in the neuromuscular synapses of invertebrates, e.g. crustaceans, where it activates channels for Cl^- ions. Acetylcholine can also be an inhibitory transmitter, which produces the curious situation that in one and the same organism the same transmitter may be both excitatory and inhibitory. In the marine mollusc *Aplysia* (sometimes called a "sea snail" but correctly a "sea slug" or "sea hare") for example, three actions of acetylcholine have been observed: a fast excitatory (e.p.s.p.) and a faster and a slower inhibitory one. Here acetylcholine receptors regulate not only the usual Na^+/K^+ channel but also a Cl^- and a specific K^+ channel.

The postsynaptic inhibition (i.p.s.p.) described here should not be confused with *presynaptic inhibition*. This is concerned with the inhibition of the transmitter *release* by hyperpolarization of the nerve ending. This hyperpolarization is exerted by a nerve axon which forms a synapse with the nerve ending of an excitatory fibre. The hyperpolarizing transmitter of this inhibitory, presynaptic synapse may again be GABA, but another candidate is an enkephalin, which as a probable "endogenous opiate" inhibits presynaptically pain transmission in the spinal cord. We will meet it again in Chapter 9.

In 1950 Fatt and Katz [3] observed a further postsynaptic process: although there was no presynaptic excitation, small depolarizations of the postsynaptic membrane occurred. They called them "**m**iniature **e**nd**p**late **p**otentials" (m.e.p.p.) (Fig. 5.10) as they only amounted to ca. 1 mV.

Their explanation of the m.e.p.p.s is still not quite proved, but numerous experiments have made it probable: a m.e.p.p. is evoked by the spontaneous release of transmitter molecules. As the m.e.p.p.s are always of about the same size or multiples of a unit size, they would appear to be derived from a standard amount of transmitter. Evidently transmitter does not diffuse arbitrarily from the presynaptic membrane but is released in packets, called *quanta*. In this theory the endplate potentials are composed of some hundreds of these m.e.p.p.s. One quantum contains about 10 000 transmitter molecules. It is thus supposed that the unit event of synaptic transmission is the release of a quantum of transmitter.

Fig. 5.10. Miniature endplate potentials (m.e.p.p.). In the region of the endplate these are small random depolarizactions, which are caused by spontaneous transmitter release [3]. They are not detectable a few mm from the endplate nor when there is presynaptic blockade by specific toxins (botulinum toxin).

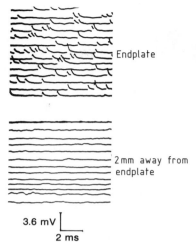

In the presynaptic nerve ending small vesicles had already been discovered, the synaptic vesicles. Today it is known that these contain transmitter, from which it has been postulated that the unit event is the exocytotic release of transmitter from one vesicle. In Chapter 9 we will discuss the mechanism and alternative hypotheses for this process.

Single channels and noise analysis: electrophysiology at the molecular level

Electrophysiological methods have long been used to investigate macroscopic processes such as the depolarization and repolarization of the whole membrane. These are the summation of a very large number of unit events, the movement of numerous ions through many ion channels. Two modern developments have made it possible to explore in detail the unit event, the single ion channel. Electrophysiological techniques have been refined so that single channels in excitable membranes can be measured (Fig. 5.11) [4]. The microelectrode, finer in dimension than a hair, can, for example, record from the acetylcholine regulated sodium/potassium channel of the neuromuscular endplate (Chapter 9) its opening and closing, its maximum conductivity, sensitivity to drugs and membrane potentials. Descriptive physiology and the molecular mechanistic biophysical and biochemical approach have thus become more closely related, no longer worlds apart, but only a few nms.

A breakthrough in this field was the development of the *patch clamp* technique [4]. In brief this method, developed by Neher and Sakmann consists of placing a microelectrode on the excitable membrane to be investigated, with an especially tight contact between electrode tip and membrane. This prevents currents leaking from the ectoplasm to the inside of the electrode which would disturb the measurement of currents through the membrane. The resistance of this contact has to be of the order of 10^{12} Ω (called a

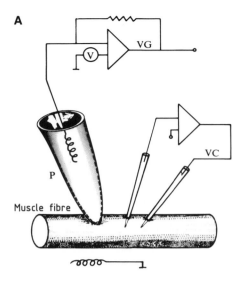

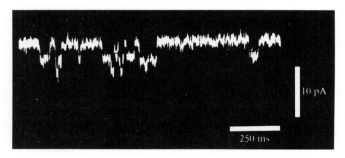

Fig. 5.11. Recording of single channels in the neuromuscular endplate (patch clamp method). (A) Above: VC, voltage clamp; P, pipette, 3.5 µm diameter, containing Ringer solution; VG, "virtual ground" electric circuit, which measures the current through the area of the membrane under the tip of the pipette. Below: Oscilloscope recording of the current through the membrane of the muscle fibre. Separate current fluctuations can be seen which are interpreted as the opening and closing of single channels. Reproduced, with kind permission, from Neher and MacMillan Journals Ltd. [4].

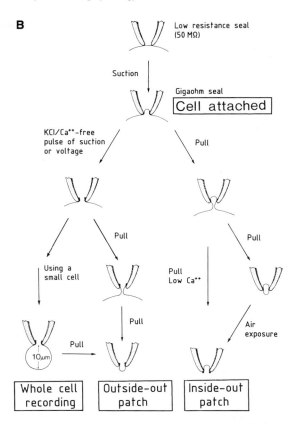

Fig. 5.11. (B) The high resistance (gigaseal) at the contact between the microelectrode and the membrane surface is achieved with an especially polished electrode tip placed on the cell surface and by suction of the membrane into the electrode. Depending on the further treatment different geometries – outside out or inside out patches – can be obtained. (The figure was kindly supplied by B. Sakmann, Göttingen).

"gigaseal"). It is achieved by special polishing and treatment of the electrode tip and of the membrane surface. Measurements are thus restricted to a small patch of membrane of 5–10 µm^2 in size which still contains many individual channels. By choosing the appropriate experimental conditions (transmitter concentration, voltage etc.) single channels can be observed by measuring the minute currents (in the pA range) flowing through them [5].

In addition the analysis of the so-called noise of excitable membranes (Fig. 5.12) has fundamentally broadened the range of methods [6-8]. This method will be briefly described here without deriving the mathematical formulae. The nerve impulse consists of ion currents through the membrane. They are conveyed through ion channels, structures which regulate the permeability of the membrane, for example for Na^+ or K^+. As we will describe in detail in Chapter 6, these channels consist of two functional components, the *selectivity filter* which determines which kind of ion is diffusible through the membrane, and the *gating mechanism* which, in response to an external signal regulates when, how fast and how long a channel opens. The opening of the channel is a statistical process, i.e. the signal (depolarization in the axonal, transmitter binding in the postsynaptic channel, see Chapter 6) increases the probability that the channel is found in the open state. This is the starting point of noise analysis: the number of open channels fluctuates statistically around a mean

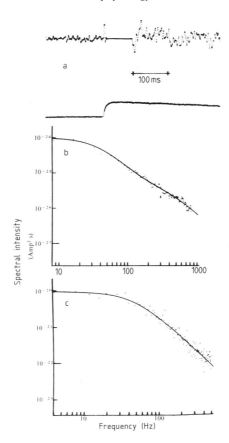

Fig. 5.12. Noise of axonal (a,b) and postsynaptic (c) membranes. (a) Current fluctuations in a node membrane (at constant voltage). (b) Frequency spectrum of the noise from the acetylcholine-activated channels of a neuromuscular endplate. Reproduced, with kind permission, from Stevens and MacMillan Journals Ltd. [6].

probability. For example, if the probability is 0.5, then of one hundred channels, on average 50 are open, but in fact there may be at one point 48, and a little later 52. This fluctuation is called noise and its analysis gives an insight into the kinetics and mode of action of the gating mechanism, since the speed with which the oscillation occurs around the mean, the *frequency*, depends on the kinetics of the unit event.

The first step in noise analysis is therefore the recording of the frequency spectrum. By mathematical analysis (Fourier analysis) of the "noise" of, for example, the potential across a membrane the contribution of the different frequencies to the whole process is determined and is plotted as the spectral density against the frequency. The second step is then the comparison of experimentally with theoretically derived frequency spectra, allowing the specific properties of the gating mechanism to be calculated. If both spectra agree, then this is good evidence for the hypothesis of the mechanism being formulated.

The method will be illustrated by one example. One ion channel can be regulated by different parameters: its conductivity can be limited either by the speed with which the signal (e.g. the transmitter molecule in the case of the postsynaptic channel) diffuses away or is degraded; or alternatively by the reaction of the channel molecules to the signal. It is accepted today that conformational changes of membrane proteins are responsible for the changes in permeability of the nerve membrane. With the help of noise analysis it can now be shown that in the case of the postsynaptic acetylcholine receptor, the closing of the channel rather than the removal and hydrolysis of the transmitter acetylcholine, is the rate-determining step for the duration of the current flow through the endplate.

We will not deal here with the theory and practice of noise analysis, it will be sufficient to give a qualitative description and to summarize the properties of ion channels which can be determined with its help. These are
1. the conductivity of the single channel
2. the mean opening time of the channel or the speed with which it closes
3. the number of states of the channel i.e. if it exists in an either open or closed state "all-or-nothing response", or if there are other states for different conductivities
4. the number of channels per unit surface area of membrane.

The value of this method for analysing ion channels is increased by the fact that the effect on these properties of various other parameters such as temperature, pH, membrane potential and nature of the transmitter can be measured.

Summary

As a summary, this is a list of the most important basic terms of electrophysiology:

Membrane potential	Potential between the inside and outside of the cell membrane due to the unequal distribution of ions.	*Refractory period*	Period after an action potential during which the membrane remains inexcitable.
Resting potential	Membrane potential before (or a long time after) stimulation of the nerve cell.	*Local potential*	Depolarization of the membrane which quickly decreases with distance from site of origin, thus does not continue to the end of the cell like the action potential. Can be summated.
Action potential	Temporary change of membrane potential following stimulation of a nerve cell; extends from site of origin in both directions. Cannot be summated, but is an "all-or-nothing" change of membrane potential. Synonymous expressions: 'nerve impulse', 'impulse', 'spike'.	*Threshold*	Limiting value of depolarization beyond which the membrane does not return to the resting potential but spontaneously developes into an action potential.

Depolarization	Equalization of ion imbalance across the membrane.	*i.p.s.p (continued)*	hibitory because it raises the threshold required for generating the action potential.
Hyperpolarization	Increase of the ion imbalance across the membrane; shift of membrane potential to more negative value.	*Noise*	Statistical slight variations (fluctuations) of the membrane potential consisting of spontaneous changes of the ion conductivity of the membrane. Interpreted as opening and closing of ion channels.
Equivalent circuit	Representation of electrical properties of a cell membrane using electrical terms: resistance, condenser, batteries.		
e.p.s.p.	**E**xcitatory **p**ost**s**ynaptic **p**otential; depolarization of the postsynaptic membrane; summatable local potential which releases an action potential on reaching a threshold.	*Ion channel*	Molecular structure in the cell membrane selectively permeable for certain ions. Exists in only two states: open and closed.
		Transmitter	Substance transmitted from one cell to the next at the synapse by a nerve impulse.
e.p.p.	**E**nd**p**late **p**otential; special expression for e.p.s.p. of the neuromuscular synapse or 'endplate'.	*Receptor*	Recognises biological signals in the postsynaptic membrane – binding site for transmitter molecules.
m.e.p.p.	**M**iniature **e**nd**p**late **p**otential; small, spontaneously occurring fluctuations of the membrane potential of the postsynaptic membrane, produced by the spontaneous (i.e. without stimulation) release of presynaptic transmitter molecules.	*Quantum*	Packet of presynaptically released transmitter molecules causing a m.e.p.p.; probably represents the content of a synaptic vesicle.
		Inactivation	Apparently spontaneous closing of ion channels, e.g. the voltage dependent sodium channel of the cell membrane.
i.p.s.p.	**I**nhibitory **p**ost**s**ynaptic **p**otential; hyperpolarization of the postsynaptic membrane by changing the conductivity of Cl^- or K^+. In-	*Activation*	Opening of ion channels e.g. by depolarization or by a transmitter molecule.

References

Cited:

[1] Katz, B.: *Nerve, Muscle, and Synapse.* McGraw-Hill, New York 1966.
[2] Aidley, D.J.: *The Physiology of Excitable Cells.* Cambridge University Press, Cambridge 1978.
[3] Fatt, P., and Katz, B., "Spontaneous subthreshold activity at motor nerve endings", *J. Physiol.* **117**, 109–128 (1952).
[4] Neher, E., and Sakmann, B., "Single-channel currents recorded from membrane of denervated frog muscle fibres", *Nature* **260**, 799–802 (1976).
[5] Sakmann, B., Bormann, J., and Hamill, O.P., "Ion transport by single receptor channels", *Cold Spring Harbor Symp. Quant. Biol.* **48**, 247–257 (1983).
[6] Stevens, C.F., "Study of membrane permeability changes by fluctuation analysis", *Nature* **270**, 391–396 (1977).
[7] McBurney, R.N., "New approaches to the study of rapid events underlying neurotransmitter action", *TINS* **6**, 297–302 (1983).
[8] Neher, E., and Stevens, C.F., "Conductance fluctuation and ionic pores in membranes", *Ann. Rev. Biophys. Bioeng.* **6**, 345–381 (1977).

Further reading:

Chang, D.C., Tasaki, I., Adelman, W.J., and Leuchtag, H.R. (eds.): *Structure and Function in Excitable Cells.* Plenum Press, New York and London 1983.
Kandel, E.R., and Schwartz, J.H. (eds.): *Principles of Neural Science.* Arnold, London 1981.
Sakmann, B., and Neher, E. (eds.): *Single Channel Recording.* Plenum Press, New York 1983.

Chapter 6

Ion Channels

Nerve cells have become specialized to conduct signals in the form of electrical impulses. The basis for this are ion currents through the cell membrane driven by a periodically formed electric potential difference between the inside and the outside of the cell (Chapter 5).

Active and passive ion transport are independent of each other

The subject of this chapter is the cell membrane structures which regulate and mediate the ion currents, particularly those of sodium and potassium (Fig. 6.1). These ion currents can, by the expenditure of energy derived from cell metabolism, be pumped from a more dilute to a more concentrated solution (active transport, ion pump). They can also be driven in the other direction by thermal movement and electrochemical potential without the expenditure of metabolic energy (passive transport). Both processes, active transport of ions against a concentration gradient, uphill as it were, and passive diffusion with the gradient or downhill from the point of view of energy expenditure, are necessary for nerve impulse conduction. Thus, in order to maintain ionic balance the passive ion currents must be compensated for by active transport. For the present we will only consider the passive phenomenon; active transport with its sodium-potassium pump driven by the hydrolysis of ATP (sodium/potassium ATPase – abbreviation: Na/K pump) will be described in Chapter 7. This division already indicates that biochemically different structures are involved. There are several lines of evidence for this:
1. As will be discussed later in detail, active and passive transport can be differentiated pharmacologically. Specific inhibitors of active ion transport like ouabain (strophanthin) do not affect passive ion currents and selective blockers of the passive sodium current, tetrodotoxin (TTX) and saxitoxin (STX) have no effect on the Na/K pump.
2. The Na/K pump works independently of membrane potential; the passive currents, however, are evoked by a depolarization and are potential-dependent.
3. Inhibitors of energy metabolism block the ion pump, but have no effect on passive ion transport. The latter remains unchanged when the cytoplasm is totally removed and therefore no metabolism is taking place. This has been demonstrated in the squid giant axon where normal action potentials are produced, even after the total extrusion of the axoplasm and perfusion with salt solution.

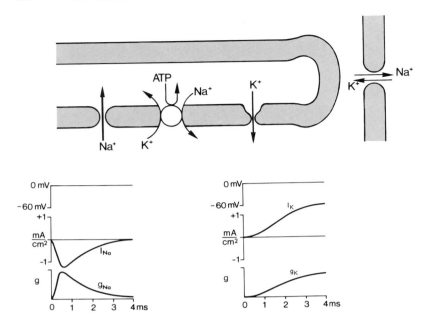

Fig. 6.1. Schematic diagram of a nerve fibre with synapse. ATP-mediated active transport and three different passive transport systems are shown. That shown above right is chemically regulated by a transmitter molecule; an example is the channel in the postsynaptic membrane of the muscle endplate which permits the passage of both Na$^+$ and K$^+$; left above the separate Na$^+$ and K$^+$ channels of the axon membrane which are opened electrically i.e. through depolarization. Below, the sequential nature of the sodium (g_{Na}) and potassium (g_K) conductances and the inwardly directed sodium (I_{Na})- and outwardly directed potassium (I_K)-currents after a depolarization of 60 mV are shown. The clearly differentiated kinetics of the two parameters (g_{Na} and g_K) demonstrates the separate molecular structure of the passive sodium and potassium transport systems.

4. The processes have different temperature dependence. Active transport is, like a chemical reaction, speeded up by a rise in temperature, while the maximum passive sodium conductivity has only a low temperature coefficient. (It is only the opening of the sodium channels which is affected by temperature and not the maximum conductivity of the open channels).

5. Further differences in the ion transport systems are evident in their ion selectivity, their maximum capacity, and their number per unit surface area of membrane (see Table 7.1, Chapter 7).

Chemical and electrical regulation of passive ion currents

The passive transport systems – from now on to be referred to as channels – are not a uniform group of functional elements in the membrane. As they are closed during the resting state they only come into action after opening. The opening or "gating" can occur electrically i.e. by a change of the membrane potential or chemically i.e. by interaction with a specific molecule. The chemical gating mechanism will be discussed in connection with the biochemistry of the synapse in Chapters 8 and 9. It will only be noted here that it also differs from other transport systems in its pharmacology, ion selectivity and kinetics.

Passive Na^+ and K^+ transport are independent of each other

We return to the electrically gated channels which mediate passive sodium and potassium ion movements. The careful analysis by Hodgkin and Huxley [1-3] of the action potential of the squid giant axon has shown that there are at least two different and physically separated channels: the sodium channel opens after depolarization of the membrane to allow an inward flow of sodium ions; after a short delay the potassium channel opens and permits potassium ions to flow in the opposite direction (Fig. 6.1). In addition to the fact that the membrane permeability for sodium and potassium ions does not rise synchronously the following two important points are further evidence that there are two separate ion channels:

1. If the depolarization of the axon membrane is artificially prolonged, the sodium channels close – in electrophysiological terms there is inactivation – but the potassium channels remain open for the total duration of the depolarization. This can be explained only by the existence of separate channels for each ion.
2. Selective inhibitors are known for the different ion currents. Tetrodotoxin and a series of other toxins specifically block the sodium channel without affecting the potassium channel. In contrast, the tetraethylammonium ion (TEA) inhibits the potassium current, but has no effect on the sodium current.

To sum up: We can conclude that in the axonal membrane (and naturally not only in this) there are the following separate ion transport systems (schematically represented in Fig. 6.1):

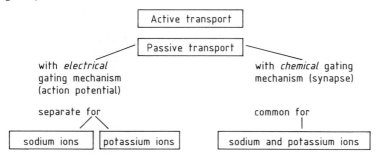

We shall go on to consider if these systems have basically different molecular structures. One of the goals of contemporary neurochemistry is their biochemical characterization [4]. Some of the confirmed experimental results and the tentative models and hypotheses arising from them will be given in the following sections.

Gate and selectivity filter are functional elements of ion channels

The aim of neurochemical research in this field is to combine biochemical analysis with electrophysiological results to produce a picture of the structure and function of the ion channel. Such a picture must include two elements:
1. The gating mechanism which regulates the ion current
2. The selectivity filter which characterizes the ion specificity.

Electrophysiological and biochemical experiments have been supplemented by electron microscopy and measurements of changes of the biophysical parameters of the axon membrane during excitation. The following account of current findings will show how far we are from a complete molecular description of ion channels and the action potential.

The sodium channel: the gating mechanism

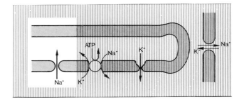

Fig. 6.1 shows the time course of the intensity of the sodium current I_{Na} measured by means of a voltage clamp during a prolonged depolarization: the inward directed current (in Fig. 6.1 below left) increases to a maximum and decreases "spontaneously" until the resting value is reattained.

The conductance of the membrane for sodium ions, g_{Na} is potential and time dependent. The following equation for the squid giant axon was proposed by Hodgkin and Huxley [3]:

$$g_{Na} = \bar{g}_{Na}\, m^3\, h$$

where $\bar{g}_{Na}$ is the maximum conductance per unit surface area of membrane and m and h are dimensionless variables, which describe the activation (m) and the inactivation (h) of the channels.

These are defined by the differential equations

$$\frac{dm}{dt} = \alpha_m(1-m) - \beta_m m$$

$$\frac{dh}{dt} = \alpha_h(1-h) - \beta_h h$$

where α and β are velocity constants for the forward and back reactions and are functions of the membrane potential.

The exponents of m and h represent the kinetics of activation and inactivation, respectively. They recall in the case of activation the well known phenomenon of enzyme cooperativity, but the molecular basis of this equation is not clear.

A basic assumption of the Hodgkin-Huxley equation is however that of an activation or gating mechanism. The ion channel must contain a component which, dependent on the membrane potential applied either permits the passage of the cations (activation in electrophysiological terminology) or blocks it (inactivation). The potential dependence indicates that the component itself carries an electrical charge (perhaps it is a dipole) and can be displaced by changes in membrane potential. This kind of charge movement in the membrane can be detected experimentally as very small positive currents (independent of sodium and potassium ions) flowing from inside to outside in less than 1 µs after a depolarization and uninfluenced by the usual inhibitors of ion currents [5-7]. They are called "gating currents" and can represent a rotation of molecular dipoles or charge displacements by changes in conformation of, for example, membrane proteins. Two current models under discussion are shown in Fig. 6.2 [8]. They show the activation of the sodium channel as the removal of a positive charge that inhibits the passage of positively charged sodium ions through rotation of the m particle (A) in Rojas' model, or in Fig. 6.2B

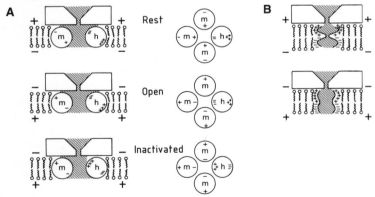

Fig. 6.2. Models for the electric gating mechanism. (A) Electric dipoles (m) are reorientated by depolarization (i.e. formation of a positive charge on the inside of the axonal membrane) from their position in the resting state (above) to that of the active state (middle). This ensures that the passage of positively charged sodium ions through the channel is not blocked by positive charges. After a short delay a further dipole (h) rotates and inactivates the channel (below). Model (B) illustrates an alternative mechanism in which the rotating dipoles are replaced by conformational changes in helical arrays of membrane molecules. Reproduced, with kind permission, from Keynes [8].

through partial extension of a helical structure. The inactivation in this model is caused by the acquisition of a positive charge by the channel through a delayed rotation of the h particle relative to the m particle (A); or alternatively through a slow further conformational change (B). Positive biochemical evidence is now required for the existence of the postulated m and h particles.

Calcium ions affect the threshold of excitation, but not m^3

A rise in calcium ion concentration in the region of the axon increases the threshold of excitation, i.e. a greater depolarization is necessary for the release of an action potential. On the other hand the kinetics of the activation of the sodium current is not essentially altered by the absence of calcium ions which are therefore not, as originally believed, involved in the gate to the channel. The calcium effect is more probably indirect: the cation affects the properties of the membrane by acting as a counter-ion to its surface charge.

The sodium channel: the selectivity filter

The second necessary functional element of an ion channel, besides the gating mechanism, is that of a selectivity filter [9]. The permeability of the sodium channel for sodium ions is about twelve times greater than for potassium ions and about the same as that of Li^+ (Table 6.1). As the permeability decreases with increasing radius of the diffusing ions the selectivity filter might be thought of as a kind of sieve. That this can be at least only partially true is shown by the fact that the K^+ channel is practically impermeable to the smaller Li^+ and Na^+ ions. Even if instead of the non-hydrated ion, whose energy of solvation and hydration shell increases as the radius decreases, the hydrated ion is taken into account inconsistencies still arise with the sieve model; the differences in the radius of the hydrated ion could explain the lower permeability of the K^+ channel to Li^+ but not the higher permeability of the Na^+ channel to it. The concept of a filter is thus only of an operational filter which tells us nothing about its structure and mechanism of action.

Table 6.1. Ion permeability of the sodium channel relative to that of sodium ions.

Ion	Permeability relative to Na^+
Li^+	1.1
Na^+	1.0
K^+	0.083
Rb^+	0.025
Cs^+	0.016

A better description of the sodium channel was derived from a series of elegant experiments by Hille [10]. By comparing the permeability of alkali metal ions with that of a series of organic cations (Table 6.2) he calculated the average size of the channel as 0.3 x 0.5 nm and postulated that it is surrounded by a ring of oxygen atoms such as those of the carboxyl groups of a membrane protein. The selectivity depended, according to Hille, not only on the dimensions of the permeating ions but also on their ability to form hydrogen bonds. The following comparison of three ions illustrates this:

Hydroxylammonium H_3N^+ — OH
Hydrazinium H_3N^+ — NH_2
Methylammonium H_3N^+ — CH_3

They are practically isosteric; the van der Waals radii of the hydroxyl-, amino- and methyl groups range from 0.37 to 0.38 nm yet the hydroxylammonium ion is about a hundred times more permeable than the methylammonium ion. This selectivity can be explained by the fact that methyl groups cannot form hydrogen bonds. The bond length between the hydrogen donor and acceptor is about 0.08 nm shorter than the sum of the van der Waals radii, so that a 0.3 nm opening allows the diffusion of a hydroxylammonium but not a methylammonium ion.

On the basis of this and other studies Hille formulated the theory that sodium ions do not permeate by simple diffusion but by means of a sequence of dehydration and binding steps to channel components [11]. In terms of energy this means the surmounting of a series of energy barriers, of which the highest represents the actual selectivity filter (Fig. 6.3). The filter is formed of oxygen atoms. The pH dependence of the sodium conductivity is evidence that carboxyl groups form part of the filter.

Channel or carrier?

Wary authors do not refer to sodium channels or pores but to the sodium system, so that they can avoid statements about the molecular structure and the mechanism of the ion transport through the membrane. Does the sodium system then act as a channel or carrier [12]? The rate of facilitated ion diffusion through a carrier is about 10^4 ions/s, about three orders of magnitude less than that through a pore. By this criterion the sodium system clearly acts as a pore. Biophysically the mechanisms are differentiated further according to the concentration dependence of their transport rate. In carrier transport a saturation value is reached in high concentrations when all carrier molecules are occupied by ions while ion diffusion through a channel is determined only by the Brownian motion of the ions and the electrochemical gradient, i.e. individual ions travel through the membrane without competing with other either similar or different ions. This principle of independence formulated by Hodgkin and Huxley is expressed in the following equation:

$$\frac{I_{Na'}}{I_{Na}} = \frac{[Na_{o'}] - [Na_{i'}] \exp.(FE/RT)}{[Na_o] - [Na_i] \exp.(FE/RT)}$$

Table 6.2. Permeability of the sodium channel for univalent inorganic ions and nitrogen-containing organic ions.

P_{ion}/P_{Na}	Ion
1.0	Sodium
0.94	Hydroxylammonium
0.59	Hydrazinium
0.33	Thallium
0.16	Ammonium
0.14	Formamidinium
0.13	Guanidinium
0.12	Hydroxyguanidinium
0.083	Potassium
0.06	Aminoguanidinium
0.056	N-Methylhydroxylammonium
0.025	Methyhydrazinium
0.007	Methylammonium
0.008	Acetamidinium
0.010	Methylguanidinium
0.007	Dimethylammonium
0.005	Tetramethylammonium
0.008	Tetraethylammonium
0.014	Ethanolammonium
0.007	Cholinium
0.007	Tris(hydroxymethyl)methylammonium
0.008	Imidazolium
0.010	Biguanidinium
0.013	Triaminoguanidinium

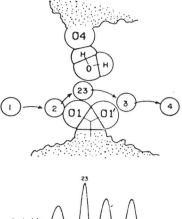

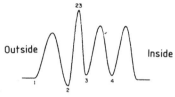

Fig. 6.3. Hille's model for ion transport through the sodium channel. The free ion (1) is bound (2) and desolvated (23); it is subsequently resolvated on the inner side of the membrane (3) and then dissociates from the channel opening (4). The diagram below shows the energy profile of this process: the rate determining step is the surmounting of the energy-barrier (23) which constitutes the selectivity filter.

I_{Na} and $I_{Na'}$ is the amount of sodium ions in the current at two different concentrations. The index o stands for outside (relative to the membrane) and i for inside. E is the potential difference $E_i - E_o$, F is the Faraday constant.

While electrophysiological measurements appeared to support the independence principle inconsistencies became evident both for the sodium and for the potassium system. That the ion channels of the excitable membrane are not simply holes is shown by the fact that there is saturation at high ion concentrations analogous to the saturation of an enzyme by a substrate and mutual competition between Na^+ and impermeable ions which block the channel. Hille's model accounts for the facts, by showing the sodium channel as taking one sodium ion at a time, its dissociation constant being 368 mmol/l. In the classical carrier model the ligand combines with the carrier molecule and diffuses with it from the outside of the membrane to the inside where the ion is released. There is no evidence for such a mechanism here. The sodium system should therefore be considered as a channel with a cation binding site (and a gating system); in contrast to the carrier it spans the membrane and is immobile.

Some physical properties of the sodium channel

The temperature coefficient (Q_{10}) of sodium conductivity is only 1.3 and is similar to the coefficient for diffusion in water. The value corresponds to an energy barrier of ca. 17 kJ/mol and together with the high dissociation constant of 368 mol/l explains the ease with which the sodium ion goes through the membrane. The rate of diffusion through a single channel is about 10^8 sodium ions per second. By noise analysis single channel conductivity in the squid axon was determined as 4 x 10^{12} Ω^{-1} which is one third of that of the potassium channel.

Biochemical characterization of the sodium channel

For the isolation and biochemical characterization of a biomolecule two things are necessary: a good starting material and a method of assay. In enzyme purification the bioassay is based on the reaction which the enzyme catalyses. In isolating ion channel components this possibility does not exist. An ideal test of activity would be provided if the isolated membrane material (i.e. the pore) could be inserted into an artificial lipid membrane (see Chapter 3), where changes in its electrical conductivity could be measured, assuming of course that the isolated material retains all the functionally important channel components. As these reconstitution experiments have proved very problematic and too complex with molecules of nerve membrane, other methods have been devised. The test of activity (i.e. ion transport) is substituted by a "binding test" with radioactively marked ligands, e.g. neurotoxins, whose selective interaction with the particular channel has been demonstrated electrophysiologically. In the case of the sodium channel tritium-labelled tetrodotoxin ([^{3}H]TTX) or alternatively saxitoxin ([^{3}H]STX) are used [13] as their dissociation constants at 10^{-9} mol/l are low enough for a binding test.

A good starting material must be available in large enough quantities and contain the channel in sufficiently high concentration. Table 6.3 shows that the number of sodium channels per unit surface area of the membrane is both very low and very variable [13]. On average 50 channels per μm^2 are each surrounded by 3 x 10^6 lipid molecules. The channels are diluted by lipid to an average distance of 140 nm.

In the electric organ of the electric eel, *Electrophorus electricus*, the model for much neurochemical research (see Chapter 12), it is possible that all the functions of the voltage dependent sodium channel are represented on one single huge polypeptide chain (see p. 119). A protein of M_r 260 000 has been isolated which binds TTX and which can be reconstituted into artificial lipid vesicles. Patch clamp experiments with this reconstituted system reveal all the expected properties of the channel: voltage dependent conduction, selectivity (for Na$^+$), inactivation, response to TTX and batrachotoxin.

Another difficulty in the isolation of sodium channels is that they are comparatively unstable outside the membrane. The following biochemical properties are known so far: the TTX-binding component of the axon membrane has a relative molecular mass of 230 000 (calculated from its inactivation by radiation) or 260 000 (determined by biochem-

Table 6.3. Sodium channels in excitable membranes [13].

Tissue	Axonal surface (cm^2/g)	Concentration (pmol/g)	Dissociation constant K_D (nmol/l)	Density of sodium channels (binding sites/µm^2)
a) Nerve				
Garfish	65 000	377	9.8	35
Lobster	7 000	94	8.5	90
Rabbit	6 000	110	1.8	110
Squid	80	–	–	550
Rabbit (node of Ranvier)	8.2	17.1	3.4	12 000
b) Muscle				
Frog	350–550	14–37	3–5	200–380
Rat (diaphragm)	700	24.5	3.8	206

ical methods), it has a sedimentation coefficient of 9.2, and it is inactivated by proteases, heating and ionic detergents (sodium dodecylsulphate). The part of the sodium channel responsible for the inhibitory binding of TTX or STX consists at least partly of protein [14]. The relative molecular mass of the sodium channel protein of brain synaptosomes is in total ca. 320 000 and consists of two small (37 000 and 39 000) and one large (260 000) polypeptide chains. It cannot be excluded that other molecules, lipids or carbohydrates are involved in part or all the sodium transport mechanism.

The inhibition of TTX-binding by proteases can be accelerated by treating the membrane first with phospholipase and neuraminidase. This is evidence that the TTX-binding component is embedded in the membrane lipid and surrounded by carbohydrate.

The participation of proteins in the sodium channel is also shown by the blocking of the sodium current by reagents [15], which combine covalently with functional groups of proteins (Fig. 6.4). Amongst these are mercury salts as for example $HgCl_2$ or p-chloromercuribenzoate (pCMB) and other SH reagents such as N-ethylmaleimide (NEM) or 5,5'-dithiobis(2-nitrobenzoic acid) (DTNB, Ellman's reagent). Stimulating the nerve increases the effectiveness of the SH reagents, perhaps because the reactive groups are only accessible after the channel has opened. The inhibition by mercury salts can be reversed by cysteine, mercaptoethanol or dithiothreitol.

Apart from essential SH groups there are apparently disulphide bridges, which are not accessible in the membrane bound sodium channel, but can be reduced through DTT after solubilization with Triton X-100. The reduction of the S-S bond blocks the binding of TTX. For further biochemical characterization of the sodium channel it is necessary to have methods that selectively affect single steps in the transport mechanism. Perfusion of the interior of the squid giant axon with the proteolytic enzyme pronase suppresses sodium channel inactivation, the h-process, from which it may be inferred that the molecule involved is a protein.

Using recombinant DNA techniques the cDNA complementary to the messenger RNA coding for the sodium channel protein of the electric eel has been cloned (see Chapter 12) and sequenced. The primary structure reveals some interesting features: it contains four repeated homology units each containing a cluster of positively charged amino acids, which may be involved in the gating structure. Other specific tools for the chemical analysis of channel components are neurotoxins, which we will deal with in detail on page 121 [16].

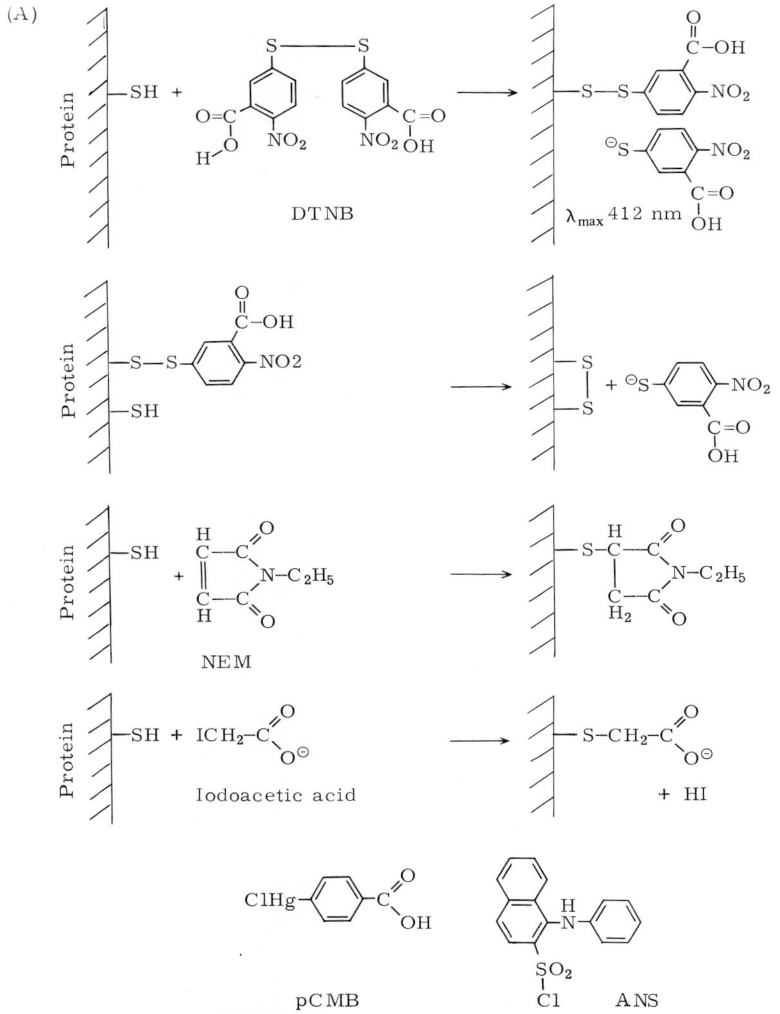

Fig. 6.4. Some examples of group-specific protein reagents commonly used for the analysis of nerve membrane proteins. (A) SH group reagents. Line 1: 5,5'-dithiobis-(2-nitrobenzoic acid) (DTNB, Ellman's reagent) reacts with SH groups by disulphide exchange and liberation of coloured thiophenolate ions (λ max = 412 nm). Line 2: If a neighbouring SH group is present, a further disulphide exchange may occur giving rise to intra- or intermolecular disulphide bridges. Other SH reagents are (line 3) N-ethylmaleimide (NEM), (line 4) iodoacetic acid, (line 5) p-chloromercuribenzoate (pCMB) and ANS, which being fluorescent serves to introduce a fluorescent group into a protein.

(B) Imidazole groups of protein histidine residues can react with diethylpyrocarbonate (DEP).
(C) There are many reagents for amino groups like (line 2) fluorodinitrobenzene (FDNB) introduced by Sanger or (line 3) the fluorescent reagent dansylchloride (5-dimethylamino-1-naphthalene sulphonylchloride).
(D) Arginine side chains can be substituted with dicarbonyl compounds like glyoxal or 1,2-cyclohexanedione.
(E) Indole groups of tryptophan residues react with 2-hydroxy-5-nitrobenzyl bromide (HNBB) introduced by Koshland.

(F)

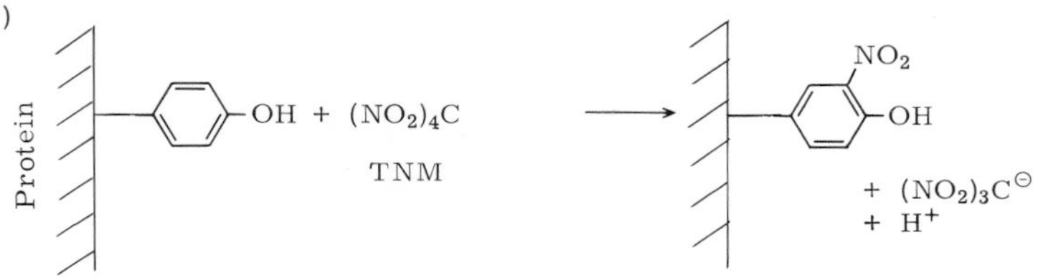

```
Protein—⟨phenyl⟩—OH + (NO₂)₄C   ⟶   Protein—⟨phenyl(NO₂)⟩—OH
                 TNM                        + (NO₂)₃C⁻
                                            + H⁺
```

(G)

```
Protein—C(=O)OH + (CH₃)₃O⊕   ⟶   Protein—C(=O)OCH₃
                   TMO
            Meerwein's reagent
```

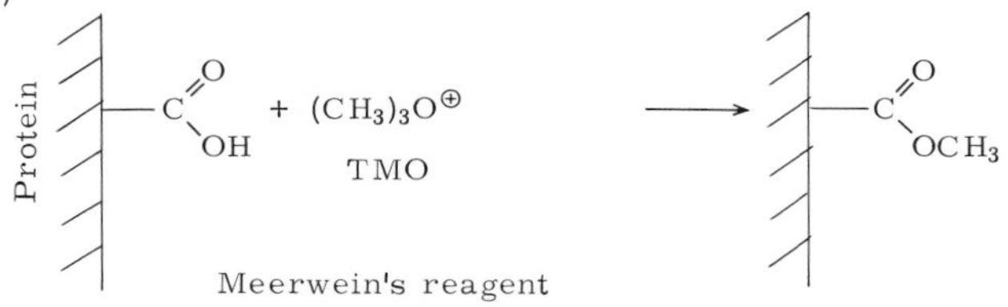

(F) Phenol groups of tyrosine residues are nitrated by tetranitromethane (TNM).
(G) Meerwein's reagent for the methylation of carboxyl groups. As most of the protein reagents react electrophilically, their group selectivity is frequently determined by the differing dissociation constants (pK) of the functional group to be substituted. For further reagents particularly affinity reagents see Chapter 9.

Drugs influence the action potential

Without attempting a comprehensive survey of neuropharmacology a few drugs will be introduced, which either contribute information about the structure and function of the excitable membrane or whose molecular mode of action is important in medicine and pharmacology.

Neurotoxins as tools for the investigation of ion channels

Tetrodotoxin (TTX). Every year some people die in Japan after consuming puffer fish (Tetrodontidae), there considered a great delicacy. The cause of these tragedies is that the ovaries and liver of these fish are not properly removed in the preparation and they contain tetrodotoxin (Fig. 6.5). This has also been found in the Californian mud puppy *Taricha torosa* and more recently in the eggs of the Costa Rican frog *Atepolus chiriquensis*. Since its discovery [17] tetrodotoxin (TTX) has become one of the most important probes for investigating the mechanism of the action potential. It specifically blocks the sodium conductivity of the axon membrane. The potassium conductivity is not affected, nor are any of the essential steps of synaptic transmission – transmitter release from the presynaptic membrane, transmitter binding and ion permeability at the postsynaptic membrane. The action of TTX has the following characteristics:

1. It only affects the outside of the axon membrane.
2. It decreases $\bar{g}_{Na}$, the maximum conductivity for sodium ions, but not the term m^3h of the Hodgkin Huxley equation, the gating mechanism.
3. It is reversible.
4. It acts on the sodium channel in the molecular ratio 1:1, with a dissociation constant of approximately 10^{-8} mol/l.

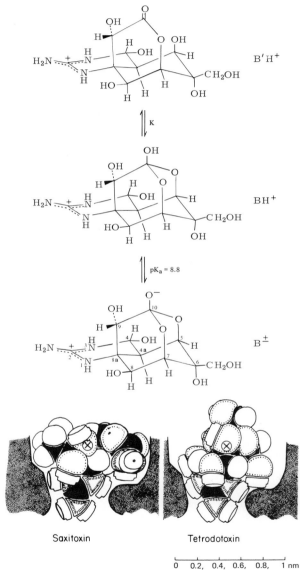

Fig. 6.5. Tetrodotoxin. Only the cationic forms (below pH 8) are toxic. Below, putative blocking mechanism demonstrated by space filling molecular models of TTX and STX [18].

5. It is independent of the permeating ion and the direction of the current.
6. It is effective below pH 8, i.e. in one of the cationic forms (Fig. 6.5) [18].
7. Slight modifications of molecular structure abolish its activity. A comparison with the structure of the almost similarly acting saxitoxin (STX) indicates that the active molecular group is the guanidinium group.
8. Competition experiments with uni- and di-valent metal ions make it probable that it is bound to a metal binding site of the ion channel.

Saxitoxin (STX) is a poison sometimes found in the Alaskan shellfish *Saxidomas giganteus* due to its ingestion of the microorganism *Gonyaulax catenella*. It is fundamentally different from TTX in structure but it also has a guanidinium group as its functional group (Fig. 6.6). It combines in the same 1:1 stoichiometric ratio with the sodium channel, its dissociation constant is 7×10^{-9} mol/l and its pharmacological and physiological action are similar to TTX.

Sea anemone toxins. While TTX and STX affect the maximum sodium conductivity $\bar{g}_{Na}$ and so block the open gate, the tentacles of sea anemones contain a poison which inhibits the inactivation of the sodium channel and keeps the gate open. Three different toxins have been isolated from *Anemonia sulcata,* anemonetoxin I, II and III (abbreviated to ATX I, ATX II, ATX III) [19]. They consist of small peptides with respectively 46, 47 and 27 amino acids whose sequences have been determined. ATX I and II show a high degree of sequence homology, but ATX III has practically no relationship to them. The tertiary structure of ATX I and ATX II is stabilized by three disulphide bridges. If these are broken by reduction the toxicity is lost.

These toxins are particularly interesting as probes for the biochemical analysis of channel structure as their binding site on the sodium channel is clearly different to that of TTX. The channel stabilized in its open state by ATX is nevertheless blocked by TTX; if the latter is then washed out, the inactivation of the sodium channel by ATX is still slowed down. If both toxins are not simply displacing each other, it must be concluded that the "gate way" (g_{Na}) and the gate (h in the Hodgkin-Huxley terminology) are either different parts of one of the channel forming molecules or totally different ones.

ATX II binds with a dissociation constant of 10^{-5} mol/l to the node of Ranvier, but to crustacean nerves with an affinity three orders of magnitude greater [20]. It is remarkable that it acts only on the outside of the membrane, whereas the pronase study (see page 119) appears to have established that the "gate" is localized on the inside [21]. Perhaps ATX penetrates deeply into the membrane; lengthy hydrophobic sequences seem to be evidence for this.

Grayanotoxin, Batrachotoxin, Veratridine. This group of alkaloid neurotoxins extends the range of actions in that it raises the conductivity of the resting membrane for sodium ions and so produces a reduction of the resting potential. The effect is neutralized by TTX, but at concentrations of an order of magnitude greater than that needed to block the sodium channel. These observations can be unified by a two state model of the sodium channel: in this [22] the channel exists in an open and a closed state. GTX, BTX and

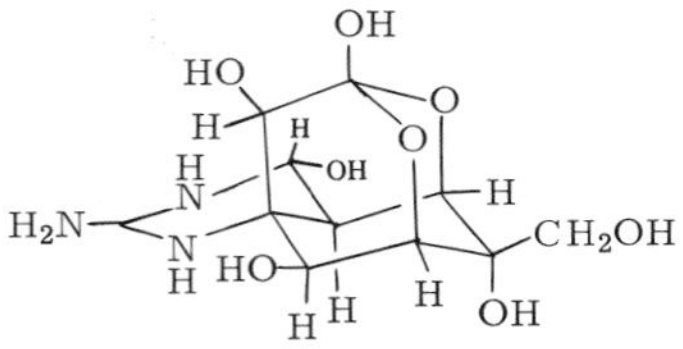

Tetrodotoxin (TTX)

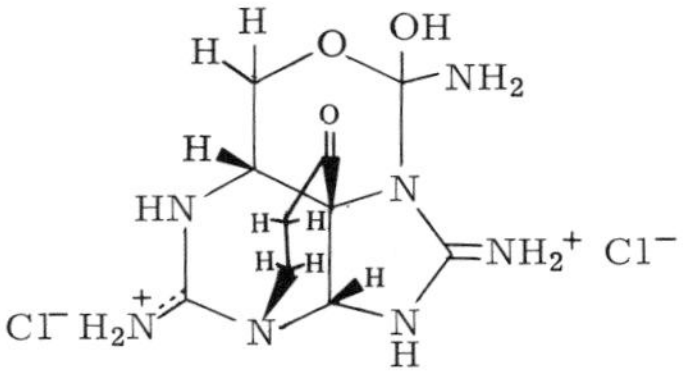

Saxitoxin (STX)

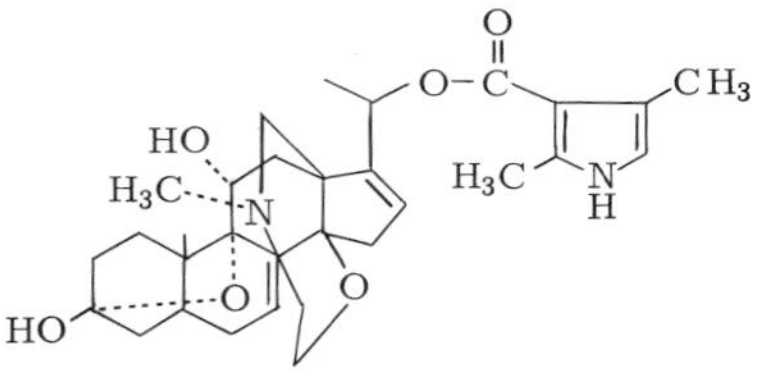

Batrachotoxin (BTX)

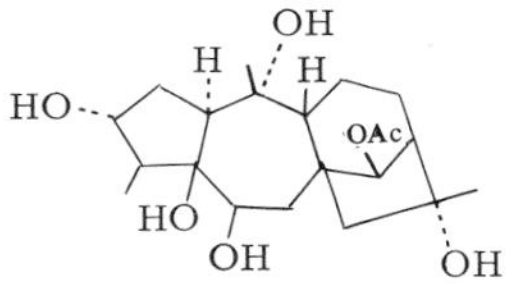

Grayanotoxin (GTX)

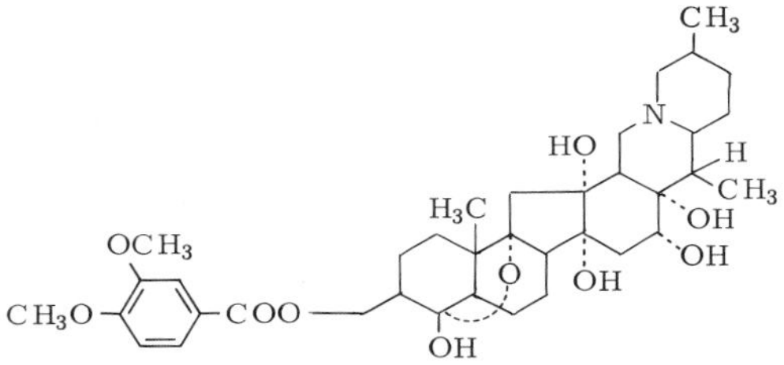

Veratridine

Gly-$\genfrac{}{}{0pt}{}{\text{Ile}^*}{\text{Val}}$-Pro-Cys-$\overset{5}{\text{Leu}}$-Cys-Asp-Ser-Asp-$\overset{10}{\text{Gly}}$-
-Pro-Ser-Val-Arg-$\overset{15}{\text{Gly}}$-Asn-Thr-Leu-Ser-$\overset{20}{\text{Gly}}$-
-Ile-Ile-Trp-Leu-$\overset{25}{\text{Ala}}$-Gly-Cys-Pro-Ser-$\overset{30}{\text{Gly}}$-
-Trp-His-Asn-Lys-$\overset{35}{\text{Lys}}$-His-Gly-Pro-$\overset{40}{\text{Thr}}$-
-Ile-Gly-Trp-Cys-$\overset{45}{\text{Cys}}$-Lys-Gln

Sea anemone toxin (ATX II)

Fig. 6.6. Some neurotoxins which affect the sodium permeability of excitable membranes.

veratridine are bound by the open channel and displace the equilibrium between the two states, towards the open configuration.

Besides this action on the excitable membrane GTX, BTX, veratridine and also ATX II stimulate chemical synaptic transmission. This is not unexpected as it has been shown electrophysiologically that transmitter release occurs when depolarization produces a rise in calcium permeability of the presynaptic membrane.

Scorpion toxins (ScTX). Scorpion venom also contains several neurotoxins. They are miniproteins with 65–66 amino acids and four disulphide bridges. Some of their amino acid sequences are known. Their action is less selective. ScTX I inhibits simultaneously the inactivation of sodium and the activation of potassium conductivity. In certain species (*Centruroides*) toxins have also been isolated some of which selectively affect the activation of the sodium channel, and one which blocks sodium conductivity [23].

Summary of neurotoxins as tools. During an action potential three processes can be pharmacologically differentiated, the activation (opening) of the canal, the ion transport through the open pore and the inactivation (closing) of the channel. The neurotoxins affecting voltage-dependent sodium channels appear to act via three distinct sites within the channel [14]: site 1 (TTX, STX), related to the ion transport, site 2 (BTX, veratridine, aconitine) regulating channel activation, and site 3 (ScTX, ATX) regulating channel inactivation (Table 6.4.).

As these toxins do not compete for a common binding site, a separate molecular structure for their site of action must be postulated. They are not small molecules and they do not interfere with each other, so it is thought that their binding sites are widely separated on one large membrane molecule or on several different molecules. With the help of radioactive toxins in a binding test these toxin-binding molecules can now be isolated and characterized biochemically.

Anaesthesia

Anaesthesia is the reversible elimination of sensitivity to pain by the blocking of nerve conduction. Anaesthetics differ from the toxins of Table 6.4 quantitatively but not qualitatively. Both classes of substances can provide useful information about the mechanism of nerve conduction, and when their mode of action is understood, other safer and more effective drugs may possibly be synthesized.

The variety and structural diversity of the substances which block the conduction of nerve impulses (Fig. 6.7, Tables 6.4 and 6.5) is so great that their mode of action cannot be described by a single molecular mechanism. As no overall classification is thus possible anaesthetics have been divided into non-specific and specific. The latter stop the conduction of the impulse at particular sites in the nervous system because their structure is only recognised by specific receptors (e.g. opiates) (Fig. 6.7 below). Non-specific anaesthetics block neuronal membrane in general due to their solubility in the membrane lipid when their concentration in the serum exceeds a definite value. However many drugs act as specific as well as non-specific anaesthetics although at differing concentrations, so that this

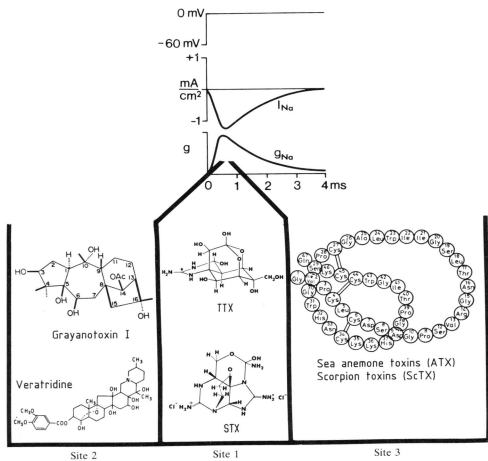

Toxin binding site	Ligands	Physiological effect
1	Tetrodotoxin Saxitoxin	Inhibition transport Alter activation and inactivation
2	Batrachotoxin, Veratridine Aconitine Grayanotoxin	Cause persistent activation
3	Scorpion toxin Sea anemone toxin	Inhibit inactivation Enhance persistent activation

Table 6.4. Neurotoxin binding sites associated with the sodium channel [14]. – Site 3 is apparently located on a part of the sodium channel which undergoes a conformational change during channel activation. This part therefore appears to sense voltage changes and may represent a gating structure.

classification is also not very useful. Within the classes there are so many possible modes of action that a systematic classification must await further research. We must be satisfied for the moment with the statement that anaesthetics can block both axonal conduction and synaptic transmission. In this chapter we shall confine ourselves to the effect on the axon.

Table 6.5. Anaesthetics.

Anaesthetic	Formula	Concentration for nerve block (mol/l)
Neon	Ne	0.268
Nitrous oxide	N_2O	0.128
Nitrogen	N_2	0.088
Ethylene	C_2H_4	0.052
Methanol	CH_3OH	2.4
Ethanol	C_2H_5OH	0.5
Propanol	C_3H_7OH	0.218
Nonanol	$C_9H_{19}OH$	6.4×10^{-5}
Chloroform	$CHCl_3$	5.0×10^{-3}
Diethyl ether	$C_2H_5OC_2H_5$	5.0×10^{-2}
Halothane	$F_3CCBrClH$	5.0×10^{-3}

The effectiveness of a substance in blocking nerve conduction depends on its solubility in the axon membrane [24]. But the basic requirement, as shown by anaesthetic gases, is that they are taken up by the extracellular medium and conveyed to the nerve. They must thus be soluble in water. It is basically the partition coefficient between plasma and membrane which decides their anaesthetising property. In addition the size of the molecule is important. Large molecules like chlorpromazine block the membrane in lower concentrations than small ones e.g. ethanol. Finally, in a yet unknown way the diameter of the nerve fibres plays a role: fibres of a small diameter are more easily blocked than those with a large one. Ethanol produces unconsciousness (general anaesthesia) in a blood serum concentration of only 0.2%, while 4–5% is needed to block nerve impulses in peripheral nerves (local anaesthesia), the fibres of the central nervous system being thinner than those of the periphery.

In Chapter 3 it has been mentioned that local anaesthetics increase the fluidity of the lipid membrane. This is accompanied by a lateral expansion of the membrane resulting perhaps in a change in the ion channels and so in a blocking of impulse conduction. Local anaesthetics selectively reduce the sodium conductivity ($\bar{g}_{Na}$) and the gating mechanism. There are various hypotheses about the mechanism of this action [25,26]. The lateral expansion could directly change the structure of the sodium channel. It could however also be an indirect consequence of the increased membrane fluidity; the ion channels are stabilized by a stiff "collar" of fluid crystalline lipid in a functional conformation which collapses when the "collar" is made more fluid by the drug.

It is certain though that local anaesthetics do not only enter the lipid phase of the membrane [15]; they also interact directly with membrane proteins [16]. This is demon-

Anaesthetic	Formula	Concentration for nerve block (mol/l)
Procaine (Novocaine)	$H_2N-\langle\bigcirc\rangle-CO-O-CH_2-CH_2-N{\overset{C_2H_5}{\underset{C_2H_5}{\diagdown}}} \cdot HCl$	4.6×10^{-3}
Tetracaine	$\overset{C_4H_9}{\underset{H}{\diagdown}}N-\langle\bigcirc\rangle-CO-O-CH_2-CH_2-N{\overset{CH_3}{\underset{CH_3}{\diagdown}}} \cdot HCl$	
Lidocaine	$\langle\bigcirc\rangle(CH_3)_2-NH-OC-CH_2-N{\overset{C_2H_5}{\underset{C_2H_5}{\diagdown}}} \cdot HCl$	
Chlorpromazine	phenothiazine-Cl, $CH_2-CH_2-CH_2-N(CH_3)_2$	1.0×10^{-5}
Cocaine	tropane-COOCH$_3$, O-CO-C$_6$H$_5$	2.6×10^{-3}
Morphine	morphine structure, N-CH$_3$	5.5×10^{-2}

Fig. 6.7. Anaesthetics. Below, anaesthetics block impulse conduction either by interaction with the lipid membrane (right) or with the membrane proteins (left) by direct blocking of the ion channels.

strated by the fact that enzymes like ATPase can still be influenced by anaesthetics even after extraction from the membrane. It is also observed that the lateral expansion of a protein-containing membrane is about ten times greater than that of a pure lipid membrane. One can only speculate on the significance of this amplification factor, perhaps it shows that protein and lipid together are responsible for the anaesthetising action of many substances.

The problem is further complicated by the fact that local anaesthetics can either release calcium ions from the axon membrane or increase their binding to them. As calcium affects the threshold of excitation of nerve and muscle membrane a description of the mode of action of anaesthetics must include a consideration of this action of calcium. The tertiary amines e.g. procaine and chlorpromazine due to their positive charge displace Ca^{2+}, while the negatively charged barbiturates and also numerous neutral anaesthetics increase the binding of Ca^{2+} to the membrane.

The potassium channel

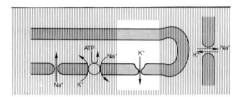

Fig. 6.1 shows the time course of the potassium current during depolarization: in contrast to the sodium current it is directed outwards and does not show the phenomenon of inactivation, i.e. the potassium channel remains open during the whole duration of the depolarization. Hodgkin and Huxley described the change in potassium conductivity g_K by the equation

$$g_K = \bar{g}_K n^4$$

$\bar{g}_K$ is the maximum conductivity per unit surface area of the membrane. By analogy with the values m and h of the equation for sodium conductivity, n represents a dimensionless variable which characterizes the kinetics of activation of the potassium channel. Thus

$$\frac{dn}{dt} = \alpha_n(1-n) - \beta_n n$$

Gating currents have not yet been measured in the potassium system, nevertheless it must be assumed that this consists also of a gating mechanism and selectivity filter. Like the sodium channel the permeability of the axon membrane is not correlated simply with the ion radius (Table 6.6).

Blocking experiments with quaternary ammonium salts have produced a more detailed picture of the structure and properties of both functions of the potassium channel. Tetraethylammonium ions (TEA) block the outward directed potassium current. In the squid giant axon this effect has only been observed when TEA is allowed to act on the

Table 6.6. Ion permeability of the potassium channel relative to the permeability of potassium ions.

P_{ion}/P_K	Ion	Ionic diameter (nm)
0.018	Lithium	0.120
0.010	Sodium	0.190
1.00	Potassium	0.266
2.3	Thallium	0.280
0.91	Rubidium	0.296
0.13	Ammonium	0.30
0.029	Hydrazinium	0.33
0.077	Caesium	0.338
0.021	Methylammonium	0.36
0.020	Formamidinium	0.36
0.013	Guanidinium	0.48

inside of the membrane by perfusing the axon with it. Therefore it follows that the channel penetrates the membrane asymmetrically. Even more interesting is the observation [27] that the blocking action of the substituted ammonium ion is increased if a hydrophobic, longer side chain is substituted for one of the ethyl groups (Fig. 6.8). These derivatives do not simply block, but they inactivate the potassium current already developed (Fig. 6.9). They appear to slip into the open channel.

Although a TEA ion has a diameter similar to a hydrated potassium ion it cannot pass through the membrane. As potassium can permeate it must therefore be concluded that it does so only in an unsolvated form. TEA cannot reduce its effective diameter like K^+ without shedding its alkyl groups. As it only blocks the permeability of potassium ions when applied to the inside of the axon it may be deduced that the channel on the inside is

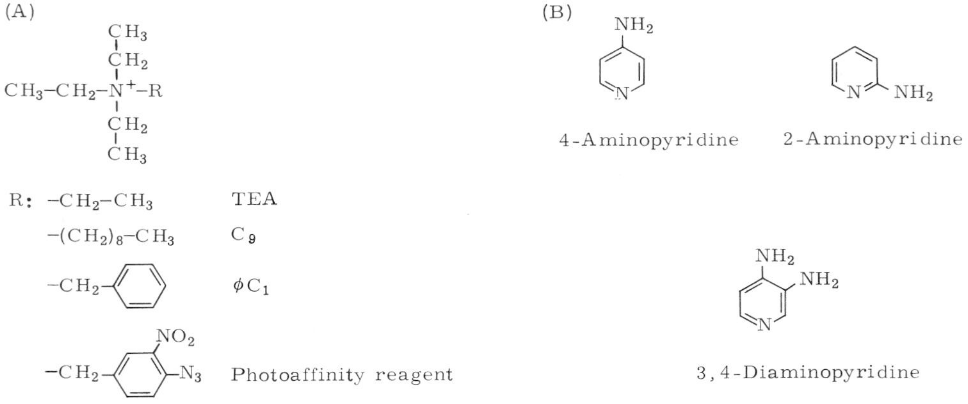

Fig. 6.8. Specific inhibitors of the potassium permeability of the excitable membrane. (A) Derivatives of tetraethylammonium (TEA), (B) Aminopyridines.

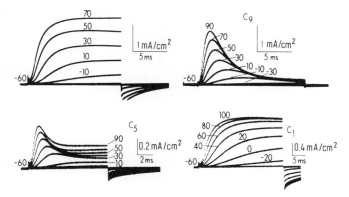

Fig. 6.9. Time dependence of the potassium current flowing through the axon membrane after depolarization at different values in mV. Above left: control without an inhibitor. Others: Inhibition of the potassium current by TEA derivatives. The blocking action is increased by the substitution of a TEA ethyl group by a hydrophobic side chain (C_9, nonyl; C_5, pentyl; C_1, methyl); the potassium current increases normally but then decreases with C_5 and C_9 derivatives (inactivation) which indicates that these reagents block by entering the opened channel.

funnel-shaped, so that both TEA ions and hydrated potassium ions can be bound there (Fig. 6.10). The funnel must have a diameter of about 0.8 nm.

As the tetraalkylammonium salt does not simply block the potassium current but causes inactivation TEA can apparently only enter into the funnel after the channel is opened by stimulation of the nerve. The gating mechanism seems to be on the inside of the membrane in order to control the funnel. Fig. 6.10 illustrates these observations. It includes the following characteristics of the potassium channel [28]:
1. The gating mechanism on the inside of the membrane.
2. The funnel which makes it possible for hydrated potassium and tetraalkylammonium ions to enter after opening of the gate (diameter ca. 0.8 nm).
3. The true selectivity filter which allows the potassium ion after dehydration to pass but not the TEA ion (diameter ca. 0.3 nm).
4. A hydrophobic region which interacts with one of the side chains of the tetraalkylammonium ions and increases its binding.

The model Fig. 6.10 describes the conditions in the squid giant axon. In contrast in the node of Ranvier of the frog sciatic nerve TEA inhibits both on the outside and on the inside of the membrane. Here there must be a funnel shaped enlargement on both sides of the potassium channel. However the phenomenon of inactivation is found only on the inside of the axon membrane, so the gating mechanism seems also to be localised on the inside in these myelinated fibres.

Other blockers, like for example, 4-aminopyridine and 3,4-diaminopyridine [29] inhibit the potassium current both on the inner and outer sides even of the squid giant axon. Aminopyridine is already bound to the closed channel.

The "hydrophobic zone" in the funnel probably consists of lipids. This zone can be radioactively marked and identified by a photo affinity reagent (Fig. 6.8), a derivative of TEA which can be activated by irradiation to form a covalent bond with its binding area.

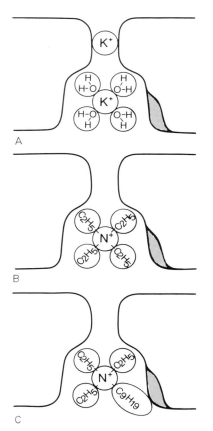

Fig. 6.10. Model of the potassium channel. (A) A funnel-shaped enlargement allows the solvated ion in; the selectivity filter allows only the desolvated ion through. (B) TEA passes into the funnel and can block it but cannot permeate the filter. (C) The blocking action is increased by a hydrophobic side chain as this interacts with a hydrophobic area (thickly outlined) in the channel funnel. The gating mechanism which closes the funnel, is omitted.

Some physical properties of the potassium channel

As with the sodium channel the temperature coefficient Q_{10} is about 1.3 which implies an energy barrier of about 17 kJ/mol. In the squid giant axon noise analysis gives a single channel conductivity of 12 x 10^{12} Ω^{-1}, i.e. it is three times greater than that of the sodium channel [30]. The same noise measurements show that there are five times more sodium than potassium channels per unit surface area of membrane. On the assumption that the opening of a channel is an "all or nothing" reaction, the duration of opening of a single channel is about 1 ms. There is probably no interaction between channels.

Biochemical characterization of the potassium channel

From the dose response curve it is concluded that TEA is bound to the potassium channel with a dissociation constant of 0.4 mmol/l. This is not enough for a binding test as a basis for the isolation and biochemical investigation of potassium channel molecules. There is therefore less information about this than about the sodium channel. One possibility perhaps will be to use TEA derivatives (Fig. 6.8) which, as they are more hydrophobic, have very much lower dissociation constants. Even better are derivatives which react covalently with components of the potassium channel, so that these can be radioactively marked and their purification and structure can be investigated. Such a derivative is the photoaffinity reagent 4-azido-2-nitrobenzyl-triethylammonium-tetrafluoroborate which reacts selectively and irreversibly with the potassium channel.

The proteinaceous nature of the components of the potassium channel was shown by their reaction with protein specific reagents such as the amino group reagent fluorodinitrobenzene (FDNB); the pH dependence of potassium conductivity suggested the participation of a group with a P_K value of 6.3 (crustacean) or 5.2 (frog); and the assertion that this group might be the imidazole ring of a histidine residue was supported by the selective blocking of potassium conductivity by the imidazole reagent diethyl pyrocarbonate (DEP) (Fig. 6.4).

Structure of the axonal membrane: biochemistry, electron microscopy, spectroscopy

The axon membrane consists of other components besides sodium and potassium channels. Lazdunski [31] estimated that in the crustacean nerve only one in a thousand polypeptide chains is a channel component. The active transport system will be dealt with in Chapter 7, here we will only note that ATPase along with acetylcholinesterase is the most notable enzymatic activity of the membrane. While ATPase has a clearly defined role as a component of the sodium potassium pump, the role of acetylcholinesterase is not clear. Because of its ubiquity Nachmansohn [32] proposed the hypothesis that the propagation of the action potential and synaptic transmission involve the same mechanism and that acetylcholine plays a central role in both since it is the transmitter at numerous synapses. This hypothesis was supported by the discovery in the axon membrane of binding sites for nicotine, a synaptic agonist. Incompatible with it, however, was the fact that axons without binding sites for nicotine were found and that typical blockers of synaptic transmission like some cobra toxins had no effect on the mechanism of action potential. In addition there are postsynaptic membranes which are only stimulated chemically and not, like the axon membrane electrically. While this hypothesis undoubtedly stimulated neurobiological research it is no longer considered tenable and the role of acetylcholinesterase in the axon membrane remains a mystery.

The action potential of the squid giant axon can be described on the basis of the Hodgkin and Huxley theory by assuming two types of voltage and time dependent channels, one each for K^+ and Na^+. But the following facts suggest that this description is incomplete: the node of Ranvier in vertebrates propagates action potentials without voltage dependent potassium channels; the theory on the other hand does not include other conductances which are clearly present in several of the excitable membranes investigated in vertebrates;

in bullfrog sympathetic neurons, (and incidently in molluscan cells) in addition to the Hodgkin and Huxley channels, a calcium-activated K^+ channel has been found; a voltage sensitive Ca^{2+} channel has also been observed (this should not be confused with the calcium current entering through the sodium channel).

Functionally similar channels do not have to be identical in different cells and organisms. Cardiac muscle cell membranes for example contain a voltage dependent sodium channel which is insensitive to TTX! For further information the reader is referred to a recent review [33].

Besides the enzymes, ATPase and acetylcholinesterase already mentioned, a large quantity of contractile proteins similar to actin, myosin and tropomyosin, and also tubulin-like molecules have been found in the axonal membrane. However the total protein content seems to be considerably less than that of other cell membranes.

There is little chance of discovering anything having directly to do with the mechanism of the action potential by studying the axon membrane molecules electronmicroscopically since channels are so scarce; as mentioned before the average distance between the fifty channels in $1\,\mu m^2$ of axon membrane is 140 nm. For every channel molecule there are thousands of other protein and more than 10^6 lipid molecules. In electronmicrographs the axon membrane shows the usual trilaminar structure about 8.5 nm thick. Using the freeze fracture technique with crayfish axons Peracchia [34] (Fig. 6.11) discovered regions containing ordered globular structures whose function is not clear.

The propagation of the nerve impulse is a dynamic process which is accompanied by structural changes in the excitable membrane. These changes are manifested as changes in the ion permeability and also in the reactivity of functional groups to specific reagents. We have mentioned the acceleration of the reaction with SH reagents by nerve excitation; further methods for detecting alterations in structure were proposed by L.B. Cohen [35] who showed that the birefringence and light scattering of the membrane were changed during excitation. At the same time the volume of the excited axon increased. However no intrinsic parameter of the axon, neither UV absorption nor fluorescence can so far be directly correlated with the action potential. Extrinsic signals such as the fluorescence of merocyanine dye incorporated into axons show the same time dependent changes as the membrane potential during the passage of a nerve impulse (Fig. 6.12) [35,36]. This method perhaps provides only indirect information about molecular processes in the excitable membrane, but optical methods may nevertheless allow a valuable extension of experimental possibilities for the measurement of membrane potentials where electrical methods are not possible.

Merocyanine belongs to the group of potential-dependent dyes. The alteration of their absorption with applied potential may be explained by two mechanisms depending on the structure of the dye used: in one the dye molecule is forced out of the membrane, a change in absorption results then from the fact that the absorption of the chromophore in the lipid environment of the membrane differs from that in the watery medium of the cell; in the other, which applies to dyes possessing a dipole momentum, the orientation of the dye and thus its absorption of incident light is altered by the applied potential.

Voltage dependent dyes can also be used for monitoring potential changes in organelles or vesicles too small to be impaled by microelectrodes. Alternatively lipophilic cations can

Structure of the axonal membrane: biochemistry, electron microscopy, spectroscopy

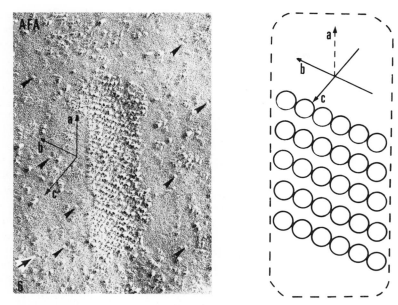

Fig. 6.11. Electron micrograph of a crayfish axon. The picture shows linear depressions in which globular particles are arranged in parallel chains. The function of these structures is unknown. They could be concerned with a) the excitation of the membrane, b) the exchange of substances between axon and its surroundings, c) intercellular adhesion. Reproduced, with kind permission, from Peracchia [34].

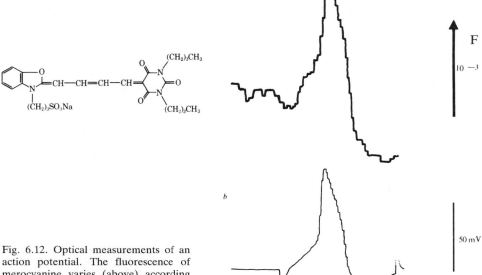

Fig. 6.12. Optical measurements of an action potential. The fluorescence of merocyanine varies (above) according to the membrane potential (below).

be utilized as, for example, tetraphenylphosphonium or triphenylmethylphosphonium ($TPMP^+$) which accumulate in membrane systems in a voltage dependent manner.

Summary

Channels for the passive transport of ions through excitable membranes consist of two functional components, the gating mechanism and the selectivity filter. The gating mechanism which opens or shuts the channel can be activated electrically i.e. by changes of the membrane potential or chemically, e.g. by a transmitter molecule at the synapse. The selectivity filter differentiates by its dimensions and constituent molecular groups which ions can pass and which are retained. Ion transport through the channel is described not simply as a diffusion but rather as a sequence of individual steps, whose activation energy characterizes the speed of transport: binding of ions to channel molecules, desolvation, solvation and dissociation of the ions from the other side of the membrane.

Numerous neurotoxins affect passive ion transport through the axon membrane. As they impair various phases of the transport they can be valuable tools for the analysis of the membrane molecules taking part in ion transport. GTX and BTX affect the activation of the sodium channel, TTX and STX reduce the conductivity of the channel, without affecting the gating mechanism, ATX and certain scorpion toxins block the inactivation mechanism thus inhibiting the closing of the sodium channel.

Anaesthetics, a heterogeneous class of substances with different modes and sites of action, can affect the conduction of the impulse by blocking the gating mechanism and the maximum conductivity $\bar{g}_{Na}$.

References

Cited:

[1] Hodgkin, A.L., and Huxley, A.F., "Current carried by sodium and potassium ions through the membrane of the giant axon of Loligo", *J. Physiol.* **116**, 449-472 (1952).
[2] Hodgkin, A.L., and Huxley, A.F., "The component of membrane conductance in the giant axon of Loligo", *J. Physiol.* **116**, 473-496 (1952).
[3] Hodgkin, A.L., and Huxley, A.F.," Quantitative description of membrane current and its application to conduction and excitation in nerve", *J. Physiol.* **117**, 500-544 (1952).
[4] Pfenninger, K.H., "Organization of neuronal membranes", *Ann. Rev. Neurosci.* **1**, 445-472 (1978).
[5] Keynes, R.D., and Rojas, E., "Characteristics of the sodium gating current in squid giant axon", *J. Physiol.* **223**, 28P (1978).
[6] Armstrong, C.M., and Benzanilla, F., "Currents related to movements of the gating particles of the sodium channel", *Nature* **242**, 459-460 (1973).
[7] Hille, B., "Gating in sodium channels of nerve", *Ann. Rev. Physiol.* **38**, 139-152 (1976).
[8] Keynes, R.D., "Organization of the ionic channels in nerve membranes". In: *The Nervous System*. Vol. 1, D.B. Tower (ed.), Raven Press, New York 1975.

[9] Rojas, E., and Bergmann, C., "Gating currents: molecular transition associated with he activation of sodium channels in nerve", *TIBS* **2**, 6-9 (1977).
[10] Hille, B., "Ionic selectivity of Na and K channels of nerve membranes". In: *Membranes*. G. Eisenman, (ed.), Marcel Dekker Inc., New York, 1975.
[11] Hille, B., "Ionic selectivity, saturation, and block in sodium channels. A four-barrier model", *J. Gen. Physiol.* **66**, 535-560 (1975).
[12] Läuger, P., "Kinetic properties of ion carriers and channels", *J. Membrane Biol.* **57**, 163-178 (1980).
[13] Ritchie, J.M., and Rogart, R.B., "Density of sodium channels in mammalian myelinated nerve fibres and nature of the axonal membrane under the myelin sheath", *Proc. Natl. Acad. Sci. USA* **74**, 211-215 (1977).
[14] Catterall, W.A., "The emerging molecular view of the sodium channel", *TINS* **5**, 303-306 (1982).
[15] Sigworth, F.J., and Spalding, B.C., "Chemical modification reduces the conductance of sodium channels in nerve". *Nature* **283**, 293-295 (1980).
[16] Narahashi, T., "Chemicals as tools in the study of excitable membranes", *Physiol. Rev.* **54**, 813-889 (1974).
[17] Narahashi, T., Deguchi, T., Urakawa, N., and Ohkubo, Y., "Stabilization and rectification of muscle fibre membrane by tetrodotoxin", *Am. J. Physiol.* **198**, 934-938 (1960).
[18] Narahashi, T., Moore, J.W., and Frazier, D.T., "Dependence of tetrodotoxin blockage of nerve membrane conductance on external pH", *J. Pharmacol. Exp. Ther.* **169**, 224-228 (1969).
[19] Beress, L., Beress, R., and Wunderer, G., "Isolation and characterization of three polypeptides with neurotoxic activity from *Anemonia sulcata*", *FEBS Letters* **50**, 311-314 (1975).
[20] Bergman, C., Dubois, J.M., Rojas, E., and Rathmayer, W., "Decreased rate of sodium conductance inactivation in the node of Ranvier induced by a polypeptide toxin from sea anemone". *Biochim. Biophys. Acta* **455**, 173-184 (1976)..
[21] Rojas, E., and Rudy, B., "Destruction of the sodium conductance inactivation by a specific protease in perfused nerve fibres from Loligo", *J. Physiol.* **262**, 501-531 (1976).
[22] Catterall, W., "Activation of the action potential Na$^+$ ionophore by neurotoxins", *J. Biol. Chem.* **252**, 866-867 (1977).
[23] Carbone, E., Wanke, E., Prestipino, G., Possani, L.D., and Maelicke, A., "Selective blockage of voltage-dependent K$^+$ channels by a novel scorpion toxin", *Nature* **296**, 90-91 (1982).
[24] Seeman, P., "The actions of nervous system drugs on cell membranes", In: *Cell Membranes*, Weissmann, G., and Claiborne, R., (eds.), H.P. Publishing Co., New York 1975.
[25] Lee, A.G., "Model for action of local anaesthetics", *Nature* **262**, 545-548 (1976).
[26] Richards, C.D., Martin, K., Gregory, S., Keightley, C.A., Hesketh, T.R., Smith, G.A., Warren, G.B., and Metcalfe, J.C., "Degenerate perturbations of protein structure as the mechanism of anaesthetic action", *Nature* **276**, 775-779 (1978).
[27] Armstrong, C.M., "Interaction of tetraethylammonium ion derivatives with the potassium channels of giant axons", *J. Gen. Physiol.* **58**, 413-437 (1971).
[28] Armstrong, C.M., "Potassium pores of nerve and muscle membranes", In: *Membranes*, G. Eisenman, (ed.), Marcel Dekker Inc., New York 1975.
[29] Kirsch, G.E., and Narahashi, T., "3,4-Diaminopyridine. A potent new potassium channel blocker", *Biophys. J.* **22**, 507-512 (1978).
[30] Conti, F., de Felice, J.L., and Wanke, E., "Potassium and sodium ion current noise in the membrane of the squid giant axon", *J. Physiol. (London)* **248**, 45-82 (1975).
[31] Balerna, M., Fosset, M., Chicheportiche, R., Romey, G., and Lazdunski, M., "Constitution and properties of axonal membranes of crustacean nerves", *Biochemistry* **14**, 5500-5511 (1975).
[32] Nachmansohn, D., and Neumann, E.: *Chemical and Molecular Basis of Nerve Activity*. Academic Press, New York 1975.
[33] Adams, P., "Voltage-dependent conductance of vertebrate neurons", *TINS* **5**, 116-119 (1982).
[34] Peracchia, C., "Excitable membrane ultrastructure: I. Freeze fracture of crayfish axons", *J. Cell Biol.* **61**, 107-122 (1974).

[35] Cohen, L.B., Salzberg, B.M., and Grinvald, A., "Optical methods for monitoring neuron activity", *Ann. Rev. Neurosci.* **1**, 171-182 (1978).
[36] Salzberg, B.M. Obaid, A.L., Senseman, D.M., and Gainer, H., "Optical recording of action potentials from vertebrate nerve terminals using potentiometric probes provides evidence for sodium and calcium components", *Nature* **306**, 36–40 (1983).

Further reading:

Chang, D.F.C., Tasaki, I., Adelman, W.J., and Leuchtag, H.R. (eds): *Structure and Function in Excitable Cells*. Plenum Press, New York and London 1983.
Reichardt, L.F., and Kelly, R.B., "A molecular description of nerve terminal function", *Ann. Rev. Biochem.* **52**, 871–926 (1983).
Hucho, F., and Ovchinnikov, Yu. (eds): *Toxins as Tools in Neurochemistry*. De Gruyter, Berlin 1983.
Marty, A., "Ca^{2+}-dependent K^+ channels with large unitary conductance", *TINS* **6**, 262–265 (1983).
Hille, B.: *Ionic Channels of Excitable Membranes*. Sinauer Associates, Massachusetts 1984.
Adams, P., "Voltage-dependent conductance of vertebrate neurons", *TINS* **5**, 115–119 (1982).
Kostyuk, P.G., "Calcium channels in the neuronal membrane", *Biochim. Biophys. Acta* **650**, 128–150 (1981).
See also ref. 7, 12, 14, 16, 24, 26, 33 and 35.

Chapter 7

Active Ion Transport

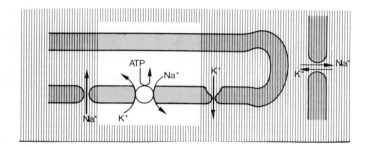

Fig. 7.1.

In Chapter 6 the sodium and potassium channels which regulate the passive ion flow during an action potential were described, but one aspect of axonal membrane function concerned with conductance of nerve impulses has not yet been covered: active ion transport. If an inflow of sodium ions followed by an outflow of potassium ions were the only processes involved, the concentration gradient between the inside and outside of the cell would soon be dissipated. Sodium ions leaking passively into the membrane during the resting state would have the same effect. The entering sodium ions must therefore be transported out again and the outwardly diffusing potassium ions taken up into the axon. Energy must be expended as the process occurs against the concentration gradient. For this the axon membrane contains ion pumps which, with the help of metabolic energy stored in ATP, takes over the active transport of ions for the maintenance of the membrane potential. The single ion movements and potential gradients are schematically summarized in Fig. 7.2. Hodgkin and Keynes [1] demonstrated active sodium transport through the nerve membrane. They showed that the outflow of radioactive sodium ions injected into a cell is inhibited by 2,4-dinitrophenol (Fig. 7.3A) which blocks the synthesis of ATP. The participation of enzymatically catalysed reactions in sodium transport was shown in further experiments by the same authors (Fig. 7.3B). Cooling the cell to 9.8°C (or even to 0.5°C) clearly slowed the sodium outflow. Passive Na^+ diffusion would not be so strongly temperature dependent.

Does this, however, refute our earlier statement that in the squid giant axon it is possible to release numerous action potentials, although the axoplasm has been replaced by simple salt solutions (Chapter 5)? Apparently the membrane potential is hardly altered by cutting off the energy supply. This paradox can be resolved by a short calculation: during an action potential ca. 5×10^{-12} moles of potassium ions go through an axon surface

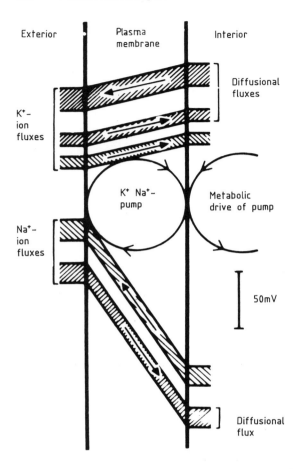

Fig. 7.2. Diagram showing ionic fluxes across the excitable membrane for K^+ and Na^+ ions under resting conditions. These are in part passive diffusional fluxes down the electrochemical gradients (Na^+ inwards, K^+ outwards), and in part active transport against these gradients (Na^+ outwards, K^+ inwards) by a specific ion pump driven by the energy of metabolism. The fluxes due to diffusion and the operation of the pump are distinguished by the cross-hatching; the respective widths of the channels indicate the magnitude of the fluxes and their slopes that of the electrochemical gradients. The diffusional fluxes are increased greatly during activity (after Eccles [16]).

of 1 cm². The potassium concentration in the axoplasm is about 0.5 mol/l. So one axon with a diameter of 0.5 mm accommodates 10 μmol K^+ under a one cm² surface area which is 2×10^6 times the amount lost by one action potential, in other words: 1000 action potentials change the potassium concentration by only about 0.05%.

Thus in the short term the neuron could function without ion pumps, but long term damage would occur, not only as a result of the depletion of potassium ions in the interior of the cell, but also from the great rise in the K^+ concentration in the narrow extracellular space. In Chapter 1 we have mentioned that an accumulation of K^+ in the immediate environment lowers the excitability of the axon and can lead to a spontaneous release of nerve impulses; perhaps this is one of the reasons for an epileptic attack and – as we have suggested in Chapter 1 – an important role for glial cells may be to buffer the extracellular K^+ concentration and thus influence the excitability of neighbouring axons.

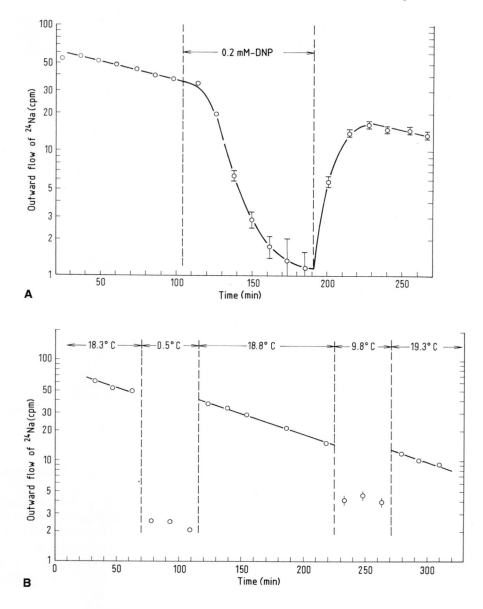

Fig. 7.3. Evidence for the active transport of Na^+ ions. (A) The transport of radioactive ions is blocked by 2,4-dinitrophenol (DNP) an inhibitor of oxidative ATP synthesis; (B) cooling to 0.5°C or 9.8°C slows down the outward Na^+ flow considerably–evidence that Na^+ transport behaves like an enzyme-catalysed reaction and not as a diffusional process which would have been less temperature dependent. Reproduced, with kind permission, from the authors and the Journal of Physiology [1].

In any case the sodium and potassium concentrations are restored by active transport. ATP, as fuel for the ion pump represents an important link between cell metabolism and nerve conduction. We will see that both Na^+ and K^+ concentrations are reestablished by a common molecular mechanism, the sodium/potassium pump.

Three examples of ATP-driven ion pumps

Active ion transport is not only significant for the maintenance of the membrane potential of excitable membranes. Equally the functional capacity of a nerve cell does not only depend on the Nernst potential (see Chapter 5) of its cell membrane. As in most other cells there are other transport processes which are energy dependent and fulfil quite different functions. The following are three examples of ion pumps: the first is the sodium/potassium pump, which takes care of the "correct" distribution of Na^+ and K^+ in the interior and exterior of cells (not only nerve cells); the second is a group of transport mechanisms which maintain the correct calcium concentration both at the excitable membrane and also in the interior of cells; the third pump transports protons. We are concerned here with two diametrically opposite physiological functions: the sodium/potassium pump and also the calcium transport mechanism act with the use of ATP ions against a concentration gradient; in contrast the proton pump in a reverse action uses the ion flux in the synthesis of ATP. The reaction

$$ADP + P_i \rightleftharpoons ATP$$

is reversible, and the two functions differ principally in that the cell in one case "utilizes" the forward and in the other case the reverse direction. In one case the electrochemical potential of an ion gradient is used for chemical synthesis (ATP synthesis) and in the other the chemical energy of ATP for the construction of ion gradients.

Energy-driven ATP synthesis does not appear at first sight to be a neurochemical problem, but there are certain similarities between signal transduction and energy coupling. On the structural level they have much in common: both processes are membrane phenomena and they are executed by proteins characteristically built into lipid membranes. The link between them will be seen clearly when we discuss the light driven proton pump of halophilic bacteria (p. 151). There a receptor analogous to neuronal receptors (Chapters 8 and 9) receives a signal from the exterior which is transduced through the plasma membrane to the interior. The external signal is the energy of a light source converted into the intracellular synthesis of ATP. Bacteriorhodopsin and the elucidation of its mechanism of coupling between photo-reception and energy driven proton transport (and finally ATP synthesis) is of particular interest in neurotransmitter receptor research.

The sodium/potassium pump

In 1957 Jens Skou discovered the Na^+,K^+-ATPase (EC 3.6.1.3), an enzyme which hydrolyses ATP only in the presence of sodium and potassium ions (and Mg^{2+}) [2]. The enzyme was membrane bound, and now it is certain that it is an integral part of the sodium/potassium pump [3]. The evidence for this is the following:

1. ATP hydrolysis and Na^+,K^+ transport show the same ion dependence; both need sodium and potassium ions (and Mg^{2+}).
2. Digitalis glycosides like ouabain (strophanthin) and digitoxin inhibit Na^+,K^+-ATPase and the pump equally.
3. Both processes show the same orientation in the membrane: they act only if ATP and Na^+ are on the inside of the membrane and K^+ on the outside. Digitalis glycosides block the enzyme and the pump only on the outside.
4. Reconstitution experiments in which highly purified Na^+,K^+-ATPase is incorporated into unilamellar liposomes shows that the pump is at least closely associated with the enzyme.
5. Extremely high sodium concentrations on the outside of the membrane cause a reverse reaction: ATPase is changed to ATP-synthetase i.e. it forms ATP from ADP and P_i.

In Chapter 6 it has been shown that passive and active transport are independent of each other in every respect. The most important evidence for this is summarized in Table 7.1.

In addition to energy dependence it can be concluded from the ion selectivity, pharmacological properties and density per unit surface area of the membrane that the two transport systems also act as separate molecular structures.

One of the most interesting problems of biochemistry concerns the direct transformation of chemical energy into work, the basis of such varied biological processes as muscle contraction, the transport of materials between neuronal perikarya and synapses and the active transport of ions and molecules across the cell membrane. It has been estimated that 30% of the energy of respiration at rest is used for the sodium/potassium pump.

The isolation and characterization of the molecules involved must precede an understanding of the molecular mechanisms. To summarize what is known about the biochemistry of sodium/potassium ATPase: it consists of several subunits; polypeptide chains with relative molecular masses of 95 000 and 45 000, a glycoprotein, have been identified. The larger chain represents the catalytic subunit and has been cloned and sequenced. The complete molecule has a relative molecular mass of 250 000 and hydrolyses ATP in a multi-step reaction in which it is transiently phosphorylated and oscillates between several conformations. The polypeptide chain with the relative molecular mass 95 000 is phosphorylated so that the phosphate group forms an acylphosphate with the carboxyl group of a glutamic acid side chain (Fig. 7.4). Fig. 7.5 shows how the ions are transported in a model which has only been partially established experimentally. This involves the following steps:

7 Active Ion Transport

Table 7.1. Comparison of passive and active sodium and potassium ion transport.

	Na^+ channel K^+ channel	Na^+/K^+ pump
Direction of ion movement	down with the electrochemical gradient	against the electrochemical gradient
Source of energy	the ion gradient	ATP
Voltage dependence of conductance	dependent on potential	independent of potential
Inhibitors	TTX blocks g_{Na^+} at 10^{-8} mol/l TEA blocks g_{K^+} at 10^{-3} mol/l ouabain has no effect	TTX and TEA have no effect ouabain blocks at 10^{-7} mol/l
Ca^{2+} (external)	increases the threshold for excitation	no effect
Selectivity	Li^+ and Na^+ are similarly permeable	Li^+ is pumped much more slowly than Na^+
Effect of temperature	speed of opening and closing channels has high temperature coefficient, but the maximum conductivity is little affected.	speed of pump greatly inhibited by cooling
Surface concentration of transport system	between 3 and 500/µm² (squid) TTX binding sites	between 750 (rabbit vagus) and 4000/µm² (squid) ouabain binding sites
Maximum speed of transport for Na^+	10^{-8} mol·cm^{-2}·sec^{-1} during the ascending phase of the action potential	6×10^{-11} mol·cm^{-2}·sec^{-1} at room temperature
Metabolic inhibitors	no effect; electrical activity is normal in an axon perfused with pure salt solution.	CN^- (1 mmol/l) and 2,4-dinitrophenol (0.2 mmol/l) block as soon as ATP supply is exhausted.

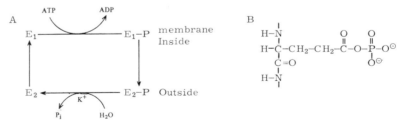

Fig. 7.4. Mechanism of action of Na^+,K^+ ATPase. (A) The enzyme goes through a phosphorylation–dephosphorylation cycle resulting in the hydrolysis of ATP to ADP and P_i. The phosphorylation step (kinase reaction) is Na^+ and Mg^{2+} dependent, the dephosphorylation step (phosphatase reaction) is activated by K^+. (B) The phosphate group forms an anhydride with a glutamyl side-chain of the protein.

1. The transport system has a higher affinity for Na^+ than for K^+ in the conformation E_1 on the inside of the membrane.
2. Phosphorylation on the inside leads to the conformation change $E_1 \rightarrow E_2$ and causes the translocation of Na^+.
3. In the conformation E_2 the transport system on the outside of the membrane has a greater affinity for K^+.
4. K^+ activates the cleavage of the phosphate group, resulting in a translocation of K^+ from the outer to the inner side.

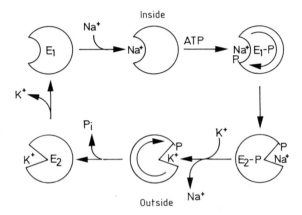

Fig. 7.5. One model of active ion transport through the membrane. It describes the sodium/potassium pump as a carrier having a higher affinity for sodium ions on the inside of the cell membrane and a higher affinity for potassium ions on the outside. The changes in affinity are caused by conformation changes brought about by phosphorylation and dephosphorylation. It is not clear how the sodium binding sites of the proteins move from the inside to the outside of the membrane. There is no evidence for the rotation suggested in the model. It is also not clear why only two K^+ are transported for three Na^+ ions.

The model in Fig. 7.5 represents kinetically the separate stages of ion transport but makes no statement about the molecular mechanism of translocation. In particular it should be realised that the rotation of the ion carrier through the membrane as suggested by the model has not yet been confirmed experimentally. On the contrary it has been shown, at least for the calcium transport system of the sarcoplasmic reticulum, that calcium is still transported when its carrier movement is blocked by antibodies [4]. Instead of rotation, a conformation change is probably responsible for ion transport somewhat in the manner suggested by S.J. Singer in his model (Fig. 7.6).

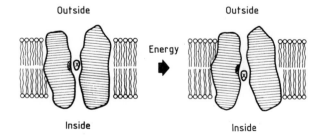

Fig. 7.6. Ion transport model proposed by Singer [17]. It avoids the assumption of the rotation of the carrier and is supported by some biochemical findings.

The detailed role played by the different polypeptide chains in ion transport is not yet known. It is important however that Na^+,K^+-ATPase is only active in the presence of phospholipids [5], above all phosphatidylserine. The model does not demonstrate another important property: Na^+ and K^+ transport is coupled, but not in ratio 1:1. For each hydrolysed ATP molecule approximately three Na^+ are transported for two K^+, a fact which is difficult to reconcile with the competition of both ions for one binding site in the molecule.

Pumps which transport charges: electrogenic pumps

The stoichiometry of ion transport has an important effect: i.e. since more sodium ions are transported outwards than potassium ions inwards a net positive outwardly directed current results. The sodium/potassium pump is therefore an electrogenic pump [6]. If the current is not equalized through the membrane by passive ion currents a hyperpolarization of the membrane is caused (Fig. 7.7). As we have seen, the membrane potential is based on the different passive permeability of the membrane for Na^+ and K^+. But if the sodium/potassium pump is blocked one observes in addition a depolarization – if only a small one – a proof of the contribution of the electrogenic pump to the membrane potential. This amounts to only 1.8 mV in the squid giant axon due to its size, but in smaller neurons or in heart muscle cells it can amount to up to 20 mV.

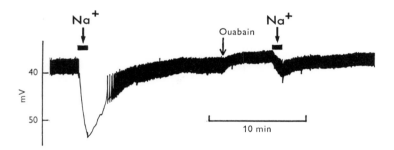

Fig. 7.7. Evidence for the electrogenic character of the sodium/potassium pump. If Na^+ is injected into the interior of the cell (arrow) the membrane potential increases and the membrane becomes hyperpolarized since more Na^+ is transported out than K^+ in. Ouabain, an inhibitor of the pump, blocks the hyperpolarizing action of the injected sodium [18].

Heart glycosides inhibit the sodium/potassium pump

Long before anything was suspected about active ion transport, the beneficial action of digitalis extract on the heart was known. Glycosides like ouabain (strophanthin) and digitoxin were identified as the active substances (Fig. 7.8). They are positive inotropes, that is they stimulate the frequency and amplitude of the contractions of the heart muscle, probably by raising the concentration of calcium which is necessary for the process of contraction. In cases of heart failure, treatment with these heart glycosides are a primary and life-saving form of therapy.

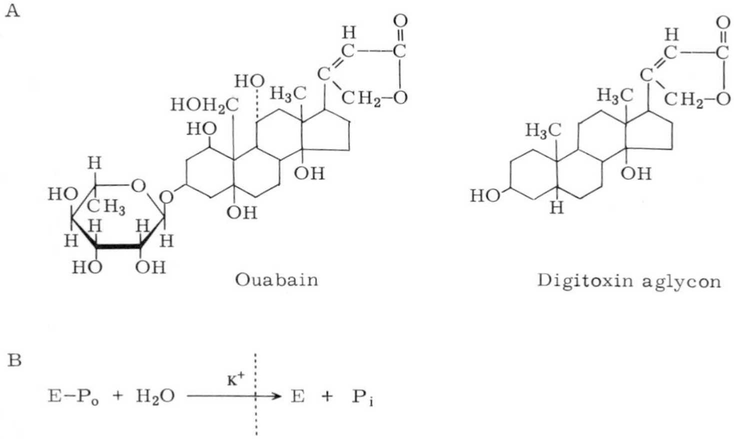

Fig. 7.8. Heart glycosides as blockers of Na^+,K^+-ATPase. (A) Ouabain and digitoxin aglycon; digitoxin is formed from the aglycon by glycosylation. The aglycon of digitoxin is related to the steroids. (B) Ouabain inhibits the K^+-dependent dephosphorylation step of ATPase and blocks the ion pump only on the outside of the membrane.

The connexion between positive inotropic effect and the sodium/potassium pump is not yet known; however it is certain that the inhibition of ATPase is responsible for the toxic side effects of digitalis preparations and also for the supply of calcium ions to the troponin of the contractile mechanism.

In most tissues ouabain concentrations of 0.1 to 1. μmol/l produce 50% inhibition of Na^+,K^+-ATPase. It is primarily the potassium dependent dephosphorylation which is blocked (Fig. 7.8) and it takes appreciably higher concentrations of the steroid to block the sodium-magnesium dependent phosphorylation of the enzyme protein. The site of action is the outside of the membrane.

Ca^{2+} transport mechanism

The functions of calcium in the cell are numerous as are the mechanisms by which its concentration is regulated. Of these functions some of particular neurochemical interest are:
1. In most crustacean muscle fibres Ca^{2+} replaces the role of Na^+ in the inwardly directed current of the action potential.
2. It affects the threshold of the action potential, the latter becomes higher as the outside concentration of Ca^{2+} increases.
3. It is the signal for the initiation of muscle contraction by an action potential. At the same time it mobilizes the energy needed for the contraction by activating phosphorylase b-kinase and so glycogenolysis and glycolysis. It thus coordinates the supply of energy through glycogenolysis for muscular contraction.
4. The entry of calcium following depolarization of the membrane is necessary for presynaptic transmitter release.

The following processes are some of those involved in the regulation of intracellular calcium concentration (Fig. 7.9):

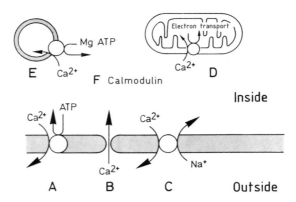

Fig. 7.9. The fate of intracellular Ca^{2+}. Passively inward-flowing calcium ions (B) are removed from the cytoplasm by active transport, (A) through the cell membrane, (D) via mitochondria, (E) via the sarcoplasmic reticulum, (C) by an outward flow coupled with sodium influx, (F) by binding to calmodulin and other Ca^{2+}-binding proteins.

1. Mitochondria can accumulate large concentrations of calcium and thus remove it from the cytoplasm [7]. Its active transport is coupled with a similarly inward directed phosphate transport. The source of energy is the electron transport chain.
2. In the muscle cell the sarcoplasmic reticulum removes the Ca^{2+} liberated by a nerve impulse by active transport. This is manifested as a Ca^{2+},Mg^{2+} activated ATPase.
3. There is a constant passive inflow of Ca^{2+} through the nerve membrane. In the rising phase of the action potential the passive calcium permeabiliby increases, so that a mechanism must be present to expel the calcium which has entered. In the squid giant axon it seems that the passive inward current of Na^+ drives the outward transport of Ca^{2+} [8]. A common carrier has been postulated, which works like a revolving door, allowing one ion in and the other out.
4. There are special Ca^{2+}-binding proteins, calmodulin and calcineurin (see Chapter 10).

As it has been particularly well characterized biochemically, the Ca^{2+}, Mg^{2+} ATPase of the sarcoplasmic reticulum will now be shortly described. It has a simple structure and consists of one polypeptide chain (relative molecular mass 100 000), possibly a proteolipid. It was shown by partial cleavage with trypsin that both functions – ATP hydrolysis and ion transport – are localized on different parts of the same polypeptide chain. One product of tryptic digestion with a relative molecular mass of 30 000 contains the site which, as in the sodium/potassium pump, is phosphorylated transiently by ATP; another with a relative molecular mass of 20 000 can be incorporated in an artificial lipid membrane where it raises selectively the calcium conductivity and may perhaps constitute the actual ionophore [9].

Here again the mechanism by which the energy released by the exothermal hydrolysis of ATP is coupled to ion transport has not been clarified. Two Ca^{2+} ions are transported for one ATP molecule hydrolysed; the phosphate group of the ATP is transferred to an asparagine residue in the ATPase. If the lipids are completely removed the ATPase loses its activity; apparently it is an essential part of the system as with the sodium/potassium pump. More complete information about the mechanism of calcium transport is obtained from reconstitution experiments: highly purified ATPase has been successfully incorporated into artificial lipid vesicles which then actively take up calcium. The basic idea here as in other reconstitution experiments is to reproduce the biological conditions using clearly defined biochemical components and thus to reconstruct the molecular events step by step. By omission or addition of single components of the biological system it is possible to establish which components of the biological membrane are necessary for a given function. Thus Racker and coworkers showed that the proteolipid associated with the 100 000 molecular mass protein molecule is necessary for ion transport but not for ATP hydrolysis [10].

The proton pump: ATP synthesis is the opposite of active transport

The proton pump is considerably more complicated than the ion pumps already described. We have already seen that its physiological function is not ion transport but the reverse, the utilization of an ion gradient for the synthesis of ATP. It is the most important source of energy in the cell. The necessary proton gradient is generated by the mitochondrial electron transport chain and is thus coupled to respiration. In certain microorganisms light can replace respiration as the energy source (see below).

Electron micrographs show characteristic particles on the inner mitochondrial membrane that are connected to it by small stalks. After prolonged controversy in which the particles were claimed to be artifacts of the electron microscopic staining methods it is now accepted that the particles themselves are an ATPase and discussion now centres on the question as to which molecules and molecular mechanisms are involved in the ion transport processes and the formation and hydrolysis of ATP and how these are related to the stalked particles.

Models of proton pumps

Three hypothetical models have been proposed of which the first is increasingly favoured [11].

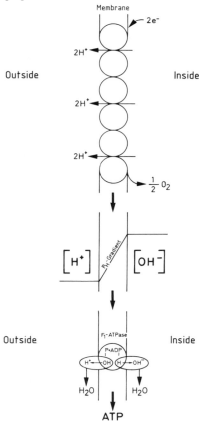

Fig. 7.10. ATP synthesis as the reverse of a proton pump. According to Mitchell during the oxidative electron transport protons are channelled through the membrane. The resulting pH gradient together with the membrane potential permits the synthesis of ATP. Conversely a pH gradient can be formed across the membrane by the hydrolysis of ATP [12]. Reproduced, with kind permission, by A.L. Lehninger.

1. The chemiosmotic hypothesis, also called the Mitchell hypothesis after its protagonist (Fig. 7.10), postulates that the underlying respiratory chain molecules are aligned vectorially in the membrane, so that the various stages of electron transfer from the substrate to oxygen are accompanied by an outwardly directed transport of protons causing a fall in pH on the outside of the membrane relative to the inside. This produces an electrochemical gradient; the inside of the membrane is more negative than the outside. Mitchell considered that the pH gradient causes in the active centre of the ATPase the production of ATP by the splitting off of water from ADP and P_i. "The H^+ is thus delivered into the sink of alkalinity generated in the matrix by electron transport and the OH^- is delivered into the sink of acidity generated on the outside of the membrane" [12].

The net effect is the formation of ATP as a consequence of the formation of a proton gradient and of an electrochemical potential.

2. The chemical hypothesis postulates an energy rich intermediate for the synthesis of ATP as a consequence of the proton flux. A prototype for this is 1,3-diphosphoglycerate which results in glycolysis from the oxidation of 3-phosphoglyceraldehyde by 3-phosphoglyceraldehyde dehydrogenase (GAPDH) and further reacts with ADP to form ATP and 3-phosphoglyceric acid.

3. The conformational hypothesis states that the energy for the synthesis of ATP originates from a transition of ATPase from an energy rich to an energy poor conformation. The electrochemical energy of the proton gradient is utilized to transform the ATPase into the energy rich conformation by releasing the ATP formed.

The three hypotheses are not necessarily mutually exclusive. Hypotheses 2 and 3 were formulated when there was still a difference of opinion as to whether proton or electron transport was the primary energy source for ATP synthesis. The question is still not definitively answered; however there is much support for the direct participation of proton transport: uncouplers like 2,4-dinitrophenol inhibit ATP synthesis but not electron transport, and these substances make the membrane permeable for protons i.e. they inhibit the formation of the pH gradient, elegant support for Mitchell's model. Further excellent evidence came from the investigation of halophilic (salt-loving) purple bacteria (*Halobacterium halobium*), which synthesize their ATP with the help of light energy [13]: here although the electron transport chain can be blocked with specific inhibitors, ATP can nevertheless be formed as long as light energy captured by rhodopsin generates a proton gradient across the cell membrane. The question as to whether ATP synthesis in these exotic bacteria from the Salt Lake in Utah, USA had any relevance for the neurons of higher organisms was resolved by the experiments of two research groups: Ephraim Racker, who worked on the isolation and reconstitution of mitochondrial ATPase, together with Walther Stoeckenius, the expert on the purple membrane of halobacteria, incorporated bacteriorhodopsin into artificial lipid vesicles. After irradiation the lipid membrane transported protons. Racker then added purified ATPase from ox mitochondria to these vesicles. On irradiation ATP was synthesized from ADP and P_i.

We will conclude this chapter with a short description of the molecular structure of one type of proton pump (Fig. 7.11). The mitochondrial system consists of numerous polypeptide chains [14]. The main component is the so called F_1-ATPase. It has a relative molecular mass of 360 000, consists of six chains of different sizes (3 α-chains, M_r 50 000, 3 β-chains, M_r 53 000) and is bound to the membrane by three polypeptide subunits, the "δ" chain, the oligomycin sensitivity-conferring protein (OSCP, relative molecular mass 8 000) and the F_6, all of which are necessary for the attachment of ATPase to the membrane.

The true pore through which the protons penetrate the membrane seems to be a proteolipid which is coupled with another factor, F_0. It interacts with inhibitors of energy transfer like N,N'-dicyclohexylcarbodiimide (DCCD). There is however another coupling factor F_2 which is less stable and less well understood, and even in the purest ATPase preparations when analysed by sodium dodecylsulphate polyacrylamide gel electrophoresis there are other unidentified bands. It is thus clear that the coupling of Mitchell's "proton motive force" with ATP synthesis requires a complex biochemical structure, and much research is still needed before we fully understand the molecular mechanisms involved.

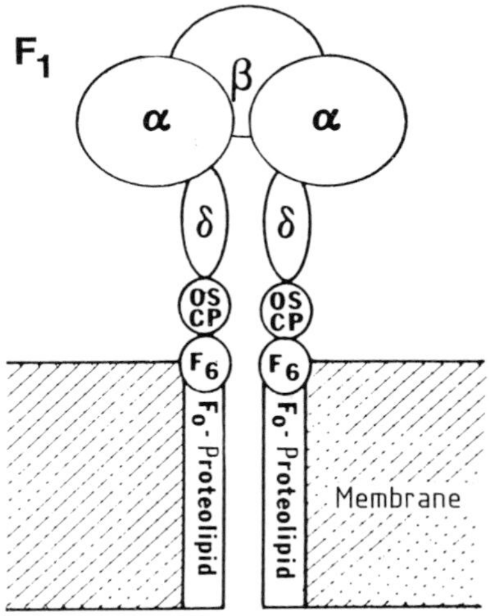

Fig. 7.11. Model for mitochondrial ATP synthetase (longitudinal section). A proton current through a membrane ionophore, the so called F_0 fraction causes the synthesis of ATP from ADP and P_i (see Fig. 7.10) in the head (F_1) of the protein complex; F_1 is a protein which in the isolated state has ATPase activity. The mechanism of this synthesis has not yet been clarified and the function of the subunits is still largely hypothetical [14]. The F_1-head has an $\alpha_3\beta_3$ quaternary structure.

What an ion pump looks like

The clearest picture of an ion pump so far has been derived from the purple membrane of halophilic bacteria. This light-driven proton pump is part of a membrane complex which includes receptors (in this case photoreceptors) converting an extracellular signal (light, the energy source of the organism) into an intracellular effect, the synthesis of ATP. It thus involves generally accepted principles described briefly below. These are shared by other model systems of basic importance in neurochemistry. These include the far more complex and as yet not fully understood mitochondrial "proton pump" which is coupled to the mitochondrial respiratory chain and the sodium/potassium pump of plasma membranes.

The purple membrane consists of 75% bacteriorhodopsin, a protein molecule of relative molecular mass 26 000, forming a Schiff base with one molecule of retinal. The rhodopsin molecules are so regularly arranged that the purple membrane can be considered to be a "two dimensional crystal". Alpha-helical regions comprise 70 to 80% of the polypeptide chain. Henderson and Unwin [15] used electron microscopy of an unstained preparation at a resolution of 0.7 nm to show the arrangement of these polypeptide chains in the membrane (Fig. 7.12).

Groups of seven rod-like α-helices span the membrane and form one molecule. This has the dimensions 2.5 x 3.5 x 4.5 nm of which the greatest extension is perpendicular to the

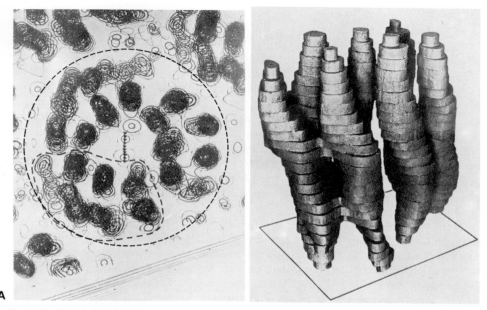

A   B

Fig. 7.12. High resolution electron micrograph of bacteriorhodopsin, the light-driven proton pump of halophilic ("salt-loving") salt water bacteria. The structure is in many respects a model for ion transport systems in other (e.g. neural) membranes. Each molecule is formed from seven helical polypeptide sequences spanning the membrane (B). As the electron density distribution shows (A) three of these molecules form a single unit in which the polypeptide chains form an inner ring of nine and an outer ring of twelve helices. The centre is not just a hole: it contains lipid. Each individual bacteriorhodopsin molecule is an active proton pump. Reproduced, with kind permission, from R. Henderson and McMillan Journals Ltd. [15].

membrane surface. Three of these molecules are arranged around a three-fold axis of symmetry in such a way that there is an inner ring of nine and an outer ring of 12 α-helices. The inner ring with a diameter of 2.0 nm, is filled with lipid (clearly visible in the electron density map Fig. 7.12A). The retinal appears to be attached to a lysine side chain and located between the three inner and four outer helices of each protein molecule. Each individual bacteriorhodopsin molecule is an active proton pump. The functional role of the trimer visualized by electron microscopy is unknown. Henderson and Unwin's model conforms to a basic concept: ion transport systems consist of integral, helical membrane-spanning polypeptide chains. Since the amino acid sequence of bacteriorhodopsin is known, there is a considerable amount of information about the structure of this ion pump. As with the rhodopsin of the retina there is also a lot known about the photochemistry of the light activation. Perhaps this biochemically accessible proton pump of halophilic bacteria will be the first system whose functional mechanism is fully clarified. This may enable inferences to be made about analogous ion transport systems in the cells of higher organisms.

Summary

Active ion transport in the nerve cell has many functions: it maintains the membrane potential in the excitable membrane (sodium/potassium pump) and regulates the intracellular Ca^{2+} concentration (Ca^{2+},Mg^{2+}-ATPase) and the energy supply of the cell (F_1-ATPase, proton pump). The sodium/potassium pump is electrogenic – for every three Na^+ ions transported in an outward direction, only two K^+ ions are conveyed inwards, thereby taking one positive charge out of the cell in each cycle. ATP supplies the energy for active transport against the ion gradient, and is thus the link between impulse conduction and the metabolism of the nerve cell. The ion transport system consists of an ATPase and an ionophore which are more or less complex membrane proteins. One of the protein components undergoes intermediate phosphorylation by ATP. Digitalis glycoside and ouabain (strophanthin) inhibit the sodium/potassium pump by inhibition of the dephosphorylation step.

The Ca^{2+} concentration controls the threshold for the release of action potentials and transmitter release at the nerve ending and is the link between the nerve impulse and muscular contraction. Ca^{2+}, Mg^{2+}-ATPases, in the mitochondrial membrane and the membrane of the sarcoplasmic reticulum regulate the cytoplasmic calcium concentration. The transport system has been isolated, biochemically characterized and its function demonstrated by insertion in an artificial lipid membrane (reconstitution).

The proton pump as opposed to other ATPases, synthesizes ATP with the help of a proton gradient. The system has been isolated from the mitochondrial membrane and partly characterized biochemically and by reconstitution. A three dimensional picture of the light driven proton pump of halophilic bacteria has been obtained by high resolution electron microscopy. This supports earlier hypotheses that such transport systems consist of α-helical polypeptide chains spanning the membrane.

References

Cited:

[1] Hodgkin, A.L., and Keynes, R.D., "Active transport of cations in giant axons from Sepia and Loligo", *J. Physiol. (London)* **128**, 28-60 (1955).
[2] Skou, J.C., "The influence of some cations on an adenosine triphosphatase from peripheral nerve", *Biochim. Biophys. Acta* **23**, 394-401 (1957).
[3] Dahl, J.L., and Hokin, L.E., "The sodium potassium adenosine triphosphatase", *Ann. Rev. Biochem.* **43**, 327-356 (1974).
[4] Dutton, A., Rees, E.D., and Singer, S.J., "An experiment eliminating the rotating carrier mechanism for the active transport of Ca^{2+} ions in sarcoplasmic reticulum membranes", *Proc. Natl. Acad. Sci. USA* **73**, 1532-1536 (1976).

[5] Priestland, R.N., and Whittam, R., "The temperature dependence of activation by phospatidylserine of the sodium pump adenosine triphosphatase", *J. Physiol.* **220**, 353-361 (1972).
[6] Thomas, R.C., "Electrogenic sodium pump in nerve and muscle cells", *Physiol. Rev.* **52**, 563-594 (1972).
[7] Bygrave, F.L., "Mitochondrial calcium transport", *Curr. Top. Bioenerg.* **6**, 259-318 (1977).
[8] Baker, P.F., Hodgkin, A.L., and Ridgway, E.B., "Depolarization and calcium entry in squid giant axons", *J. Physiol.* **218**, 709-755 (1971).
[9] Shamoo, A.E., Ryan, T.E., Stewart, P.S., and MacLennan, D.H., "Localization of ionophore activity in a 20 000 dalton fragment of the adenosine triphosphatase of sarcoplasmic reticulum", *J. Biol. Chem.* **1**, 4147-4157 (1976).
[10] Racker, E., and Eytan, E., "A coupling factor from sarcoplasmic reticulum required for the translocation of Ca^{2+} ions in a reconstituted Ca^{2+} ATPase pump", *J. Biol. Chem.* **250**, 7533-7534 (1975).
[11] Boyer, P.D., Chance, B., Ernster, L., Mitchell, P., Racker, E., and Slater, E.C., "Oxidative phosphorylation and photophosphorylation", *Ann. Rev. Biochem.* **46**, 944-1026 (1977).
[12] Lehninger, A.L.: *Biochemistry*, 2nd edition. Worth Publishers, Inc., New York 1975.
[13] Stoeckenius, W., and Bogomolni, R.A., "Bacteriorhodopsin and related pigments of halobacteria", *Ann. Rev. Biochem.* **52**, 587-616 (1982).
[14] Racker, E., "Structure and function of ATP-driven ion pumps", *TIBS* **1**, 244-247 (1976).
[15] Henderson, R., and Unwin, P.N.T., "Three-dimensional model of purple membrane obtained by electron microscopy", *Nature* **257**, 28-32 (1975).
[16] Eccles, J.: *The Understanding of the Brain*. McGraw Hill, 1973.
[17] Singer, S.J., "The molecular organization of membranes", *Ann. Rev. Biochem.* **43**, 805-833 (1974).
[18] Thomas, R.C., "Membrane current and intracellular sodium changes in snail neuron during extrusion of injected sodium", *J. Physiol.* **201**, 495-514 (1969).

Further reading:

Tanford, Ch., "Mechanism of free energy coupling in active transport", *Ann. Rev. Biochem.* **52**, 379–409 (1983).
Amsel, L.M., and Pedersen, P.L., "Proton ATPases: Structure and mechanism", *Ann. Rev. Biochem.* **52**, 801–824 (1983).
Stoeckenius, W., and Bogomolni, R.A., "Bacteriorhodopsin and related pigments of halobacteria", *Ann. Rev. Biochem.* **51**, 587–616 (1982).
Campbell, A.K.: *Intracellular Calcium*. John Wiley & Sons, New York 1983.
Joergensen, P.L., "Mechanism of the Na^+, K^+ pump", *Biochim. Biophys. Acta* **694**, 27–68 (1982).
Klee, C.B., and Vanamann, T.C., "Calmodulin", *Adv. Protein Chem.* **35**, 213–303 (1982).
Goldin, S.M. Moczydlowski, E.G., and Papazian, D.M., "Isolation and reconstitution of neuronal ion transport proteins", *Ann. Rev. Neurosci.* **6**, 419–446 (1983).

Chapter 8

The Synapse, Part 1

Neuron and synapse, historical remarks

In 1891 Waldeyer formulated his neuron theory. It stated that the nervous system did not consist of a continuum of cytoplasm but many individual cells – the neurons, whereas in the now discarded reticular theory, the nervous system was thought to be a syncitium, a union of cells uninterrupted by membranes. The neuron theory was supported predominantly by the work of Ramón y Cajal and was finally generally accepted. There then remained the vital question of the mode of communication between these single neurons. In 1897 Sherrington introduced the term synapse for the hypothetical region specialized for the exchange of signals between cells. A quarter of a century later the synapse had become a well accepted concept in neurobiology and it is still one of the most interesting subjects of biological research.

Why is the synapse so interesting?

In Chapter 5 we commented briefly on the mode of action of the synapse from the electrophysiological point of view and introduced two alternative synaptic mechanisms, corresponding to the two types of synapses – the excitatory and the inhibitory. At excitatory synapses the nerve impulse is transferred from one cell to another; at inhibitory synapses the transmission prevents the excitation of the receiving cell. This indicates a basic characteristic of the synapse: it is the site of modulation of the nerve impulse. Series of action potentials conducted along the axon to the nerve ending can at the synapse be reinforced, reduced or transmitted unchanged to the next cell.

The synapse as the site of internal regulation of the nerve impulse is also thought to be the site of temporary or permanent changes in the storage of information (learning and memory) or in behaviour as a result of external stimulation (adaptation or habituation). This is called the *plasticity* of the synapse.

The synapse is also the site of action of external regulation i.e. the influence of drugs and toxins. A classical example is the Indian arrow poison *curare* which was shown by Claude Bernard in the last century to block the transmission of the nerve impulse to the muscle fibre. Numerous anaesthetics and psychopharmacological drugs affect synaptic transmission.

Synapses are the site of a wide range of *pathological disorders*. We will discuss as examples Parkinsonism, myasthenia gravis and psychiatric disorders like schizophrenia (Chapter 9).

In the network of cell communication the synapse often has an important function as a rectifier or valve: it is polar and thus conducts an impulse in one direction only.

Finally the synapse is the key to the specificity of a neural junction. In the functioning of the nervous system, the wiring diagram of the neural network is determined not by chance but genetically and probably by other internal and external factors. These define which cells combine together in the ontogeny of a synaptic contact. The formation of synapses is specific (e.g. axons find specific targets and innervate them) – this is the result of the cell's genetic programme.

Specificity, polarity, modulation, plasticity, pharmacology, and pathological disturbances are key phenomena whose molecular correlates are central subjects of neurochemistry and which we will discuss in detail in the following chapters.

Electrical and chemical synapses

Signals can be transmitted between cells either by the direct conduction of action potentials (electrical synapses) or by special molecules of a transmitter substance (chemical synapses). These have very different structures specialized for their specific function. At a chemical synapse the distance between the cells is about 20–40 nm; the synaptic cleft between the cells is part of the extracellular space and contains fluid of low resistance, so that an electric signal becomes dissipated before it reaches the next cell. Electrical transmission, in contrast, only occurs at special structures (gap junctions) where the cells are about 2 nm apart and are connected by conducting channels. Here there is in fact something close to the postulated syncitium or multicellular cytoplasmic continuum, and it is an irony of scientific history that the discovery of the electrical synapse by Furshpan and Potter in 1959 came at a time when Waldeyer's neuron theory had finally superseded the reticular theory.

Properties of the electrical synapse

Electrical synapses are comparatively rare, and their significance in the central nervous system of higher organisms is still not understood. Furshpan and Potter discovered them in the crab abdominal nerve and since then they have been observed in numerous organisms: molluscs, arthropods and mammals. In contrast to the chemical synapse where there is a delay due to the release and diffusion of transmitter substance, the transfer of the signal across an electrical synapse is fast. This rapid coupling of specific cells may denote its physiological significance.

There has been as yet no biochemical research into the structure and mode of action of electrical synapses. However not only nerve cells are connected by gap junctions, they occur in epithelial, muscle and liver cells among many others. From these materials, membrane fragments clearly retaining their intercellular contact areas have been isolated and characterized biochemically and electron microscopically. The electron micrographs show structures of ordered particles (called *connexons* by Goodenough [1]) forming channels between the cells 2 nm apart. Two polypeptides with relative molecular masses of 25 000 and 35 000 have been isolated from these membranes and are called *connexins*. Conceivably two connexons from neighbouring cells could form a channel by dimerization (Fig. 8.1). It has been shown that this channel permits the passage not only of alkali metal ions but of molecules up to a relative molecular mass of 1000 to 2000. They thus provide in addition to electric coupling the possibility of exchange of metabolites between cells. The permeability of the channels can be regulated by Ca^{2+}.

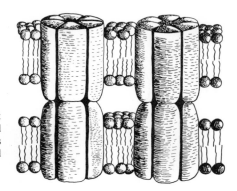

Fig. 8.1. "Gap junction". Model of the area of contact between cells. A similar structure has been proposed for electrical synapses. Protein subunits form channels which span the lipid bilayer. Reproduced, with kind permission, from Goodenough [1] and Cold Spring Harbor Laboratory.

The chemical synapse, control site of the nervous system

Figure 8.2 shows a schematic diagram of a chemical synapse. It consists of a nerve ending on the presynaptic side and the specialized area of a receiving cell on the postsynaptic side. The pre- and postsynaptic membranes are separated by a distance of 20-40 nm. The synaptic cleft is probably filled with oligosaccharide-containing connective tissue, the *basal lamina*, acting as a supporting structure for the two cells. The subsynaptic membrane, the area of the postsynaptic membrane directly opposite the presynaptic nerve ending, is recognizable electron microscopically [2] as a thickening (Fig. 8.3). In nerve muscle synapses, the neuromuscular endplate of vertebrates, it is highly invaginated. The presynaptic nerve ending contains mitochondria and in addition, characteristic vesicles, the synaptic vesicles, in which transmitter is stored.

While electrophysiologists work with the intact synapse, biochemists try to isolate the functional substructures by suitable disruption procedures and subsequent fractionation by density gradient centrifugation. Examples of these are *synaptosomes* (Fig. 8.3A) – pinched-off and sealed nerve endings detached from the axon –, synaptic vesicles, presynaptic membranes, *synaptic complexes* (Fig. 8.3B) – in which the postsynaptic

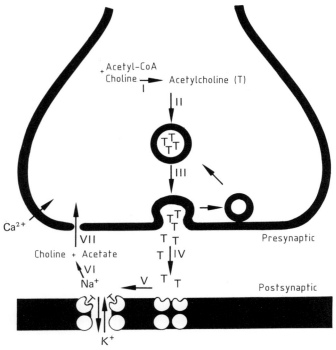

Fig. 8.2. Synapse: Schematic diagram of a nicotinic cholinergic synapse. The presynaptic nerve ending contains the constituents for the synthesis of the transmitter (here acetylcholine). After synthesis (I) the transmitter is packaged in vesicles (II). The synaptic vesicles fuse (probably temporarily) with the presynaptic membrane (III) and the transmitter is thus released into the synaptic cleft. The transmitter diffuses to the postsynaptic membrane and is bound to a specific receptor there (IV). As a result of the formation of the transmitter-receptor complex the postsynaptic membrane becomes permeable to cations (V) i.e. it is depolarized. (If the depolarization is great enough, an action potential is released i.e. the chemical signal is reconverted into an electrical nerve impulse). Finally the transmitter is "inactivated", i.e. either enzymatically degraded (VI) or taken out of the synaptic cleft by a specific "uptake mechanism". In the above diagram only one cleavage product of the transmitter, choline, is taken up by the nerve ending (VII) and recycled. The basal lamina, a diffuse structure seen in the electron microscope, in the synaptic cleft (Fig. 8.3A) is not shown.

membranes or densities remain adherent to a portion of the presynaptic membrane – and the separated postsynaptic densities. Synaptosomes are an especially useful subcellular fraction of the nervous system. Their purification and application as pioneered by V.P. Whittaker has become routine in many neurochemical laboratories. They retain after preparation most of the properties of the original nerve terminal. Among these are membrane excitability, uptake of metabolites, transmitters or transmitter precursors, and Ca^{2+}-dependent transmitter release upon depolarization. They are also a valuable starting point for further subfractionation. Upon osmotic shock they release their cytoplasm and intraterminal organelles (including the functionally important synaptic vesicles); under

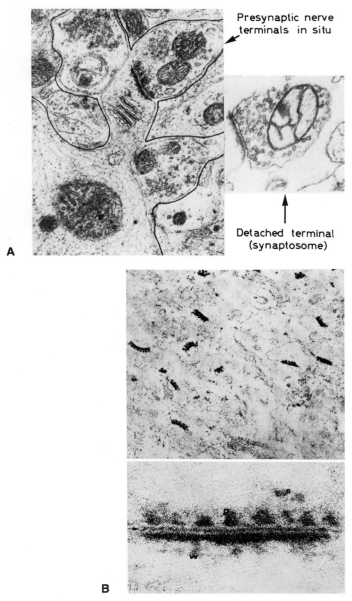

Fig. 8.3. (A) Synapse: Electron micrographs. Left: Nerve terminals *in situ* (outlined); Right: A detached and sealed terminal (synaptosome). The large dark organelles are mitochondria, the small light ones are synaptic vesicles (from V.P. Whittaker [27]). (B) Synaptic complexes which can be isolated from homogenates of nerve tissue consisting of paired pre- and postsynaptic membrane. The arrow indicates the basal lamina, which may bind the pre- and postsynaptic membrane. Magnification above × 33 000, below × 218 000 [2].

appropriate conditions preparations enriched in presynaptic and/or postsynaptic membranes including synaptic complexes and postsynaptic densities can be obtained. Synaptosomes prepared from brain tissue are heterogeneous with respect to transmitter type because they originate from the whole variety of neurons present in this tissue but they are valuable for many investigations of neuronal function especially synaptic function, and are an experimental system with many applications in neurochemistry, neuropharmacology and neurotoxicology.

The cholinergic synapse, peripheral and central

There is no one cholinergic synapse. It is a group of structurally, functionally and pharmacologically very different synapses all of which utilize acetylcholine as transmitter. Of special significance is the neuromuscular junction at which the nerve impulse is transmitted to the muscle fibre and there gives rise to a muscle contraction. There is, however, much evidence that the cholinergic synapse besides this peripheral function, has fundamental functions in the central nervous system [3,4] and takes part in basic phenomena like behaviour, consciousness, affect, learning and memory. Evidence for this is derived from biochemical investigations of acetylcholine metabolism and the associated enzymes in the central nervous system and also from experiments with psychopharmacological substances which show effects on cholinergic synapses. Acetylcholine is also an important transmitter in the autonomic nervous system. In all the ganglia of the sympathetic and parasympathetic system there are cholinergic synapses. In postganglionic nerves i.e. those from the ganglion to the target organ, the nerve impulse is transmitted by acetylcholine at all parasympathetic synapses (e.g. in eyes, heart, lungs, stomach, intestine) and at some sympathetic (e.g. sweat glands).

Two classes of cholinergic synapses, the muscarinic and nicotinic

The pre- and postganglionic cholinergic synapses are not identical. They are distinguishable electrophysiologically and pharmacologically. The release of acetylcholine from the nerve ending of the preganglionic axon produces a short, excitatory (i.e. releasing an action potential) postsynaptic potential (e.p.s.p) whereas the postganglionic axon releases slow excitatory (e.p.s.p) or inhibitory (i.p.s.p) postsynaptic potentials. Synapses of the preganglionic type can be stimulated by nicotine as well as acetylcholine (Fig. 8.4): these are called nicotinic cholinergic synapses. In the postganglionic type the action of acetylcholine can be simulated not by nicotine but by muscarine, the toxin of Fly Agaric (*Amanita muscaria*). They are called muscarinic cholinergic synapses. Both types are also selective for inhibitors: nicotinic synapses are inhibited by curare or its main active substance d-tubocurarine, while muscarinic synapses are inhibited by atropine the toxin of deadly nightshade (*Atropa belladonna*) (Fig. 8.4). Further pharmacological and biochemical differences will be discussed in the section on receptors. The neuromuscular endplate belongs to the nicotinic cholinergic type.

Individual stages of chemical synaptic transmission

When an action potential arrives at the nerve ending, it causes the release of the transmitter by depolarization. The transmitter diffuses across the synaptic cleft to the postsynaptic membrane, producing a change in ion permeability in the membrane and thus a change in membrane potential (see Chapter 5). This in its turn can cause the release of an action potential.

We shall now briefly describe the complex sequence of molecular events analysed into single steps. A transmitter molecule, for example acetylcholine, goes through the following stages:
1. Synthesis of the transmitter.
2. Loading of the transmitter into the vesicle. In the case where stages 1 and 2 take place in the perikaryon, axoplasmic transport of the vesicle to the nerve ending.
3. Fusion of the vesicle on depolarization with the presynaptic membrane and the release of transmitter (*exocytosis*).
4. Diffusion to the postsynaptic membrane.
5. Recognition and binding to a specific receptor e.g. a membrane protein of the postsynaptic membrane, combined with the effect of the transmitter on the ion channel or an enzyme of the postsynaptic membrane.
6. Inactivation of the transmitter, so that the presynaptic signal is of limited duration. The inactivation occurs by enzymic degradation or by reuptake of the transmitter by the presynaptic membrane.

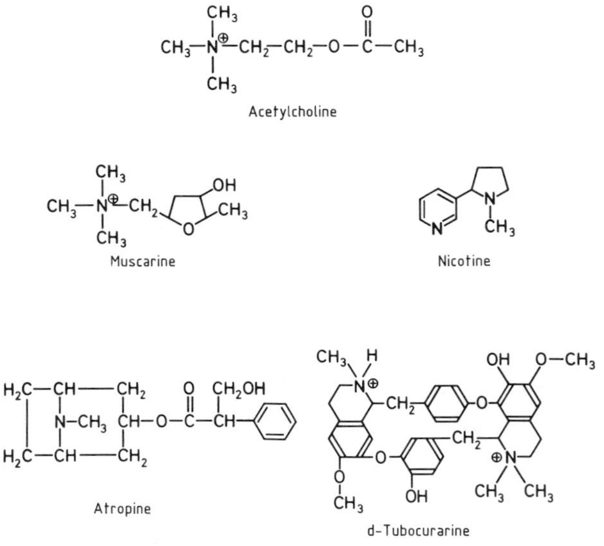

Fig. 8.4. Cholinergic ligands. Above, the agonists acetylcholine, muscarine and nicotine, below, an antagonist for the muscarinic and nicotinic cholinergic synapses respectively.

The cycle of the transmitter is complete and reverts to step 2 (or to step 1 if the degradation product of the inactivated transmitter has been taken up into the cell).

This is a general outline of synaptic function; for the biochemistry of specific structures and molecular processes it is best to take particular examples. The most fully researched synapse is the cholinergic, we will therefore deal next with the acetylcholine cycle. This will be followed by a description of other transmitters, their biosynthesis, biological function and pharmacology and then a section on receptors. The conclusion will be a discussion on the plasticity of synapses and the specificity of synaptic connections.

The nicotinic cholinergic synapse

The first stage in the acetylcholine cycle is the synthesis of acetylcholine from acetyl-coenzyme A (acetyl-CoA) and choline (Fig. 8.2). Acetyl-CoA is the end-product of glycolysis and originates in the mitochondrion by oxidative decarboxylation of pyruvate catalysed by the multienzyme complex pyruvate dehydrogenase. As acetyl-CoA can not permeate the mitochondrial membrane, it must be transported indirectly into the cytoplasm where acetylcholine is synthesized. It is not clear whether the same process takes place in nerve tissue as for example in fatty tissue. There acetyl-CoA reacts with oxaloacetate forming citrate; this is transported out of the mitochondrion and is acted on in the

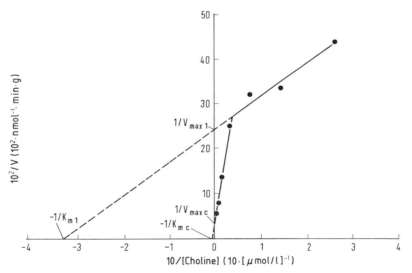

Fig. 8.5. Choline "uptake". The concentration dependence of the rate with which choline is taken up by the synaptosomes shows two mechanisms, a "high affinity" (K_M = 2.4 μmol/l) and a "low affinity" (K_M = 167 μmol/l). The high affinity mechanism is concerned in transmitter synthesis [28]. Reproduced, with kind permission, from Whittaker and Cold Spring Harbor Laboratory.

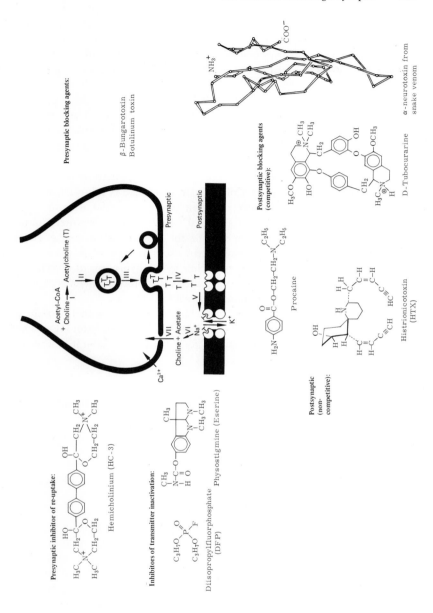

Fig. 8.6. Neuroactive drugs and toxins, which affect the individual stages of nicotinic cholinergic synaptic transmission.

cytoplasm by ATP-citrate-lyase reforming acetyl-CoA and oxaloacetate. Experiments with ^{14}C-labelled citrate fail to support this. They show that neither citrate nor acetate are utilized as the source of acetylcholine in nerve tissue, so the question of the origin of acetyl-CoA remains unanswered. Probably there is in any case a separate acetyl-CoA pool [5].

The second constituent of acetylcholine is choline. It is not synthesized in the nerve ending itself, but comes from the extracellular choline pool. There it originates mainly in the liver from phosphatidylcholine which is formed from phosphatidylethanolamine by series of methylations (see Chapter 2). Choline is also the degradation product of lipids and is present in the brain in concentrations of about 20 µmol/l.

The mechanism by which choline is taken up from the extracellular space is unknown [6]. If the concentration dependency of the rate of uptake is followed in synaptosomes (Fig. 8.5) evidence is obtained for two different transport mechanisms: a high affinity uptake with a Michaelis constant (K_M) of about 10^{-6} mol/l. and a low affinity uptake with a K_M of about 10^{-4} mol/l. The latter is not specific for nerve cells but occurs generally. The high affinity transport mechanism is stimulated by sodium ions and is specific for nerve endings where choline produced in the synaptic cleft by the hydrolysis of acetylcholine is taken up. There is as yet no evidence of an uptake mechanism for the intact acetylcholine, as there is for example in the case of γ-aminobutyric acid (GABA).

There are specific inhibitory drugs for choline transport as there are for every individual step of the transmitter cycle (Fig. 8.6). Here it is hemicholinium HC-3 which inhibits high affinity transport about 500 times more ($K_I = 0.1$ µmol/l) than it inhibits the low affinity.

Choline reacts with acetyl-CoA forming acetylcholine

The reaction

$$CH_3-\overset{CH_3}{\underset{CH_3}{\overset{|+}{N}}}-\overset{H}{\underset{H}{\overset{|}{C}}}-\overset{H}{\underset{H}{\overset{|}{C}}}-OH + CH_3-\overset{O}{\overset{\|}{C}}-SCoA \longrightarrow CH_3-\overset{CH_3}{\underset{CH_3}{\overset{|+}{N}}}-\overset{H}{\underset{H}{\overset{|}{C}}}-\overset{H}{\underset{H}{\overset{|}{C}}}-O-\overset{O}{\overset{\|}{C}}-CH_3 + CoASH$$

is catalyzed by the enzyme choline acetyltransferase (ChAT) discovered by D. Nachmansohn. It is a protein of relative molecular mass 65 000 found in the soluble cytoplasm and also more or less tightly bound to the membranes of perikarya, axons and nerve endings with some species variation. It is synthesized in the cell body and conveyed to the nerve ending by "fast axonal transport" with a speed of about 190 mm per day (rabbit vagus nerve). The long half-life of the enzyme of 12-20 days suggests that it stays there and is not released by nerve excitation. Both ChAT and acetylcholine are distributed throughout the neuron but in a particularly high concentration in the nerve ending. The extent to which axonal transport is also responsible for the supply of transmitter has not yet been established. Transport of acetylcholine and also of vesicles (see below) has been demonstrated (see Chapter 10).

Acetylcholine is packaged in vesicles

In 1955 De Robertis and Bennett discovered spherical structures in the nerve ending, the so-called synaptic vesicles (Fig. 8.3). They assumed that they acted as organelles containing stored transmitter which as Castillo and Katz concluded from their study of miniature endplate potentials (m.e.p.p.s) is released in discrete quanta on nerve excitation and also spontaneously in the resting state. Postsynaptic potentials are always a specific multiple of a unit, the quantum (see Chapter 5). Synaptic vesicles were isolated and their acetylcholine content identified simultaneously in the laboratories of Whittaker and of De Robertis in 1963. The question remained as to whether the transmitter was released directly into the synaptic cleft or indirectly via the cytoplasm. We shall return to this when we discuss the mechanism of transmitter release; next we shall describe how acetylcholine is taken up into the storage vesicles.

There is information about this from the biochemical analysis of vesicles and their contents. They contain a high concentration of ATP and acetylcholine in a molecular ratio of 1:5. The ATP content seems to indicate an active transport uptake mechanism for acetylcholine. However this has not yet been demonstrated. In the vesicle membrane of the electric ray *Torpedo* there is a magnesium-dependent ATPase. As it is stimulated by acetylcholine it is possibly the "transmitter pump". An active transport mechanism is clearly necessary since the acetylcholine concentration in the vesicle varies from 0.2 to 0.6 mol/l according to species.

How does acetycholine get into the synaptic cleft?

In an elegant experiment, molecules of acetylcholinesterase, the acetylcholine-hydrolysing enzyme, were incorporated into synaptic vesicles. This caused an inhibition of synaptic transmission, showing that transmitter stored in the vesicles was involved in the process of presynaptic transmitter release. Subsequently acetylcholinesterase was injected into the axoplasm of cholinergic neurons of *Aplysia,* where again synaptic transmission was inhibited. This time acetylcholine was involved which was stored not in the vesicles but in the cytoplasm. The interrelationship of cytoplasmic and vesicular acetylcholine pools has not yet been completely clarified. It is however generally accepted that transmitter release, both spontaneous and following stimulation, occurs directly from the vesicles themselves which fuse with the presynaptic membrane and liberate, depending on the tissue, between about 2000 (guinea pig cortex) and 200 000 (electric organ of *Torpedo*) molecules of acetylcholine per vesicle [7]. This hypothesis has been supported by electron micrographs of fixed preparations which show many stages in the association and fusion of vesicles with the presynaptic membrane (Fig. 8.7). A question remaining to be answered is the discrepancy between the number of acetylcholine molecules in a vesicle and the size of the m.e.p.p. It is estimated that invertebrate muscle requires a quantum of 10 000 to 12 000 molecules.

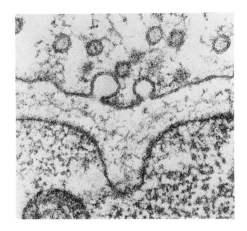

Fig. 8.7. Electron micrograph of the fusion of a synaptic vesicle with the presynaptic membrane, possible evidence for exocytotic transmitter release. From *Sci.Am.* **236**, 107 (1977); Reproduced, with kind permission, from the publisher.

Exocytosis – endocytosis

Fusion of the vesicle does not imply a permanent incorporation of the vesicle membrane into the presynaptic membrane since even after prolonged stimulation i.e. after the fusion of a great number of vesicles, the surface area is only temporarily increased. If peroxidase is present in the synaptic cleft during stimulation, it is soon found in synaptic vesicles – evidence that by endocytosis the fused vesicle membrane is rapidly taken up to form vesicles trapping or reinternalizing extracellular material.

The role of calcium

The cause of the release of acetylcholine is the depolarization of the nerve ending by the incoming action potential. There is no effect however when there are no calcium ions in the extracellular space. We have already mentioned that the threshold of the action potential is affected by Ca^{2+} ions. They are now seen to have a key role in chemical synaptic transmission. The depolarization of the nerve ending increases the membrane permeability for calcium and thus the intracellular calcium concentration. The calcium entering the nerve terminal must however be got rid of again if the stimulation of the synapse is to be transitory. There is a great deal of evidence that intracellular calcium is regulated by the mitochondria and proteins such as calmodulin and calcineurin (see Chapter 7). The mitochondria have a very efficient calcium pump; also inhibitors of mitochondrial function cause an increase in the number of m.e.p.p.s, evidence that the absorption of the calcium influx by the mitochondria is being inhibited. It is not clear where the calcium is transferred to by the mitochondria in order that they themselves shall not become saturated with ions. Still less is known about the molecular site of Ca^{2+}-stimulation of transmitter release. Speculations can be made about the contribution of an actomyosin-like complex but there is as yet no experimental proof. The concentration dependence of the calcium effect shows that several, perhaps four ions, cooperatively activate the liberation of a quantum. Mg^{2+} competes with Ca^{2+} at this site of action, i.e. it inhibits the Ca^{2+}-stimulated transmitter release without being able itself to function as an activator.

The comparatively slow removal of the calcium which has flowed in during depolarization probably accounts for phenomena like "facilitation" and "post-tetanic potentiation", when after one action potential a subsequent one causes a greater postsynaptic potential due to the raised Ca^{2+} concentration already present (see Chapter 11).

Acetylcholine is bound by the postsynaptic membrane

How does the transmitter molecule released from the presynaptic membrane reach the postsynaptic membrane? The superficially correct answer is by diffusion. But an explanation is needed of how it diffuses past the numerous molecules of acetylcholinesterase present in the synaptic cleft which would theoretically be able to hydrolyse many times the quantity of released transmitter and would prevent it from reacting with the postsynaptic membrane. It is postulated that either the substance in the synaptic cleft, the basal lamina, excludes this by its structure perhaps by forming channels, or the enzymatic activity of the esterase is temporarily inhibited perhaps by interaction with the postsynaptic membrane or by saturation with its substrate. It has also been suggested that the esterase is not present in the cleft i.e. in the diffusion path of acetylcholine but is in the postsynaptic membrane, but there is no evidence for this model [8].

The next step is the interaction of the transmitter with the postsynaptic membrane. A drastic change in ion permeability of the membrane occurs due to transmitter binding which operates the gating mechanism of ion channels. Although it sounds straightforward it is a complicated biochemical process. We will discuss two of the complex topics involved:
1. The specificity,
2. The molecular mechanism of the permeability change.

The specificity of the transmitter action has two aspects: substrate specificity – only molecules of specific structure can act as transmitters; and the specificity of action – a transmitter can open or close pores for K^+ or Na^+ or Cl^- – or for both K^+ and Na^+ according to the function of the particular synapse (Table 8.1). We will deal here predominantly with the neuromuscular synapse, in which acetylcholine opens common channels for Na^+ and K^+, but we will also mention other actions of this transmitter.

Acetylcholine produces a local depolarization at the neuromuscular endplate, the endplate potential. Substances which have the same action as the natural transmitter are called *agonists*; those which inhibit the action of agonists are called *antagonists* (Fig. 8.8). The postsynaptic membrane must possess structures which can accurately identify and differentiate these substances. A binding or receptor protein, like an enzyme, binds the low molecular mass ligand at a special active site with high affinity and selectivity. The binding is reversible i.e. transmitter and receptor are in an association-dissociation equilibrium.

The binding protein of the transmitter at the neuromuscular synapse, the acetylcholine receptor, is biochemically the most fully investigated receptor (see Chapter 9). This is because organisms occur in nature which contain this receptor in unusually high concentrations: these are the electric fish and they have been of great value to neurochemists. The electric organs of the electric eel (*Electrophorus*) and the electric ray (*Torpedo*) are developmentally derived from muscle tissue, are uniformly cholinergically innervated and contain nicotinic acetylcholine receptor protein in milligram quantities. These have been used as a model system (see Chapter 12) for very many fundamental experiments [9]. The molecular properties of the acetylcholine receptor will be covered in detail in the section on "receptors"; here they will be described only as far as they are necessary for our presentation of the transmitter cycle and the functions of the synapse listed at the beginning of this chapter.

Table 8.1. Specificity of ACh [29].

Cell	Effects of ACh	Change in electrical ion conductivity	Type of AChR	Effect of α-Bungarotoxin
Vertebrates				
Skeletal muscle	d	↑Na, K	n	+
Electroplaque	d	↑Na, K	n	+
Heart	h	↑K	m	−
Smooth muscle	d	↑Na (K, Ca)	m	−
Sympathetic ganglion:				
Fast EPSP	d	↑Na, K	n	−
Slow EPSP	d	↓K or ↑Na, Ca	m	−
Slow IPSP	h	↓Na		
Cortical neurons	d	↓K	m	
Invertebrates				
Snail neurons:				
Fast EPSP	d	↑Na, K	n	−
	d	↑Cl		
Fast IPSP	h	↑Cl	n	+
Slow IPSP	h	↓K		−
Leech neurons	d	↑Na	n	−
	h	↑Cl	m	−
Leech muscle	d		n	−+
Insect neurons	d	↑Na	n, m, n/m	−

d Depolarization, h Hyperpolarization;
↑ Increase of conductivity, ↓ Decrease of conductivity;
n Nicotinic AChR, m Muscarinic AChR; n/m AChR with nicotinic and muscarinic properties;
+ Inhibition, − No inhibition.

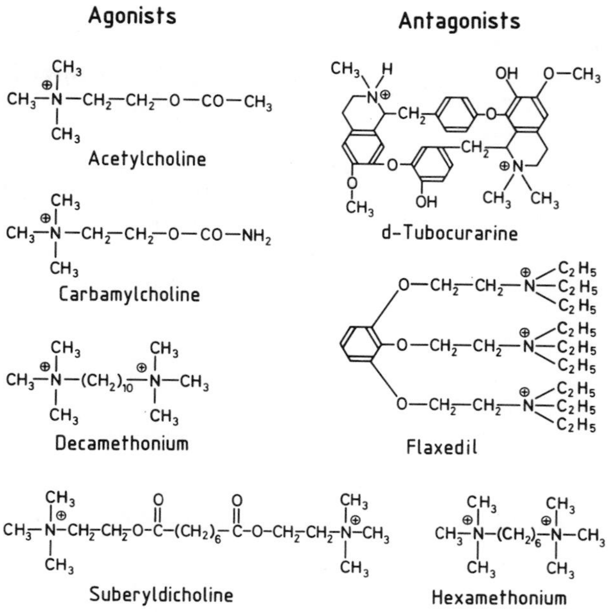

Fig. 8.8. Agonists and antagonists of the nicotinic acetylcholine receptor.

The acetylcholine receptor is an integral membrane protein, which is asymmetrically oriented in the postsynaptic membrane: when agonists or antagonists are injected intracellularly they produce no response, thus the binding sites must be oriented to the outside i.e. towards the synaptic cleft. There they are shown by electron microscopy to be so densely packed that there is hardly space for other proteins. There are 10 000 to 30 000 binding sites per μm^2; the higher number is found in the exposed area of the invaginated membrane. Outside the region of the synapse the receptor density is only 1% as compared to the subsynaptic area. After denervation by cutting the nerve however the density there clearly increases.[*]

The molecular mass of the polypeptide chain containing the binding site is 40 000 ($\pm$ 10%). It combines with other polypeptide chains of unknown function to form a receptor complex of relative molecular mass 260 000.[**] The acetylcholine receptor of higher vertebrates seems to have similar molecular properties [10].

---

[*] The hypersensitivity caused by denervation (see Chapter 9) has been observed in vertebrate muscles but not yet in *Torpedo*.
[**] These figures refer to molecular masses as determined by SDS-polyacrylamide gel electrophoresis. Data obtained by sequencing the cDNA (see Chapter 12) yielded somewhat larger numbers (50 200 for the agonist binding chain and 268 000 for the total receptor complex).

By an interaction between transmitter and receptor a pore is opened for a period of 1 ms allowing the passage of about 5×10^4 ions of Na^+ and K^+. The mechanism of this pore opening is the subject of much research. The following are two possible models, the first being the most likely in the case of the acetylcholine receptor:

1. The ligand-binding causes a conformational change in the receptor protein and a resultant change in ion permeability. This effect can be direct i.e. the receptor can itself contain the pore which is opened by the conformational change; or it can be indirect, i.e. the binding protein interacts with a further ionophore-containing protein to which the conformational change is transmitted.

2. The receptor is the regulatory subunit of an enzyme e.g. of a kinase, which by phosphorylation alters the electrostatic properties and thus the permeability of the membrane.

Conformational changes of the receptor protein have been shown in connection with changes in fluorescence of tryptophan residues after ligand binding. Phosphorylation and dephosphorylation of proteins of the postsynaptic membrane (cholinergic and other) have also been observed. Neither the observed conformational changes nor the phosphorylation reaction can as yet be correlated with the changes in ion conductivity and the basic connection remains to be shown. Obviously the models are not mutually exclusive.

Transmitter inactivation by enzymatic hydrolysis

The postsynaptic potential lasts for only a few milliseconds, if it is not reinforced by the release of further transmitter molecules. The acetylcholine concentration in the synaptic cleft is decreased by diffusion and hydrolysis. The transmitter is inactivated by an esterase, acetylcholinesterase (EC 3.1.1.7), which was crystallized to purity by Nachmansohn [8] and is one of the enzymes with the greatest turnover number.

Acetylcholinesterase is a peripheral membrane enzyme, i.e. it is extracted from the membrane by solutions of high ionic strength. It is uncertain from which membrane it is derived. Its localization in the subsynaptic region of the postsynaptic membrane seems contradictory; firstly there would appear to be inadequate space for an equivalent quantity of enzyme due to the high receptor density and secondly, in the subfractionation of synaptic membranes by sucrose gradient centrifugation it is found in a different membrane fraction from the acetylcholine receptor. This therefore indicates that it is not bound to the postsynaptic membrane but to the basal lamina between the pre- and postsynaptic membranes. It consists of one triple helical tail and one head formed of three tetramers. The tail is removed by trypsin, and also interestingly, by collagenase; the sedimentation coefficient of the molecule thereby decreases from 18 S to 11 S. Collagenase cleaves specifically only at the carboxyl side of proline, and the amino acid analysis of the tail protein shows up to 15% of proline and the rare amino acids hydroxylysine and hydroxyproline. This is evidence of a collagen-like structure and a similarity with the basal lamina which is derived from collagen fibrils.

The enzyme consists of four probably identical polypeptide chains of relative molecular mass 80 000. Each chain has an active centre and one or more peripheral, probably regulatory binding sites. The active centre contains an anionic site which binds the positively charged quaternary ammonium group of the substrate and a reactive site which catalyses the hydrolysis of the ester link of acetylcholine. The functional group responsible for this is an OH-group of a serine residue, whose nucleophilic properties are increased by the neighbouring imidazole ring of a histidine residue. The nucleophilic attack of the serine-OH group leads to the acetylation of the enzyme. The rate-determining step of this multi-stage reaction is the deacylation of the protein with formation of the free enzyme.

Inhibitors of acetylcholinesterase can either interact with the active site or with the peripheral binding sites. As a serine esterase it is irreversibly inactivated by the potent nerve gas diisopropylfluorphosphate (DFP) (Fig. 8.9).

This, like the other organophosphates developed as chemical weapons or as pesticides, blocks the active site by covalently reacting with the reactive serine residue. Parathion, also

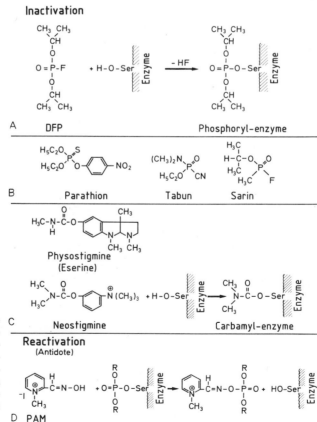

Fig. 8.9. Inactivation and reactivation of acetylcholinesterase. Inactivation: (A) By diisopropylfluorphosphate (DFP), (B) Other organophosphates which cause a similar reaction, (C) By eserine, a carbamylation reagent. Reactivation: (D) By an antidote of the aldoxime type.

known as E 605, a DFP-related compound should also be mentioned. Reactivation of the enzyme occurs by means of antidotes first developed in Nachmansohn's laboratory containing the reactive group -CH=NOH (e.g. pyridinaldoxime, 2 PAM) especially those of the bispyridine type like Toxogonin.

A further group of inhibitors such as eserine and neostigmine inactivate the active site by carbamylation of the serine-OH group, a slowly reversible reaction. The compound formed in this case is an ester of carbamic acid. These anticholinesterases are used experimentally for temporary inhibition of the esterase to produce a prolongation or increase of the action of acetylcholine.

Besides these covalently reacting inhibitors there is another group of effectors of acetylcholinesterase, which react not with the esteratic site but, due to their quaternary ammonium group, with the anionic centre (Fig. 8.10). Their binding sites whether in the active centre (competitive inhibitors) or on the periphery (uncompetitive inhibitors) often depend on the concentration of the inhibitor and on the ionic strength. A substance relatively selective for the active site is edrophonium – known pharmacologically as Tensilon; Flaxedil (gallamin) is specific for the peripheral binding sites (see Fig. 8.8) as is Propidium which is a fluorescent reporter-ligand for its binding site.

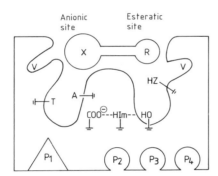

Fig. 8.10. Model of acetycholinesterase. It contains an active centre and probably many peripheral binding sites for different effectors ($P_1 - P_4$). The active centre includes an anionic group (-COO$^-$) which reacts with the quaternary ammonium group of acetylcholine and a serine-OH-group which is involved in the esterase action itself. The latter is activated by a "charge relay system" which transports a proton during the catalysis via a histidine imidazole residue. V, hydrophobic region; HZ, acid group; X-R, substrate. According to [11].

Acetylcholinesterase should be distinguished from the so-called pseudo-cholinesterases (EC 3.1.1.8). These are usually only quantitatively different in substrate and inhibitor specificity. They are found in serum where their physiological function is unknown; they do not however take part in synaptic transmission.

The cycle of acetylcholine is completed by the inactivation of the transmitter by hydrolysis and the presynaptic uptake of the resulting choline. Further details will be given when the receptors and their possible role in the plastic changes in the synapse are considered.

Inhibitors of the individual steps in synaptic transmission

Selective inhibitors are available for each step in chemical synaptic transmission; they are summarized in Table 8.2 and Fig. 8.6. With their help it is possible to isolate *in vivo* a single step of synaptic transmission, so that the synapse can, as it were, be dissected pharmacologically.

Table 8.2. Steps of the transmission cycle in cholinergic synapses and their inhibitors.

Steps of chemical synaptic transmission	Inhibitor
1. Transmitter synthesis	Styrylpyridine derivatives
2. Presynaptic release	Botulinum toxin
	β-Bungarotoxin
	Mg^{2+}
3. Binding to receptor	α-Bungarotoxin
	Curare etc.
4. Ion transport	Local anaesthetics
	(e. g. Procaine, Lidocaine)
	Histrionicotoxin
5. Hydrolysis of transmitter	DFP or other organophosphates
	Eserine
6. Choline uptake	Hemicholinium

Substances acting presynaptically include inhibitors of transmitter release, β-bungarotoxin (see below) and botulinum toxin, and an inhibitor of choline uptake, hemicholinium HC-3. The inactivation of the transmitter by hydrolysis is inhibited by acetylcholinesterase inhibitors. Substances acting postsynaptically can be divided into two groups. One blocks the binding of the transmitter to the receptor, for example receptor antagonists of the curare type and snake venom α-neurotoxin from cobra and krait; the other seems to interfere with the events following the formation of the transmitter receptor complex, for example local anaesthetics like procaine (Fig. 8.11) and animal toxins like histrionicotoxin, isolated from the skin of certain Columbian frogs. They act as *noncompetitive antagonists* blocking ion transport through the postsynaptic membrane, without competing with the transmitter at its binding site on the receptor.

Inhibitors are important for the neurochemical characterization of synaptic molecules *in vitro*. Typical examples are the snake venom neurotoxins. Following the work of Lee, the isolation of pure α-bungarotoxin in quantity and the availability of radioactive ^{125}I-labelled toxin has enabled significant progress to be made in our understanding of postsynaptic mechanisms.

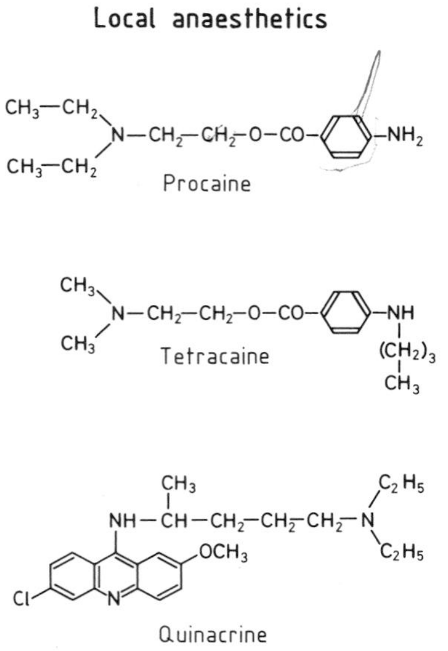

Local anaesthetics

[Structure: Procaine — (CH₃—CH₂)₂N—CH₂—CH₂—O—CO—C₆H₄—NH₂]

Procaine

[Structure: Tetracaine — (CH₃)₂N—CH₂—CH₂—O—CO—C₆H₄—NH—(CH₂)₃—CH₃]

Tetracaine

[Structure: Quinacrine — chloro-methoxy-acridine with NH—CH(CH₃)—CH₂—CH₂—CH₂—N(C₂H₅)₂ side chain]

Quinacrine

Fig. 8.11. Local anaesthetics. Above: Two representatives used in medicine, Procaine and Tetracaine. Below: Quinacrine which due to its strong fluorescence can be used as a "probe" for the investigation of the binding site and the mechanism of action of local anaesthetics on excitable membranes.

Neurotoxins from snake venoms

Snake venoms contain a whole range of toxic substances. We are interested here less in the cardiotoxins, which because of their lytic properties cause local lesions, cardiovascular changes or lysis of red blood cells, than in the neurotoxins which in the last few years have provided a valuable tool for the biochemical characterization of receptors. The neurotoxins of the cobra (e.g. *Naja naja, N. naja siamensis*) and the krait (*Bungarus multicinctus, B. ceruleus*), both Elapidae, and of the sea snakes (e.g. *Laticauda semifasciata*) Hydrophiidae, have been particularly well investigated.

Their value for the analysis of molecular processes at the synapse depends on their specificity for a particular site of action. They are classified according to their action on the nicotinic cholinergic synapse as follows:

α-neurotoxins which, like curare, inhibit postsynaptic receptors,

β-neurotoxins which inhibit the presynaptic release of the transmitter molecule.

α-neurotoxins are subdivided according to their primary structure into type I – which are polypeptides with 61-62 amino acids and a molecular mass of approx. 6 800 – and type II with 71-74 amino acids and a molecular mass of 7 800. The tertiary structure of type I is stabilized by four disulphide bonds, type II by five. When these are reduced the peptide loses its toxicity. The neurotoxins of Hydrophiidae belong to type I and have a closely similar amino acid sequence to the type I toxins of Elapidae. Their tertiary structure is also very similar.

The primary structures of many α-neurotoxins are known [12]: some X-ray crystallographic structures are also available; the tertiary structure of the crystallized erabutoxin of the sea snake *Laticauda semifasciata* (Fig. 8.12) [13,14] was the first determined. It shows that the polypeptide chain is folded into three loops, which are stabilized by the disulphide bonds and also by hydrogen bonds. The larger type II cobra toxin is folded in a similar way [15].

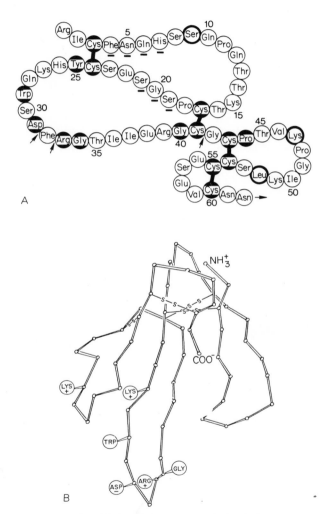

Fig. 8.12. α-neurotoxin from snake venom. The erabutoxin shown here is very similar in its structure to α-bungarotoxin and cobra α-neurotoxin. (A) Primary structure with the four disulphide bonds essential for its toxic action; (B) tertiary structure derived from X-ray crystallography [13,14]. Reprinted, with kind permission, from Elsevier North-Holland.

One of these loops contains all the amino acids essential for the toxic action, and is probably also responsible for the binding to the postsynaptic membrane. It may form as it were the portion of the key which fits snugly into the lock, the postsynaptic membrane.

The molecular mechanism of these neurotoxins has not yet been determined. α-neurotoxins bind to the acetylcholine receptor protein with high affinity ($K_D < 10^{-9}$) and thus inhibit the binding of agonists. Krait α-bungarotoxin is practically irreversibly bound to the nicotinic acetylcholine receptor ($K_D = 10^{-11}$). Many non-covalent interactions are involved; such multipoint binding seems to be responsible for the low K_D and the toxicity [16].

Other transmitters: criteria and classification

Cholinergic synapses are not the end of the story and acetylcholine is not the only transmitter; the sum total of substances which might be called transmitters is not yet known. The established transmitter substances are the catecholamines dopamine, adrenaline and noradrenaline; the amino acids γ-amino-butyric acid (GABA) and glycine, perhaps also glutamate and aspartate, and in addition serotonin (5-hydroxytryptamine, 5-HT) and histamine. Recently interest has been increasingly focused on certain peptides like substance P and the enkephalins which seem to be promising candidates as transmitters. Proline, taurine and purine nucleotides (like for example ATP) are as yet doubtful. Where doubts exist the term *putative transmitter* or *transmitter candidate* is used. Many substances modulate synaptic transmission without being transmitters. It is not a sufficient criterion that they are released presynaptically and act on the postsynaptic membrane. To characterize a substance as a transmitter the following conditions should be fulfilled:

1. It (and its metabolic precursors) must be contained in the neuron concerned, and in the nerve ending in a higher concentration.
2. The enzymes for its synthesis and the mechanism for its release from the nerve ending must be present in the neuron.
3. It must be released by stimulation of the neuron and proved to be present in the extracellular space.
4. A system for its inactivation must exist, either an enzyme or an uptake mechanism.
5. Specific receptors must be shown postsynaptically and in increased concentration subsynaptically.
6. Substances applied externally must show a similar reaction as stimulation; in particular the action of drugs, e.g. specific antagonists must be the same whether the transmitter is externally applied or arises from the nerve.

Only about half a dozen substances fulfill all these criteria (Fig. 8.13); the others remain as yet putative (Fig. 8.14). To summarize the definition of a transmitter: it is a substance, synthesized and stored in a neuron, released by nerve action and specifically bound by the postsynaptic membrane where it causes stimulation or inhibition of the postsynaptic cell by depolarization and hyperpolarization respectively.

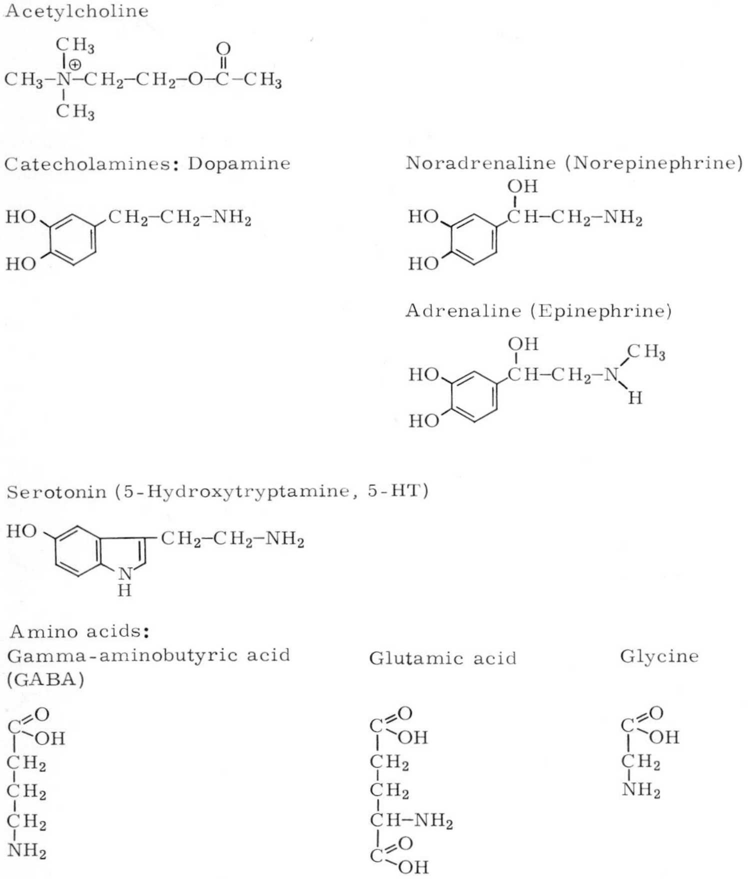

Acetylcholine

$$CH_3-\overset{\overset{CH_3}{|}}{\underset{\underset{CH_3}{|}}{N^+}}-CH_2-CH_2-O-\overset{\overset{O}{\|}}{C}-CH_3$$

Catecholamines: Dopamine

HO—C6H3(OH)—CH$_2$—CH$_2$—NH$_2$

Noradrenaline (Norepinephrine)

HO—C6H3(OH)—CH(OH)—CH$_2$—NH$_2$

Adrenaline (Epinephrine)

HO—C6H3(OH)—CH(OH)—CH$_2$—NH(CH$_3$)H

Serotonin (5-Hydroxytryptamine, 5-HT)

HO—(indole)—CH$_2$—CH$_2$—NH$_2$

Amino acids:
Gamma-aminobutyric acid (GABA)

$$\begin{array}{l} C(=O)OH \\ CH_2 \\ CH_2 \\ CH_2 \\ NH_2 \end{array}$$

Glutamic acid

$$\begin{array}{l} C(=O)OH \\ CH_2 \\ CH_2 \\ CH-NH_2 \\ C(=O)OH \end{array}$$

Glycine

$$\begin{array}{l} C(=O)OH \\ CH_2 \\ NH_2 \end{array}$$

Fig. 8.13. Transmitters.

H$_2$N—CH$_2$—CH$_2$—S(=O)$_2$—OH

Taurine

(imidazole)—CH$_2$—CH$_2$—NH$_2$

Histamine

also:

ATP

Aspartate

Proline

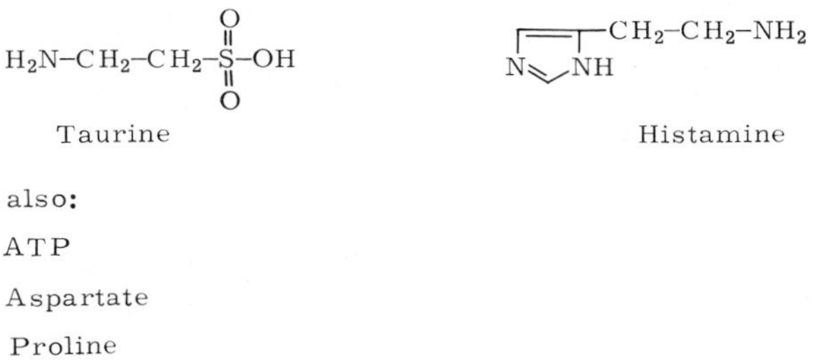

Fig. 8.14. Putative transmitters.

Neuropeptides: transmitters and hormones

It is as yet not known from the available evidence which of the neuropeptides (listed in Fig. 8.15) are transmitters and which are modulators. But they are particularly interesting as they seem to have both functions; as transmitters at specific synapses and modulators or regulators at other sites in the organism. We have defined the organs of communication within the organism as the nervous system together with the hormonal system. Neuropeptides appear to operate in both systems. The following figure will show some examples before we return to the "classical" neurotransmitters.

Met-enkephalin
Tyr-gly-gly-phe-met

Leu-enkephalin
Tyr-gly-gly-phe-leu

Substance P
Arg-pro-lys-pro-gln-gln-phe-phe-gly-leu-met-NH_2

Neurotensin
p-glu-leu-tyr-glu-asn-lys-pro-arg-arg-pro-tyr-ile-leu

ß-Endorphin
Tyr-gly-gly-phe-met-thr-ser-glu-lys-ser-gln-thr-pro-leu-
    val-thr-leu-phe-lys-asn-ala-ile-val-lys-asn-ala-his-
    lys-lys-gly-gln

ACTH (Corticotropin)
Ser-tyr-ser-met-glu-his-phe-arg-tyr-gly-lys-pro-val-gly-
    lys-lys-arg-arg-pro-val-lys-val-tyr-pro-asp-gly-ala-
    glu-asp-glu-leu-ala-glu-ala-phe-pro-leu-glu-phe

Angiotensin II
Asp-arg-val-tyr-ile-his-pro-phe-NH_2

Oxytocin
Cys-tyr-ile-gln-asn-cys-pro-leu-gly-NH_2

Fig. 8.15. Neuropeptides [30]. (A selection. Many neuropeptides have been identified, and their number is growing steadily.)

Vasopressin

Cys-tyr-phe-gln-asn-cys-pro-arg-gly-NH$_2$

Vasoactive intestinal polypeptide (VIP)

His-ser-asp-ala-
 gln-met-ala-val-phe-thr-asp-asn-tyr-thr-arg-leu-arg-lys-
  val-lys-lys-tyr-leu-asn-ser-ile-leu-asn-NH$_2$

Somatostatin

Ala-gly-cys-lys-asn-phe-phe-phe-trp-lys-thr-phe-thr-ser-cys

Thyrotropin-releasing hormone (TRH)

p-glu-his-pro-NH$_2$

Luteinizing-hormone-releasing hormone (LHRH)

p-glu-his-trp-ser-tyr-gly-leu-arg-pro-gly-NH$_2$

Bombesin

p-glu-gln-arg-leu-gly-asn-gln-trp-ala-val-gly-his-leu-met-NH$_2$

Carnosine

Ala-his

Cholecystokinin-like peptide

Asp-tyr-met-gly-trp-met-asp-phe-NH$_2$

Fig. 8.15. (continued) Neuropeptides [30].

A transmitter can have several functions

It is not reasonable to classify transmitters as stimulatory or inhibitory, as their function depends on the particular synapse and postsynaptic receptor involved. Acetylcholine, for example, is a stimulatory transmitter at the neuromuscular endplate, and is inhibitory at the synapse between the vagus nerve and the heart muscle fibre. We have already mentioned the difference between nicotinic and muscarinic acetylcholine receptors, but it was discovered in *Aplysia* that transmitter function can be even more complex. In this there are at least three types of cholinergic synapses or acetylcholine receptors, two inhibitory and one excitatory. The inhibitory synapses are distinguished by their ion specificity: at one postsynaptic membrane acetylcholine raises the potassium permeability and at another that of the chloride ion, in each case causing a hyperpolarization of the membrane. At the excitatory synapse acetylcholine causes a depolarization by opening sodium channels.

A similar dual function is described for the transmitters dopamine and serotonin. It can therefore just be said that acetylcholine and glutamate are most typically stimulating transmitters and glycine, GABA and noradrenaline most typically inhibitory.

In the majority of cases, possibly in all, each neuron utilizes only one transmitter. This statement has been known incorrectly as Dale's Principle.*)

The development of modern analytical techniques like autoradiography and immunofluorescence have shown however that putative neuropeptide transmitters may be present as neuromodulators in an aminergic neuron. This, although thought to be incompatible with Dale's Principle is seen to be in no way in conflict with its correct rendering.

Catecholamines

The catecholamines dopamine, noradrenaline and adrenaline (known in the American literature as norepinephrine and epinephrine respectively) are among the well established transmitters. We will now deal briefly with their biosynthesis, structure, function and the influence of neuropharmacological drugs on their action.

Synthesis is tightly regulated

The name catecholamine derives from catechol the structure of which is common to all three compounds and consists of a benzene ring with two ortho substituted -OH groups; i.e. o-dihydroxybenzene. The starting compound for their biosynthesis (Fig. 8.16) is the amino acid tyrosine, obtained either from the hydroxylation of phenylalanine, or directly from food. The first and rate-determining step, decisive for the regulation of their metabolic path, is the hydroxylation of tyrosine to **di**hydr**o**xy**p**henyl**a**lanine (DOPA) by the enzyme tyrosine hydroxylase (EC 1.14.16.2). DOPA is finally decarboxylated to dopamine by means of DOPA-decarboxylase (EC 4.1.1.26) and converted to noradrenaline by dopamine β-hydroxylase (EC 1.14.2.1). The last of the three catecholamines results from the methylation of noradrenaline by means of the enzyme phenylethanolamine-N-

---

*) The notion that a neuron uses the same transmitter at all its terminals was termed by Eccles "Dale's Principle". Later it became popular to refer to Dale's Principle as the postulate that each neuron releases only one transmitter, a postulate never put forward by either Dale or Eccles. It would not be tenable today considering the ample evidence that in many neurons more than one transmitter has been found and shown to be released. At present no example is known where the same neuron uses at one terminal a given transmitter and a different one at another although this should be possible in principle. The designation "Dale's Principle" should be restricted to the hypothesis that a neuron releases the same transmitter at all its terminals. (For a discussion of this problem which is not only of historical interest, see Whittaker, V.P.: "What is Dale's Principle?" In: *Dale's Principle and Communication Between Neurons.* Osborne, N.N., (ed.), p. 1-5. Pergamon Press, Oxford and New York 1983.)

Fig. 8.16. Catecholamine synthesis.

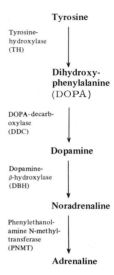

methyltransferase (PNMT) using S-adenosylmethionine (SAM) as methyl donor. As has been mentioned the most interesting step in a metabolic sequence is the rate-determining step, in this case the one catalysed by tyrosine hydroxylase (Fig. 8.17). As the *"pace-maker enzyme"* it is subject to a variety of control mechanisms.

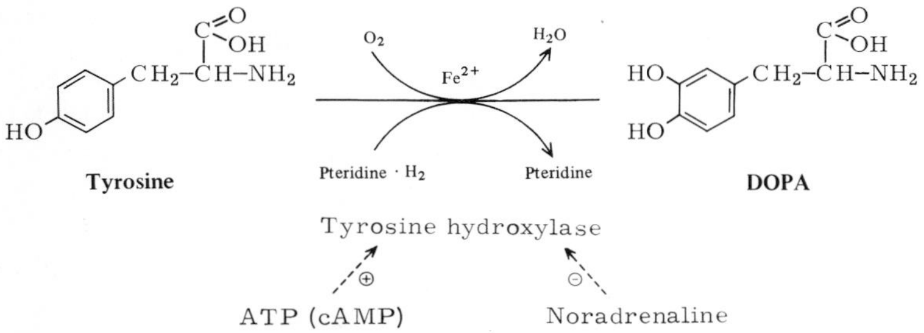

Fig. 8.17. Tyrosine hydroxylase reaction and regulation.

It is found only in the adrenal gland and in catecholaminergic neurons, predominantly in the nerve endings. It needs as cofactors a pteridine (tetrahydrobiopterine), Fe^{2+} and molecular oxygen and is subject to feedback regulation by the end-product noradrenaline. Besides these allosteric controls a regulation by phosphorylation and dephosphorylation has been found both in the enzyme of the adrenal gland and in various regions of the brain. A cAMP-dependent protein kinase evidently takes part. The result of phosphorylation is a lower K_M for the cofactor pteridine and a decrease in the affinity for the feedback inhibitor, and thus an activation of the enzyme. The phosphorylation is a result of neuronal activity.

In addition to allosteric regulation and phosphorylation the activity of tyrosine hydroxylase and a number of enzyme proteins is controlled by induction on the chromosomal level. The reader is referred to the section on *transsynaptic regulation* in Chapter 11 for further information.

The enzyme concerned in the second step, DOPA decarboxylase (DDC) also decarboxylates the precursor of serotonin, 5-hydroxytryptophan, but it is not yet certain whether it is identical with the more ubiquitous aromatic amino acid decarboxylase. It requires pyridoxal phosphate (Vitamin B_6) as a coenzyme and is also found in non-nervous tissue like kidney and liver.

Dopamine β-hydroxylase (DBH) by contrast is only found in adrenergic cells and can be considered as their specific marker enzyme. It is a copper-containing enzyme which is mainly localized in the membrane of the *chromaffin granule*, the storage vesicle of catecholamine (see below). Both DBH and DDC are present in excess. Even a 95% inhibition of DDC (e.g. by methyl DOPA or α-methyl-meta-tyrosine, α-MMT) does not affect the brain catecholamine concentration.

We have only described so far the main path of catecholamine synthesis. A more complete metabolic scheme is shown in Fig. 8.18, which also illustrates the connection with other substances classified as *putative transmitters* like octopamine. Some of these trace amines like for example phenylethylamine when present in abnormal concentrations in the CNS may perhaps be linked to psychiatric disorders.

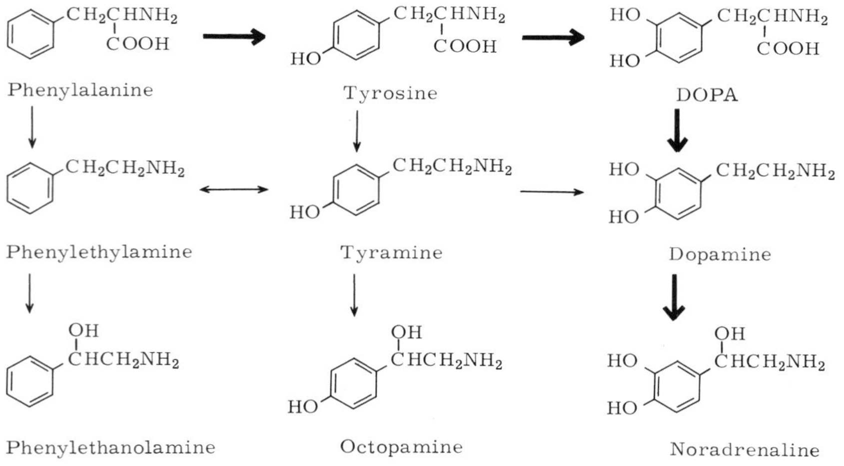

Fig. 8.18. Pathways of catecholamine metabolism (main pathway emphasized by arrows).

Catecholamines are also packed into vesicles

Like acetylcholine, catecholamines are stored in vesicles and, as the transmitters of sympathetic nerves, are released from the varicosities of these nerves. Two populations of vesicles have been observed by electron microscopy. They have an electron dense core, and are distinguished by their size as *large* and *small dense core vesicles*. The latter predominate. The physiological significance of these two different populations is as yet unknown. The larger may originate in the cell body and reach the nerve ending via axonal transport. It is estimated that each vesicle contains about 10 000–15 000 molecules of noradrenaline.

The biochemistry of the catecholamine storage vesicles found in the adrenal gland has been more fully investigated [17]. These are called *chromaffin granules* and it is assumed that their cycle is similar to that of neurotransmitter storage vesicles. They contain densely packed substances of high and low molecular masses. In addition to catecholamines there is ATP (and smaller amounts of ADP and AMP) and Ca^{2+}. Nucleotides and catecholamines are transported by separate carriers into the vesicles, by means of a pH gradient produced from an electrogenic pump driven by ATPase. The macromolecular components of the chromaffin granule have been identified as a group of acid secretory proteins, the *chromogranins*; and a smaller quantity of dopamine β-hydroxylase, which is the predominant constituent of the vesicle membrane. Both the soluble secretory and the membrane-bound protein contain mucopolysaccharides.

In vivo, dopamine seems to be taken up by the vesicles and there converted into noradrenaline by dopamine β-hydroxylase. Some of the noradrenaline leaves the vesicle and
is converted into adrenaline in the cytosol. Adrenaline is then taken up by the granules.

Release and binding

Catecholamines, like acetylcholine, are released presynaptically by exocytosis and bind postsynaptically to receptor proteins. These receptors do not seem to be coupled directly with ion channels as is the case with the nicotinic acetylcholine receptor; instead they interact with the enzyme adenylate cyclase the product of which, the second messenger cAMP, indirectly regulates among other cellular functions, the ion permeability of the postsynaptic membrane. This coupling can be either stimulatory or inhibitory causing an increase or decrease of the cAMP concentration in the target cell.

In Chapter 9 we will deal in further detail with the interaction between receptor and enzyme and the other biochemical properties of catecholamine receptors. Here it will be enough to mention that like the acetylcholine receptors in the peripheral nervous system, catecholamine receptors are divided into two classes, α and β. They were distinguished from one another about thirty years ago and defined by their pharmacological properties. Adrenaline has many sites of action in the sympathetic nervous system (see below) and there are a number of "sympathomimetic substances" which show a remarkable selectivity (Fig. 8.19).

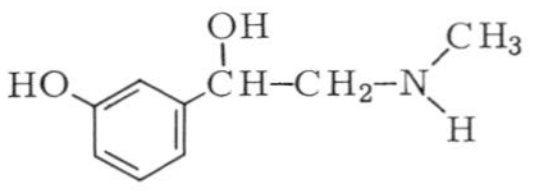

Phenylephrine
(α-sympathomimetic)

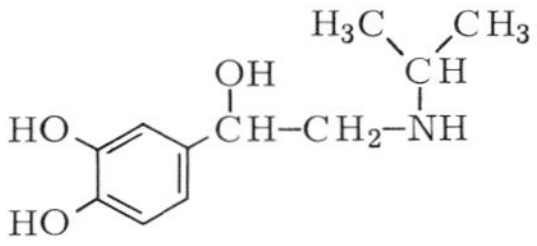

Isoprenaline (Isoproterenol)
(β-sympathomimetic)

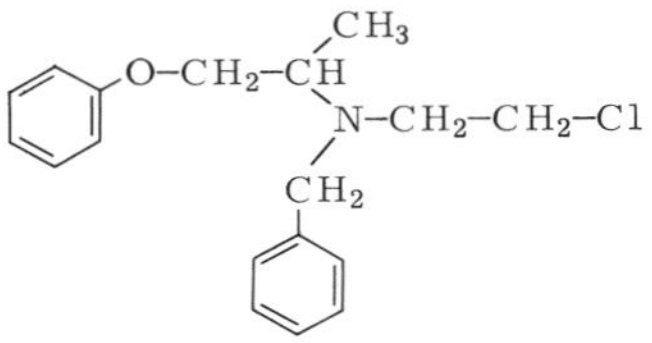

Phenoxybenzamine
(α-blocker)

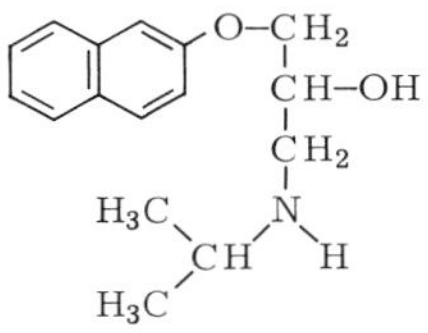

Propanolol
(β-blocker)

Fig. 8.19. Effectors of α- and β-receptors.

Phenylephrine, for example, causes stimulation predominantly of α-adrenergic receptors, isoprenaline of β-adrenergic receptors. The selectivity of antagonists is even more marked: phenoxybenzamine is an almost pure α-blocker and propanolol, alprenolol and pindolol are specific β-blockers. Among the α-blockers are also the ergot alkaloids derived from lysergic acid.

Multiple receptors assure variability of transmitter effects

As we shall see in Chapter 9 specificity and variability of effects caused by the relatively small number of transmitter substances available in the nervous system are considerably increased by the device of *multiple receptors*; thus α and β-receptors are subdivided into α_1 and α_2, β_1 and β_2 receptors respectively. There are D_1 and D_2-receptors in dopaminergic and H_1 and H_2 in histaminergic terminals. The muscarinic and nicotinic acetylcholine receptors and the μ-, $\varkappa$- and δ-opiate receptors are further examples of this device.

Various types of adrenergic effects

Adrenaline can be both an excitatory and an inhibitory transmitter (Table 8.3), but this dual action has nothing to do with the subdivision into α and β-receptors. Thus adrenaline, by causing the contraction of heart muscle fibres, accelerates the heart rate. The receptor concerned has been identified as a β_1-receptor. However adrenaline acts on β_2-receptors in the iris of the eye, the smooth muscle of the bronchi and blood vessels causing relaxation of the muscle. α-receptors are concerned in general with the contraction of involuntary (smooth) muscle, but as an exception produce relaxation of the smooth muscle of the digestive tract.

The relationships in the central nervous system are basically obscure and it is as yet not clear whether the classification into α and β receptors can be applied there [18,19].

Table 8.3. Types of receptor in various organs and their response to adrenergic substances [32].

Organ or organ system	Receptor type	Response	Action of NA	A	I
Heart	β	Increase of heart frequency and contractility, decrease of conduction velocity	3	2	1
Blood vessels in general	α	Constriction	1	2	$\emptyset$
muscle, liver and heart only	β	Dilatation	3	2	1
Bronchi, smooth muscle	β	Relaxation	3	2	1
Viscera					
muscle, longitudinal, circular	α, β	Relaxation	2	1	$\emptyset$
sphincters	α	Contraction	2	1	3
Bladder					
Detrusor vesicae	β	Ralaxation	3	2	1
Trigonum vesicae (sphincter)	α	Contraction	2	1	3
Uterus	α, β	Contraction/Relaxation			
Eye					
M. radialis (dilator)	α	Contraction	2	1	$\emptyset$
M. ciliaris	β	Relaxation	3	2	1
Salivary glands					
secretory cells	β	Secretion			
myoepithelial cells	α	Contraction			

NA Noradrenalin; A Adrenalin; I Isoproterenol; Action: strong (1); intermediate (2); weak (3); no effect ($\emptyset$).

Inactivation of catecholamines by uptake and degradation

The adrenergic synapse utilizes its transmitter more economically than the cholinergic synapse for although the adrenergic synapse has the machinery to degrade catecholamines (cf. cholinergic synapses which degrade acetylcholine) most of the released catecholamine is, in fact, recycled. It is reabsorbed from the synaptic cleft into the nerve ending by a specific uptake mechanism and packaged in the vesicles. A portion is also certainly

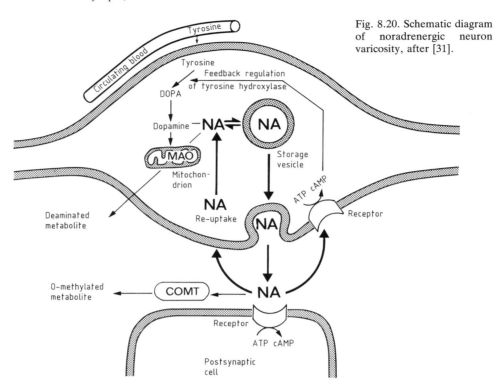

Fig. 8.20. Schematic diagram of noradrenergic neuron varicosity, after [31].

metabolized: the characteristic amino group is removed by oxidative deamination and one of the OH groups of the catechol ring is methylated either before or after deamination. The oxidative deamination takes place on the mitochondrial membrane where the flavoenzyme monoamine oxidase occurs (MAO, EC 1.4.3.4). The cytoplasmic enzyme catechol-O-methyltransferase (COMT, EC 2.1.1.6) performs the methylation utilizing S-adenosylmethionine (SAM) as methyl donor and Mg^{2+} as cofactor.

MAO is not only found in high activity in nerve tissue. Isoenzymes, MAO A and MAO B exist which differ in substrate and inhibitor specificity. Type A deaminates noradrenaline and serotonin preferentially and is sensitive to the inhibitor clorgyline. Type B deaminates the trace amine phenylethylamine and benzylamine and is inihibited by deprenyl. MAO A and B are evidently products of different genes as their subunits are of different size with relative molecular masses of 60 000 and 55 000 and they give different peptide patterns after proteolysis [21].

In addition to the endogenous regulators of the metabolism of catecholamines there are exogenous factors which can cause stimulation and inhibition. Table 8.4 (see also Fig. 8.20) shows some of these substances. Clinically they are used to alter the concentration or action of catecholamines. For example treatment with MAO blocker antidepressives causes a rise in serotonin and catecholamine concentration. We will return to some specific examples of this.

Table 8.4. Individual steps in the transmitter recycling at noradrenergic synapses and their inhibitors.

Step in chemical synaptic transmission	Inhibitor	
1. Transmitter synthesis (Tyrosine hydroxylase)	α-Methyl-p-tyrosine	
2. Storage	Reserpine Tetrabenazine	*Reserpine*
3. Presynaptic release	Amphetamine (stimulants)	*Amphetamine*
4. Receptor binding	α-Blocker (Phenoxybenzamine)	*Phenoxy benzamine*
	β-Blocker (Propanolol)	*Propranolol*
	Ergotalkaloid (Ergotamine)	*Ergotamine* R^1: H_3C- R^2: ⌬—CH_2-
	Haloperidol Phenothiazine (Dopamine receptor-blocker)	*Haloperidol*
5. Re-uptake	Imipramine Amphetamine Cocaine	*Imipramine*
6. Degradation (MAO)	Clorgyline Deprenyl	*Clorgyline*
7. Inactivation (COMT)	Tropolon	

Serotonin

5-hydroxytryptamine (serotonin, 5-HT) is classified with the catecholamines as an aminergic transmitter. There is little in common between these however apart from the NH_2 group. The best evidence that serotonin is a transmitter was derived from electrophysiological investigations of snails, but it probably also acts as a transmitter in the central nervous system of higher organisms including mammals. In man it is found in various areas of the brain (brainstem, pons, raphé nuclei) and in even higher concentrations in the intestine, blood platelets and mast cells. Its function is not known. It seems to take some part in sleep regulation, and its involvement in the neural basis of consciousness is indicated by its interaction with the hallucinogen lysergic acid diethylamide (LSD). LSD acts by binding to serotonin receptors antagonistically, but this does not fully describe its mode of action as it simultaneously inhibits the synthesis of serotonin. Both actions occur as a result of a similarity in molecular structures: they both contain an indole ring system (Fig. 8.21). Like other ergot-derivatives LSD is also a dopaminergic agonist.

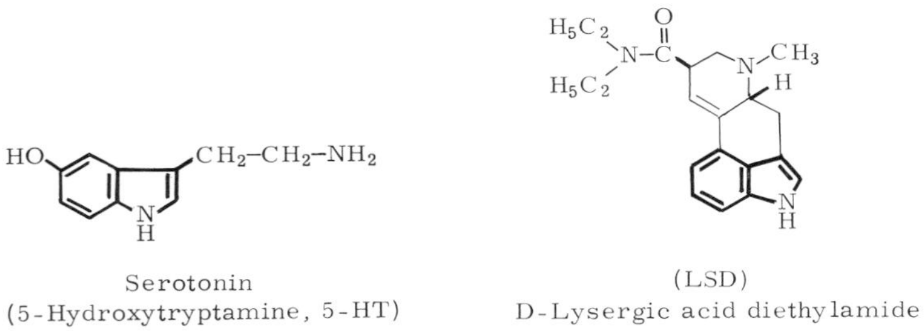

Serotonin
(5-Hydroxytryptamine, 5-HT)

(LSD)
D-Lysergic acid diethylamide

Fig. 8.21. Formulae of serotonin and LSD (showing their common indole ring structure).

Synthesis from tryptophan, degradation by MAO

Serotonin is formed from the amino acid tryptophan by hydroxylation of position 5 and subsequent decarboxylation (Fig. 8.22). The first step is catalyzed by the enzyme tryptophan hydroxylase (EC 1.14.16.4) which like tyrosine hydroxylase needs molecular oxygen and reduced pteridin as cofactors. The decarboxylation to indolamine is catalyzed by the same enzyme we have met already in the decarboxylation of DOPA. Its lack of specificity would suggest that "aromatic amino acid decarboxylase" is a better name for it than 5-HT decarboxylase (EC 4.1.1.28).

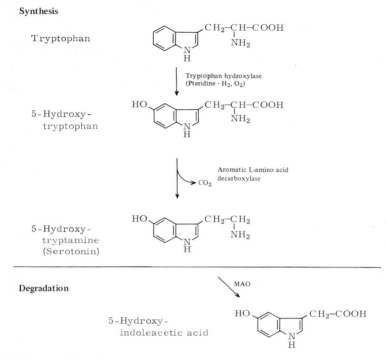

Fig. 8.22. Serotonin: synthesis and degradation.

The serotonin cycle is analogous to other transmitter cycles

Serotonin, at least in the central nervous system of higher organisms, belongs to the group of putative transmitters, therefore its metabolic cycle cannot be described with the same certainty as that of acetylcholine and the catecholamines. In preparations of nervous tissue most serotonin is found bound to particles, for example, packed in synaptic vesicles. It is released by stimulation of the appropriate brain regions and binding studies have given evidence of specific receptors (see Chapter 9). As free serotonin inhibits the further release of transmitter it is thought that there are not only postsynaptic receptors but also presynaptic receptors which may regulate exocytosis. There is also evidence of a Na^+-dependent *high affinity uptake* which is responsible for the inactivation of the transmitter by removing it from the synaptic cleft. The pharmacology of the individual stages of the serotonin cycle has been little investigated. We have already dealt with the mode of action of LSD. p-Chlorophenylalanine (pCPA) is a potent inhibitor of tryptophan hydroxylase and due to its specificity it is utilized to determine the participation of serotonin in particular behaviour patterns.

Amino acids as transmitters: GABA, glycine and others

Amino acid transmitters are divided into two groups: the excitatory acidic transmitters glutamate and aspartate, and the inhibitory neutral amino acids γ-aminobutyric acid (GABA), glycine, β-alanine and taurine. According to our criteria of transmitters (see p. 178) glycine and GABA are the best established representatives of this class [22] (Fig. 8.23).

$CH_2-CH_2-CH_2-COOH$ $\|$ NH_2 GABA (γ-Aminobutyric acid)	$HOOC-CH-CH_2-CH_2-COOH$ $\|$ NH_2 Glutamic acid
CH_2-COOH $\|$ NH_2 Glycine	$HOOC-CH-CH_2-COOH$ $\|$ NH_2 Aspartic acid
CH_2-CH_2-COOH $\|$ NH_2 β-Alanine	
$CH_2-CH_2-SO_3H$ $\|$ NH_2 Taurine	
Neutral amino acids (mainly inhibitory)	Acidic amino acids (excitatory)

Fig. 8.23. Amino acid transmitters.

GABA, an inhibitory transmitter

The identification of GABA as a neurotransmitter was due to Bazemoore, Elliott and Florey, who demonstrated an inhibitory efferent component in the sensory nerve from the crustacean stretch receptor, and produced an extract of mammalian brain having a similar action in which it was later possible to identify GABA as the active principle. In vertebrate brain GABA has been most clearly identified as a transmitter of the Purkinje cells of the cerebellum and also of many inhibitory interneurons of e.g. the neostriatum, spinal cord and cortex. GABA is derived from L-glutamic acid by decarboxylation. It occurs in the cell bodies of inhibitory neurons in a concentration of more than 10 mmol/l, and in their axons

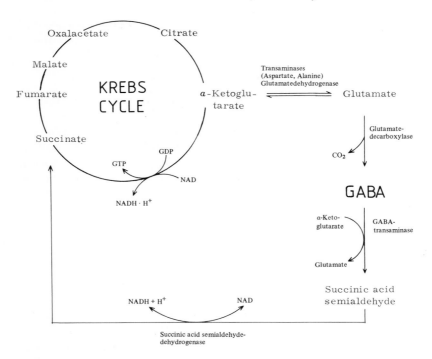

Fig. 8.24. GABA metabolism (GABA shunt).

in a concentration of 100 mmol/l. GABA metabolism therefore constitutes a considerable task for the cells. The main part of the GABA precursor glutamate is derived from α-ketoglutarate by transamination, and so originates from the citric acid cycle. After decarboxylation by the pyridoxal phosphate-requiring enzyme L-glutamate decarboxylase (EC 4.1.1.5) the carbon skeleton of GABA is redirected into the Krebs-cycle. GABA is degraded by transamination in which the NH_2 group is transferred to another molecule of α-ketoglutarate to form glutamate. The remaining succinic acid semialdehyde is oxidized to succinate. GABA synthesis and degradation is thus a supplementary loop of the Krebs cycle, which is called the GABA *shunt*. From the point of view of energy metabolism it involves a loss since the oxidation to succinate occurs without the formation of GTP. In the "normal" Krebs cycle this is formed by the conversion of α-ketoglutarate to succinate by substrate-level phosphorylation.

In Fig. 8.24 the connection between GABA metabolism and the amino acids L-glutamate, L-aspartate and L-alanine and the Krebs cycle is shown. It is interesting however that GABA synthesis and degradation do not seem to take place in the same compartment. Thus if animals are given a potent inhibitor of both glutamate decarboxylase (synthesizing) and GABA transaminase (degrading) e.g. hydroxylamine, it is the degradation which is inhibited and the GABA concentration rises. The GABA shunt seems to be a bridge between two compartments in one of which, probably the nerve endings, GABA is synthesized; after release it is degraded in the other compartment, which may be localized in the glial cells [23].

GABA regulates chloride channels

GABA is classified as an inhibitory transmitter both in crustacea and vertebrates. Inhibition involves the prevention of the depolarization of the postsynaptic membrane necessary for the release of an action potential, either by hyperpolarization or at least by stabilization of the resting potential. GABA achieves this by raising the conductivity for Cl^-; GABA receptors like glycine receptors seem to be associated with chloride channels (see Chapter 9).

GABA antagonists are convulsants

If the inhibitory action of GABAergic neurons is removed this leads to uncontrolled activity in the pathways concerned. GABA antagonists like for example picrotoxin and bicucullin (Fig. 8.25) are therefore powerful convulsants. On the other hand not all substances which act as convulsants are GABA antagonists, as they can interfere with other inhibitory transmitters (see below). Substances enhancing the inhibitory effect of GABA are relaxants and tranquillizers. Most prominent among these are the diazepines (e.g. Valium[R], see Chapter 9).

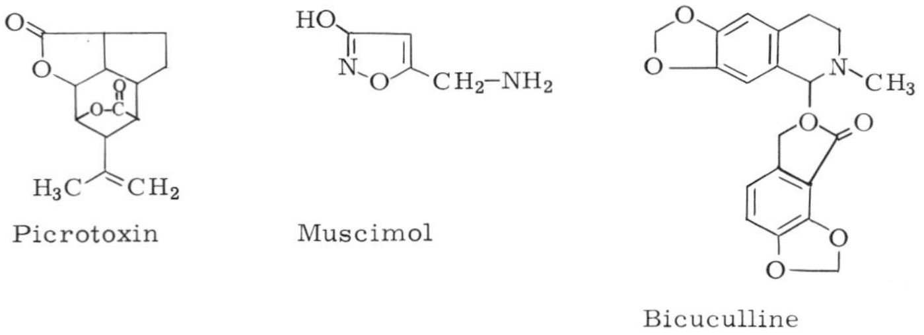

Picrotoxin   Muscimol

Bicuculline

Fig. 8.25. GABA antagonists (convulsants) and the agonist muscimol, a poison of the Fly Agaric fungus.

Glycine, another inhibitory transmitter

Glycine has a similar inhibitory action as GABA. It stabilizes the resting potential by raising the Cl⁻ conductivity of the synaptic membrane. Both these amino acids are anatomically localized to different but sometimes overlapping regions of the central nervous system. Glycine seems to be the predominant inhibitory transmitter of the spinal cord and brain stem while GABA is found more in the brain. The Cl⁻ channels of the two receptors also do not appear to be identical. The convulsant strychnine, a non competitive antagonist of the glycine synapse, interacts directly with Cl⁻ channels (see Chapter 9), but has no effect in similar concentrations on the GABA regulated Cl⁻ conductivity.

There is little known about the biochemistry and pharmacology of glycine as a transmitter, although there is some evidence for stimulus-induced presynaptic release, binding to specific receptors, a high affinity sodium-dependent uptake system and enzymes for its synthesis and degradation.

Glutamate and aspartate: excitatory transmitters

The evidence for the function of the amino acids glutamate and aspartate as excitatory transmitters in vertebrates and arthropods is also circumstantial though that for glutamate is better substantiated. Powerful non-physiological agonists for them are also known such as kainic acid, a neurotoxin from a Japanese alga, which is a structural analogue of glutamate (see Fig. 8.26). There is also a markedly high concentration of glutamate throughout the central nervous system. It is thought to be the most frequently occurring excitatory transmitter.

Fig. 8.26. Structural analogy between glutamic acid and kainic acid.

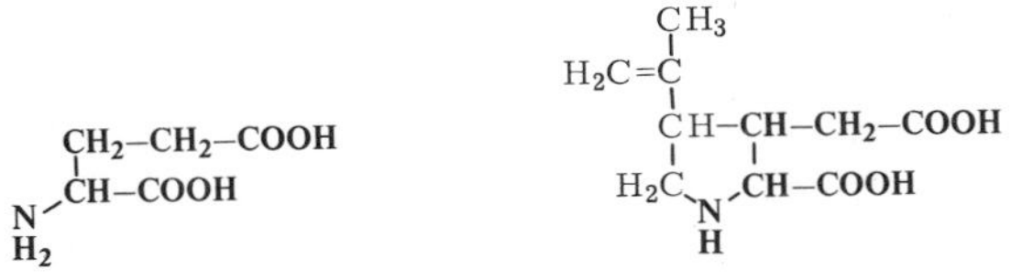

Glutamic acid

Kainic acid

Kainic acid and the method of chemical lesions

Kainic acid is an even more powerful agonist than glutamate, although it does not seem to interact directly with glutamate receptors. It is neurotoxic and causes a degeneration of nerve cells, where it selectively damages only glutaminergic neurons but not axons of other neurons in the area of the central nervous system into which it is injected. It can therefore be used in neuroanatomy and neurophysiology as a valuable method for producing defined and localized lesions. Lesions can be caused by damaging tissue, e.g. by operative procedures or thermal coagulation, but in these methods the extent of the damage is difficult to control, especially since the site of the operation can be traversed by axons which come from a long way away and have perhaps nothing to do with the function of the brain region concerned. The chemical lesion does not affect such axons. As a result of the kainic acid injection one can trace particular behavioural changes to the damage to cell bodies at the injection site and thus find out their function. Analogous lesion experiments were carried out with 6-hydroxydopamine, a specific neurotoxin for catecholaminergic neurons. Chemical lesions have increasing significance for the charting of brain functions.

Enkephalins and other neuropeptides: neuromodulators and suspected neurotransmitters

Opiates are drugs which have an analgesic and euphoric effect. Specific receptors for these drugs were found in the brain, and indeed, as already mentioned, several types of opiate receptors are now known to exist. It was assumed that endogenous ligands for these receptors must exist, but it was not until 1974 that Kosterlitz and Hughes, and Terenius and Wahlström discovered peptides in brain extracts that have the properties expected of *endogenous opiates*. It is not yet clear whether these peptides are true neurotransmitters or merely neuromodulators; the evidence will be reviewed shortly.

Two similar pentapeptides called *enkephalins* were isolated and their amino acid sequence determined:

Tyr-gly-gly-phe-met Met-enkephalin,
Tyr-gly-gly-phe-leu Leu-enkephalin.

Some fundamental transmitter criteria, but not all, are fulfilled: with the help of fluorescent antibodies opiate peptides have been histochemically detected among other tissues in the dorsal spinal cord, and thus in the area which is responsible for the conduction of pain signals. They are found in small interneurons and not in the main pathways whose transmitter is another neuropeptide, substance P (see below). This leads one to speculate that the opiate peptides may inhibit presynaptically the release of substance P (see Chapter 9). They also occur in high concentration in another area, a part of the limbic system, which is known to take part in the regulation of emotional processes. Other transmitter criteria that have been satisfied are: release by electrical stimulation and the demonstration that the effect of externally applied peptide is similar to that of electrical stimulation. The contraction of the guinea pig ileum or the vas deferens of the mouse are used as test systems; they are inhibited both by endogenous and exogenous opiates.

Opiate receptors have already been well characterized as specific transmitter receptors (Chapter 9). In binding studies opiate peptides compete with synthetic opiates.

What little is known about the biosynthesis of the opiate peptides (transmitter criterion no. 2, p. 178), may be summarized as follows.

The sequence of Met-enkephalin is the same as the sequence of positions 61-65 of β-lipotropin, a polypeptide of the pituitary to which no function has as yet been attributed apart from a weakly expressed lipotropic hormone action. Perhaps it serves as a precursor for the enkephalins or at least for Met-enkephalin. The sequence of Leu-enkephalin is not however present in β-lipotropin, but is contained in dynorphin, another pituitary peptide. A further complication is the presence in the pituitary of a large precursor protein. (Fig. 8.27) whose primary structure not only contains β-lipotropin but also ACTH

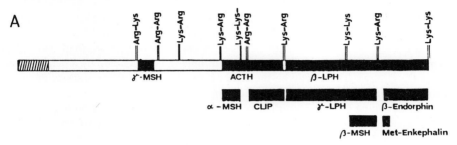

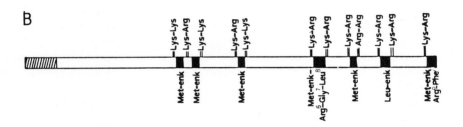

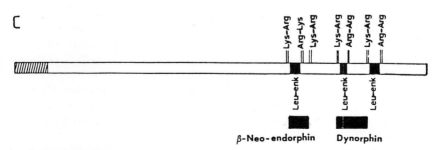

Fig. 8.27. The three precursors of opioid peptides: (A) Proopiomelanocortin, (B) proenkephalin, and (C) prodynorphin. The shaded area to the left symbolizes the signal peptide; for the other symbols and for explanation see text.

(**A**drenoc**o**rticotropic **h**ormone); this hormone regulates the growth of the adrenal cortex and the formation and secretion of cortisone. The mechanism of formation of Met-enkephalin from the precursor is not clear. Beside Met-enkephalin, other fragments of β-lipotropin have been found to have opiate properties. Among these so-called endorphins, β-endorphin (position 61-91 of lipotropin) is notable for its marked analgesic action.

With the advent of gene cloning and especially of cDNA techniques it has become possible to identify and sequence a whole set of precursor proteins comprising all hitherto known opioid peptides [24]. Three precursors have been found (Fig. 8.27): proopiomelanocortin, proenkephalin, and prodynorphin. The amino acid sequences deduced from the corresponding cDNA sequences show the following common features: all precursors contain several biologically active peptides, some of them in more than one copy (closed bars). The sequences of these peptides are terminated at each end by a pair of basic amino acids (Lys, Arg). These may be the sites of action for specific peptidases involved in the processing of the precursor. In addition there are several open bars representing cDNA sequences coding for non-identified peptides of unknown function and finally at the amino terminus there are sequences thought to be signal peptides, which are typical for secretory peptides and have to be removed during post-translational processing.

As can be seen in Fig. 8.27 proopiomelanocortin contains one copy each of the hormones ACTH, β-lipotropin (β-LPH), α-, β-, γ-MSH, the ACTH-like peptide CLIP, γ-lipotropin (γ-LPH) and β-endorphin, in addition Met-enkephalin and the putative signal peptide (making the protein shown in the figure a "pre-pro-opiomelanocortin"). The pre-enkephalin contains four copies of Met-enkephalin, one Leu-enkephalin and one each of the extended enkephalins Met-enkephalin-Arg, Phe, and Met-enkephalin-Arg, Gly, Leu. Prodynorphin, which has been identified in the hypothalamus, contains three copies of Leu-enkephalin and one each of β-neodynorphin and dynorphin. Dynorphin and β-neodynorphin both have a Leu-enkephalin-sequence at their N-terminus. Proenkephalin is found in the adrenal gland as well as in brain, and proopiomelanocortin in the pituitary.

A system for the inactivation of these putative transmitters has not yet been clearly identified (transmitter criterion no. 4, p. 178). An uptake mechanism does not seem to be present, but there is evidence of a proteolytic degradation which begins with the cleavage of the terminal tyrosine.

The relationship between structure and function can be readily investigated for the enkephalins as derivatives are easily synthesized [25] (Fig. 8.28). For example if the pentapeptide is shortened at the N-terminal tyrosine or the C-terminal methionine, its activity is largely lost. Lengthening the peptide chains has the same effect (exception: β-endorphin, whose increased action depends upon the fact that it is protected against proteolytic degradation).

It can be further shown that the free terminal carboxyl group, the OH group of tyrosine, the terminal amino group and the phenyl side chain are all necessary for activity, as is the correct distance between Phe and Tyr. The activity increases only when there is substitution of a glycine by D-alanine.

Activity means in this case the analgesic effect *in vivo*. It is determined not only by the efficiency of the peptide at the site of action, i.e. the affinity to the receptor but also by the resistance to enzymatic degradation, and the rate at which the peptide is transported to the

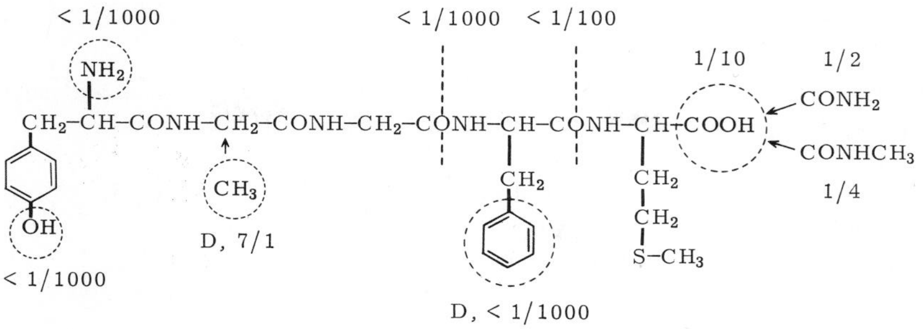

Fig. 8.28. Relationship between structure and activity of Met-enkephalin. The quotients give the change of activity after the respective modification. The circles of interrupted lines denote the cleavage or substitution of the respective functional group, for phenylalanine the substitution by D-Phe, for gly by D-Ala [25].

site of action. The instability of most enkephalin derivatives has made it difficult to do binding studies *in vitro* and to come to conclusions about receptor-ligand interactions.

Enkephalins are found concentrated in synaptosomes and they probably serve as inhibitory transmitters. Postsynaptic and presynaptic mechanisms and interactions with adenylatecyclase will be discussed in Chapter 9. Here we shall just note that the enkephalins like exogeneous opiates produce dependence and tolerance (habituation), the "natural" substance being medically no more useful than the synthetic drug.

Substance P, the longest-known neuropeptide

Another peptide which belongs to the group of putative transmitters is substance P, which has been known for a half century as an undecapeptide with the sequence

Arg-pro-lys-pro-gln-gln-phe-phe-gly-leu-met.

It was discovered in 1931 by von Euler and Gaddum in extracts of the brain and digestive tract, and was found to cause contraction of the smooth muscle of the gut and dilatation of blood vessels, an effect which occurs in man with an intravenous dose of only a few nanograms. Later, it was purified to homogeneity predominantly from the hypothalamus. The physiological role of substance P is not yet quite clear. The release of the peptide after sensory stimulation has been shown in the dorsal horn of the spinal cord. This process is inhibited by enkephalin, as has been mentioned in the previous section. Substance P acts by depolarizing the postsynaptic membrane and is thus a transmitter of sensory stimulation. The P in its name was originally the abbreviation for powder, but has come to assume the interpretation of *pain* [26].

Summary

Synapses are sites of communication between nerve cells. Chemical and electrical synapses are distinguished according to the mechanism of transmission.

Functions of the chemical synapse:
1. Nerve impulses are only conveyed in the direction presynaptic – postsynaptic; thus synapses are rectifiers for conductivity purposes.
2. They are modulators and integrators of nerve impulses.
3. "Nonspecific" action potentials are transformed into specific signals, i.e. a specific transmitter with a postsynaptic effect.
4. They are sites of action for endogenous and exogenous factors, drugs and toxins. The go-between at the synapse is the transmitter.

Criteria for the definition of a substance as a transmitter:
1. It (and its metabolic precursors) must be contained in the neuron.
2. The enzymes for its synthesis, and devices for its release must be present.
3. It must be released by stimulation of the neuron, i.e. shown to be in the extracellular space.
4. A system for its inactivation must be available (enzyme or uptake mechanism).
5. Its effect when externally applied must be similar to that when it is released by stimulation, in particular there must be the same interaction with drugs.

Transmitter cycle: 1. Synthesis, 2. uptake into vesicles, 3. if 1 and 2 occur in the perikaryon: axoplasmic transport to nerve endings, 4. presynaptic release by depolarization into the synaptic cleft (exocytosis), 5. diffusion to the postsynaptic membrane, 6. recognition and binding by a specific receptor e.g. a membrane protein (to produce "gating" of the postsynaptic membrane), 7. inactivation.

Examples of transmitter substances: definitely identified, the amines acetylcholine, dopamine, noradrenaline and serotonin (5-HT); almost definitely identified, the amino acids GABA, glutamic acid and glycine; "putative" transmitters or neuromodulators, histamine, purine nucleotides, enkephalins, and other neuropeptides. The site of action is a receptor protein in the postsynaptic membrane (and sometimes also in the presynaptic membrane) and the mechanism of action is the change in ion conductivity of the excitable membrane.

References

Cited:

[1] Goodenough, D.A., "The structure and permeability of isolated hepatocyte gap junctions", *Cold Spring Harbor Symp. Quant. Biol.* **40**, 37-47 (1976).
[2] Elfvin, L.G., "The ultrastructure of neuronal contacts", *Prog. Neurobiol.* **8**, 45-79 (1976).
[3] DeFeudis, F.V.: *Central Cholinergic Systems and Behaviour.* Academic Press, London, New York, San Francisco 1974.

[4] Davis, K.L., Berger, P.A., Hollister, L.E., and Barchas, M.D., "Cholinergic involvement in mental disorders", *Life Sci.* **22**, 1865-1872 (1978).
[5] Tucek, S.: "The synthesis of acetylcholine". In: *Handbook of Neurochemistry*, Lajtha, A., (ed.) 2nd edition, Vol. 4, p. 219-249. Plenum Press New York and London 1983.
[6] Kuhar, M.J., and Murrin, L.C., "Sodium-dependent, high affinity choline uptake", *J. Neurochem.* **30**, 15-21 (1978).
[7] Zimmermann, H.: "Biochemistry of the isolated cholinergic vesicles". In: *Neurotransmitter Vesicles*. Klein, R.L., Lagercrantz, H., and Zimmermann, H., (eds.), p. 271-304. Academic Press London, New York 1982..
[8] Nachmansohn, D., and Neumann, E.: *Chemical and Molecular Basis of Nerve Activity*. Academic Press, New York 1975.
[9] Conti-Tronconi, M.B., and Raftery, M.A., "The nicotinic cholinergic receptor. Correlation of molecular structure with functional properties", *Ann. Rev. Biochem.* **51**, 491-530 (1982).
[10] Fambrough, D.M., "Control of acetylcholine receptors in skeletal muscle", *Phys. Rev.* **59**, 165-227 (1979).
[11] Rosenberry, T.L., "Acetylcholinesterase", *Adv. Enzymol.* **43**, 103-219 (1975).
[12] Tu, A.T., "Neurotoxins of animal venoms: snakes", *Ann. Rev. Biochem.* **42**, 235-258 (1973).
[13] Low, B.W., Preston, H.S., Sato, A., Rosen, L.S., Searl, J.E., Rudko, A.D., and Richardson, J.S., "Three-dimensional structure of erabutoxin b neurotoxic protein: Inhibitor of acetylcholine receptor", *Proc. Natl. Acad. Sci. USA* **73**, 2991-2994 (1976).
[14] Tsernoglou, D., and Petsko, G.A., "The crystal structure of a postsynaptic neurotoxin from sea snake at 2.2. Å resolution", *FEBS Lett.* **68**, 1-4 (1976).
[15] Walkinshaw, M.D., Saenger, W., and Maelicke, A., "Three-dimensional structure of the 'long' neurotoxin from cobra venom", *Proc. Natl. Acad. Sci. USA* **77**, 2400-2404 (1980).
[16] Tsetlin, V.I., Pluzhnikov, K., Karelin, A., and Ivanov, V.: "Acetylcholine receptor interaction with neurotoxin II. photoactivable derivatives". In: *Toxins as Tools in Neurochemistry*. Hucho, F., and Ovchinnikov, Y.A. (eds.) p. 159–169, De Gruyter, Berlin, New York (1983).
[17] Winkler, H., and Westhead, E., "The molecular organisation of adrenal chromaffin granules", *Neuroscience* **5**, 1803-1823 (1980).
[18] Moore, R.Y., and Bloom, F.E., "Central catecholamine neuron systems: Anatomy and physiology of the norepinephrine and epinephrine systems", *Ann. Rev. Neurosci.* **2**, 113-168 (1979).
[19] Moore, R.Y., and Bloom, F.E., "Central catecholamine neuron systems: Anatomy and physiology of the dopamine system", *Ann. Rev. Neurosci.* **1**, 129-169 (1978).
[20] Kebabian, J.W., and Calne D.B., "Multiple receptors for dopamin", *Nature* **227**, 93-96 (1979).
[21] Cawthon, R.M., and Breakefield, X.O., "Differences in A and B forms of monoamineoxidase revealed by limited proteolysis and peptide mapping", *Nature* **281**, 692-694 (1979).
[22] Lajtha, A., (ed.): *Handbook of Neurochemistry*, 2nd edition, Vol. 3. Plenum Press, New York and London 1983.
[23] Siesjö, B.K.: *Brain Energy Metabolism*, p. 168-172. J. Wiley and Sons, New York, Toronto 1978.
[24] Akil, H., Watson, S.J., Young, E., Lewis, M.E., Khachaturian, H., and Walker, J.M., "Endogenous opioids: Biology and function", *Ann. Rev. Neurosci.* **7**, 223-255 (1984).
[25] Fredrickson, R.C.A., "Enkephalin pentapeptides: a review of current evidence for a physiological role in vertebrate neurotransmission", *Life Sci.* **21**, 23-41 (1977).
[26] Kerr, F.W.L., and Wilson, P.R., "Pain", *Ann. Rev. Neurosci.* **1**, 83-102 (1978).
[27] Whittaker, V.P., "The biochemistry of synaptic transmission", *Naturwiss.* **60**, 281-289 (1973).
[28] Dowdall, M.J., Fox, G., Wächtler, K., Whittaker, V.P., and Zimmermann, H., "Recent studies on the comparative biochemistry of the cholinergic neuron", *Cold Spring Harbor Symp. Quant. Biol.* **40**, 65-81 (1976).
[29] Magazanik, L.G., "Functional properties of postjunctional membrane", *Ann. Rev. Pharmacol. Toxicol.* **16**, 161-175 (1976).

[30] Iversen, L.L., "The chemistry of the brain", *Sci. Am.* **241**, 118-129 (1979).
[31] Cooper, J.R., Bloom, F.E., and Roth, R.H.: *The Biochemical Basis of Neuropharmacology*, p. 140, 2nd edition. Oxford University Press, New York, London, Toronto 1974.
[32] Schmidt, R.F., and Thews, G.: *Einführung in die Physiologie des Menschen*, S. 121, 18. Auflage. Springer-Verlag, Berlin, Heidelberg, New York 1976.

Further reading

Siegel, G.J., Albers, R.W., Agranoff, B.W., and Katzman, R. (eds.): *Basic Neurochemistry*, 3rd edition. Little, Brown and Co., Boston 1981.
Klein, R.L., Lagercrantz, H., and Zimmermann, H. (eds.): *Neurotransmitter Vesicles*. Academic Press, London, New York 1982.
Cooper, J.R., and Bloom, F.E.: *The Biochemical Basis of Neuropharmacology*, 4th edition. Oxford University Press, New York, London, Toronto 1982.
Lajtha, A. (ed.): *Handbook of Neurochemistry,* 2nd edition, Vol. 3: "Metabolism in the Nervous System". Plenum Press, New York and London 1983.
Reichardt, L.F., and Kelly, R.B., "A molecular description of nerve terminal function", *Ann. Rev. Biochem.* **52**, 871–926 (1983).
Hucho, F., and Ovchinnikov, Y.A. (eds.): *Toxins as Tools in Neurochemistry*. De Gruyter, Berlin, New York 1983.
Whittaker, V.P.: "The synaptosome". In: *Handbook of Neurochemistry*. Lajtha, A. (ed.), 2nd edition, Vol. 3, p. 1–41. Plenum Press, New York 1984.

Chapter 9

The Synapse, Part 2: Receptors

Receptors, defined as physiological sites of action

The term "receptor" is used today in two different senses. It refers firstly to the primary receiver of the sensory stimuli – light, touch, temperature and pain. In this sense it describes an organ consisting of one or more cells, for example the rods and cones of the retina are the photoreceptors. Secondly it describes in molecular terms the binding site for a low molecular active substance. Here too the definition is not precise: many authors define it as any site that binds a ligand specifically whether the latter is endogenous or exogenous. Neurochemists mean exclusively the target sites of endogenous effectors like hormones, prostaglandins and neurotransmitters. According to this definition the term "receptor" should not include the binding sites for neurotoxins in the axonal ion channels or on gangliosides of the neural membrane. The term is reserved mainly for pre- and postsynaptic receptors; these are always proteins which bind the presynaptically released transmitter – the first stage of chemical excitation of a membrane. This does not exclude the fact that such a receptor, as in the case of the opiate receptor, is detected and characterized by exogenous drugs, especially in cases where the endogenous transmitter is not yet known.

Neurotransmitter receptors have a key importance in molecular neurobiology, as they play a central role in the transmission of nerve impulses and are the site both of important regulatory processes and of changes which occur in certain neurological diseases. Since they are also the site of action of many neuropharmacological drugs they are of practical as well as theoretical interest. The biochemical characterization of receptor molecules is therefore one of the currently most active areas of neurochemical research.

Three criteria define a binding site as a receptor

The identification of a receptor is based on the following premises: a ligand – endogenous or exogenous – must be present which can be radioactively labelled; has an affinity for the receptor high enough to be measured accurately in a binding test; and the binding must be specific for the receptor concerned. For this specificity there are three criteria:

1. *Saturation* of the binding site should be in a physiologically significant concentration range.
2. *Localization* i.e. binding should only occur in the tissue in which the biological action is observed.
3. *Selectivity* i.e. this ligand should be the only one binding to the site or it should be able to compete effectively with other effectors of the same receptor.

Only by strict observation of all these criteria can artifacts be excluded. The following two examples have proved that fulfilment of a single criterion is not enough. Cuatrecasas found that insulin is bound to talcum with the same affinity as to the cell membrane *in vivo*. Snyder showed that certain glass filters bind opiates with the same stereospecificity as the biological opiate receptor.

Three levels of receptor research: the data must be correlated

Receptors are investigated on three levels, in the intact tissue or intact single cells, in membrane suspensions derived from the tissue, and in molecules isolated from it. Biochemical investigations are only physiologically relevant when they give consistent results at all levels. A breakthrough in the biochemical research into receptor molecules was therefore only achieved when these molecules could be freed from their surrounding membrane components while retaining their distinctive properties. The most important means for this has become the use of non-ionic detergents like Triton X-100, Emulphogen or Brij. They solubilize the membrane but stabilize the hydrophobic membrane protein in water, and due to their amphophilic property take the place of the lipid on the protein surface. They thus prevent aggregation and precipitation, and there is no denaturation of the protein as occurs with the ionic detergent sodium dodecylsulphate (SDS) (see Chapter 3).

The following properties of a receptor are of particular interest in neurochemistry: chemical composition, i.e. whether it consists of protein, carbohydrate, glyco- or lipoprotein; molecular mass and quaternary structure; amino acid composition and sequence; carbohydrate sequence; spatial arrangement of the molecular components; number and dissociation constants of the ligand binding sites; independence or cooperativity of the binding sites, interaction of the receptor both with its environment, (i.e. the membrane lipids, and other membrane proteins), and with the extra- and intracellular space. From these data an attempt can be made to formulate a model of the functional mechanism of the receptor.

Receptor models

Until now only one receptor, the nicotinic acetylcholine receptor, has been extensively purified and biochemically characterized, but this has not yet led to a general theory about the relationship between receptor structure and function. However it is obvious that receptors have a double function: they receive and recognise specific signals, and simultaneously initiate the first step in the cellular response to the signal. Neurotransmitter receptors regulate, among other functions, the ion permeability of the postsynaptic membrane (see Fig. 9.1). The binding of the transmitter is thus coupled with the response of opening ion channels. The question of the functional mechanism of receptors is a question of the mechanism of this coupling.

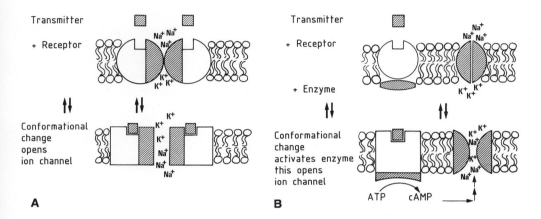

Fig. 9.1. Receptor models. Receptors regulate the ion permeability either directly (A) or indirectly (B).

The opening of ion channels however, is not the sole response to transmitter binding. In the catecholamine receptors for example, the primary response consists of the production of the "second messenger" cAMP, which by means of a protein kinase regulates not only the ion permeability of the excitable membrane but also the energy metabolism and protein biosynthesis in the cell. Receptors, defined as binding molecules for endogenous ligands are thus components of membrane complexes consisting of molecular species that bind the ligand and others that are functionally active in the membrane. The method by which the ion permeability of a cell membrane is regulated can be seen in the model developed for axonal ion channels (Chapter 6). The receptor is the gating mechanism which can be activated chemically or, in the axon, electrically, by a change in potential. It forms a unit with the selectivity filter which determines the passage of the appropriate ion through the membrane. This analogy does not apply where the receptor exerts its activity through one or more enzymatic steps. However in these cases we can speak of receptor complexes, especially when – as the β-adrenergic receptor will demonstrate – these may include regulatory proteins. The term receptor in the narrow sense is often used synonymously with the term receptor complex, i.e. implying that there are effector molecules in addition to the ligand-binding site. We shall now return to the theory of the interaction of transmitter and receptor.

In many cases the action of the transmitter can be correlated with the number of receptors occupied by it [1]. If this "occupancy" is reversible we can write:

R + L $\longrightarrow$ RL $\longrightarrow$ $\longrightarrow$ biological response, where
R is the receptor,
L is the ligand,
RL is the receptor-ligand complex.

The "occupancy theory" states that the cellular response depends on the concentration of the occupied receptors [RL]. This is formally similar to the Michaelis-Menten theory of enzyme kinetics and the overall kinetics are analysed similarly in both theories. Hyperbolic dose-response curves are obtained, and the transmitter concentration causing half the maximal effect, the EC_{50} (**effector concentration 50%**), is like the Michaelis constant K_M of enzymology interpreted in simple cases as the dissociation constant K_D of the protein-ligand complexes:

$$K_D = \frac{[R][L]}{[RL]}$$

The ratio of the biological response Q at a given ligand concentration to the maximal possible response Q_{max} varies as the ratio of the concentration of the occupied receptors [RL] to the total concentration of receptors $[R_{tot}]$:

$$\frac{Q}{Q_{max}} = \frac{[RL]}{[R_{tot}]}.$$

From this can be derived:

$$Q = \frac{Q_{max}[L]}{K_D + [L]}$$

Thus a Michaelis-Menten type equation is obtained in which the reaction velocities V and V_{max} are replaced by the responses Q and Q_{max} and the substrate concentration by the ligand concentration [L].

Q is derived from measurements of e.g. the depolarization of the receptor – containing membrane. From the double reciprocal plot $1/Q$ versus $1/[L]$ Qmax and K_D can be determined graphically in the same way as in the Lineweaver-Burk analysis used in enzyme kinetics. K_I, the inhibitor constant of an antagonist, can also be obtained analogously from the double reciprocal plot or from the Dixon plot $1/Q$ versus $[I]$, where $[I]$ equals the concentration of the inhibitor (antagonist).

Frequently, the dose response curve is not a hyperbola but has a sigmoid shape. The simple law of mass action which is the basis of the occupancy theory no longer applies and a more complex relationship between ligand binding and biological effect must be postulated. Originally the theory of "spare receptors" was put forward which were supposed to be functional only at high ligand concentrations. However here again an analogy with the enzyme-substrate interaction, this time with allosteric enzymes may be the most promising interpretation.

In this formulation a neurotransmitter (or hormone) receptor exists in at least two states, an inactive or resting (I) and an active (A) state (see Fig. 9.2). The equilibrium between the two is affected by the transmitter or hormone. Substances, which like the transmitter favour the active state, are called agonists, and those which displace the equilibrium towards the inactive state, antagonists. We shall now present and discuss various models for the quantitative analysis of results and the relationship between dose and response for transmitters and other pharmacologically active substances:

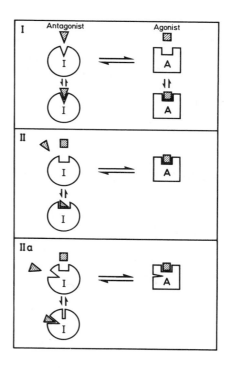

Fig. 9.2. Two state model for transmitter and hormone receptors. I, inactive state; A, active state.

1. The states (I) and (A) already exist before a ligand interacts with the receptor. They are in a "preformed equilibrium" where (I) strongly predominates. The agonist is bound only by (A); it stabilizes the active state and activates the receptor by displacing the equilibrium from (I) to (A). The antagonist is preferentially bound by (I) and inhibits the displacement. This Model I corresponds to the Monod-Wyman-Changeux model for allosteric proteins.
2. Alternatively the agonist may be already bound by (I) and in this binding causes a conversion of the receptor into configuration (A). Antagonists can either be bound to the same binding site without triggering activation to (A) (Model II) or they can inhibit agonist binding via a separate binding site (Model IIa). In the first case the antagonist would be a competitive, in the second case a non-competitive inhibitor. (The analogy to Koshland's "induced fit" model is evident).

It is difficult for anyone in the field – whether he is primarily concerned with enzymes or with receptors – to distinguish definitively between these models. In the case of neurotransmitter action the interpretation of experimental data is made even more difficult as the antagonist by no means always inhibits the binding of the agonist. It can also inhibit one of the steps subsequent to binding as for example the ion flux through the opened channel (behaving like a stopper in a tube) or the coupling between transmitter binding site and ion channel i.e. the opening of the channel. The former mechanism seems to apply for many local anaesthetics, while the latter applies to some effectors of adrenergic receptors (see below).

The ion channel may be a part of the receptor itself or a separate molecule, through which the message detected by the receptor is transmitted. We have already mentioned that transmission can be effected enzymatically or by conformational changes and protein-protein interaction. We shall meet the allosteric receptor-ion channel mechanism postulated by Nachmansohn in 1955 [3] and later by Belleau, Karlin, Changeux and others [2,4,5] in the section dealing with the nicotinic acetylcholine receptor; the enzymatic mechanism is exemplified by the catecholamine receptors (see Fig. 9.1).

Binding studies: binding does not equal effect

From what has been said we can see that the dose-response curve need not be identical with the binding curve. For example it is possible that the formation of the receptor ligand complex obeys the law of mass action and gives a hyperbolic substrate saturation curve, but that two or more binding sites must be occupied before an effect is triggered. A discrepancy of this kind between binding and dose-response curves provides a key to the understanding of the mode of action of neurotransmitters (or hormones). Attempts to characterize a receptor therefore always begin with binding studies.

Methods familiar from enzymology and protein chemistry can be utilized here i.e. equilibrium dialysis, ultracentrifugation, gel chromatography and ultrafiltration. The differentiation between specific and non-specific binding presents fundamental problems in particular in binding studies involving membrane preparations and partially purified protein fractions. In the central nervous system receptor concentrations may only amount to a few pmol/(g of tissue) and neurotransmitters in virtue of their polar properties readily interact, if only weakly, with polar membrane constituents outside the receptor area. Even if this non-specific binding has an affinity several orders of magnitude lower than the specific binding to the receptor (i.e. a much higher K_D), it is still an important factor to be taken into account because of the compensatorily large number of non-specific binding sites present. Discovering a receptor in the central nervous system by binding studies is thus only possible if there is an extremely high affinity ligand available with very high specific radioactivity (disintegrations $> 185 \cdot 10^{10} \mathrm{s}^{-1} \mathrm{mmol}^{-1}$).

A binding experiment is shown in Fig. 9.3. Curve A shows the total amount of bound ligand as a function of ligand concentration. It includes specific and non-specific binding. As evidence for this curve B shows the binding of the radioactive ligand in the presence of

Fig. 9.3. Binding of a ligand to a receptor containing preparation.
Curve A: dependence of binding on ligand concentration. Binding of radioactive ligand.
Curve B: The same, in the presence of a hundred-fold excess of non-radioactive ligand. This non-displaceable part of the binding of the radioactive ligand represents the nonspecific binding.
Curve C: difference between the two curves, giving the saturable, i.e. the specific binding.
Curve D: double reciprocal plot of data from curve C with extrapolation to K_D (intersection with the abscissa)$^{-1}$ and the maximal binding (intersection with the ordinate)$^{-1}$.

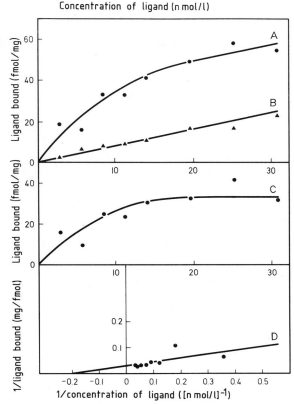

a hundred-fold excess of the "cold" unlabelled ligand. This displaces the "hot" ligand from the specific, saturable receptor (criterion 3, p. 204); the linear shape of curve B shows the non-saturable character of the residual nonspecific binding. The difference between curves A and B gives a hyperbola C, i.e. a saturation curve which expresses the specific binding of the ligand to the receptor. The data of curve C are transformed into a double reciprocal plot (curve D) and K_D and the maximum binding are graphically determined. Dissociation constants and the number of binding sites are deduced from a Scatchard diagram (ordinate is concentration of bound ligand/concentration of free ligand; abscissa is concentration of bound ligand). Table 9.1 shows some currently used radioactive ligands and compounds applicable in the displacement test (curve B in Fig. 9.3) for a range of central nervous system neurotransmitter receptors.

Tab. 9.1. Ligands for transmitter-receptor binding sites [28].

Receptor	Ligand Agonist	Ligand Antagonist	Substituting ligand	K_D (nM)
α-adrenergic	[^{3}H] Noradrenaline		(−)-Noradrenaline	26
		[^{3}H] Clonidine	(−)-Noradrenaline	5
		[^{3}H] WB 4101	(−)-Noradrenaline	0.6
		[^{3}H] DHE	(−)-Noradrenaline	1.6
β-adrenergic	[^{3}H] HBI		(±)-Propranolol	1.2
		[^{3}H] DHA	(−)-Alprenolol	1.3
		[^{125}I] HYP	(±)-Propranolol	1.5
Dopamine	[^{3}H] Apomorphine		(+)-Butaclamol	1-2
	[^{3}H] ADTN		(+)-Butaclamol	6.0
		[^{3}H] Spiroperidol	(+)-Butaclamol	0.3
		[^{3}H] α-Flupenthixol	(+)-Butaclamol	3.8
Serotonin	[^{3}H] Serotonin		Serotonin	8-15
		[^{3}H] LSD	Serotonin	5-8
Histamine Histamine H$_1$	[^{3}H] Histamine		Histamine	9.4
		[^{3}H] Mepyramine	Promethazine/Triprolidine	1.7
Histamine H$_2$		[^{3}H] Cimetidine	Cimetidine/Histamine	42
GABA	[^{3}H] GABA		GABA	16.0
	[^{3}H] Muscimol		GABA	4-5
		[^{3}] Bicuculline	GABA	380.0
Glutamate		[^{3}H] Kainic acid	Glutamate	12.0
Glycine		[^{3}H] Strychnine	Strychnine/Glycine	
muscarinic cholinergic	[^{3}H] Oxotremorin-M		Atropine	2.9
		[^{3}H] QNB	Atropine/Oxotremorin	0.06
		[^{3}H] Dexetimide	Dexetimide	0.65
nicotinic cholinergic		[^{125}I] α-Bungarotoxin	Nicotine	0.11
Opiate		[^{3}H] Naloxon	Levallorphan	1-4
	[^{3}H] Leu-Enkephalin		Leu-Enkephalin	0.4
Benzodiazepines	[^{3}H] Diazepam		Diazepam/Clonazepam	
	[^{3}H] Flunitrazepam		Chlordiazepoxide	4-6
Angiotensin II	[^{125}I] Angiostensin II		Angiotensin	0.23
TRH	[^{3}H] TRH		TRH	50
Neurotensin	[^{3}H] Neurotensin		Neurotensin	1-2
Bombesin	[^{125}I] Tyr^4-bombesin		Tyr^4-bombesin	3.0
VIP	[^{125}I] VIP		VIP	36.0

Methodological remarks: irreversible binding, affinity labelling

Binding studies in tissue, cells or even membrane preparations frequently produce data which are not reproducible with receptors in solution or purified receptors. Sometimes the binding capacity of a receptor is completely destroyed when it is freed from its environment in the membrane. For investigations of a receptor in these cases the method of affinity labelling can be used: a ligand is selected with the structural features for specific binding, including a suitable label (which may be a radioactive atom or a strongly fluorescent or light-absorbing substituent) and a reactive group which enables it to react covalently with its receptor. The receptor is then labelled *in situ* with the affinity reagent and identified after solubilization from the membrane by its radioactivity, fluorescence or light absorbance. Some typical neurochemical affinity reagents are shown in Fig. 9.4. Frequently they react spontaneously as when the reactive group confers upon them the properties of an alkylating agent; reagents requiring activation by light (the so-called photoaffinity reagents) are of particular value as the time and duration of the reaction can be accurately controlled. Fig. 9.5 shows as an example the method by which a peptide hormone, transmitter or toxin is coupled covalently to the receptor: the peptide is treated with 4-fluoro-3-nitrophenylazide, incubated subsequently in the dark with the receptor-containing tissue and activated by irradiation. This method, developed for the insulin receptor, has been used successfully to identify neurotoxin binding sites in axonal and postsynaptic membranes.

In some cases photoaffinity labelling does not require an azido-derivative of the ligand which may be difficult to obtain and pharmacologically different from the natural ligand. UV-irradiation of receptor-ligand complexes can result in covalent incorporation of the label. For example the glycine receptor was labelled in this way with [^{3}H]strychnine, the benzodiazepine receptor with [^{3}H]fluonitrazepam and the nicotinic acetylcholine receptor with a variety of radioactive noncompetitive antagonists.

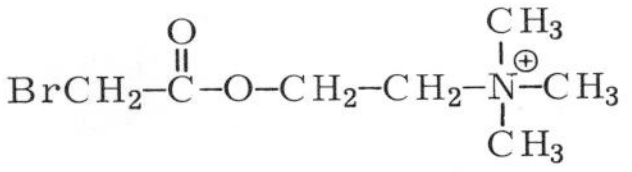

Bromoacetylcholine
alkylating derivative of a nicotinic cholinergic agonist

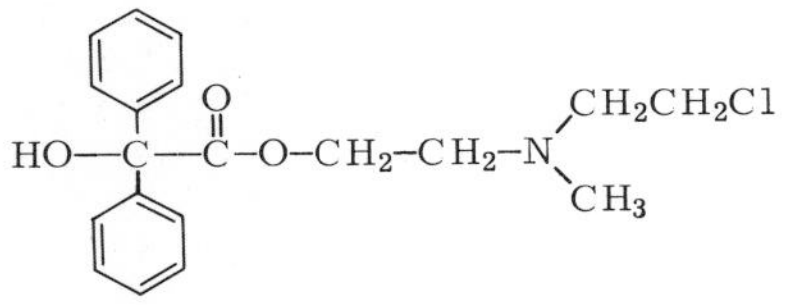

Benzilylcholinemustard
irreversible alkylating muscarinic cholinergic antagonist

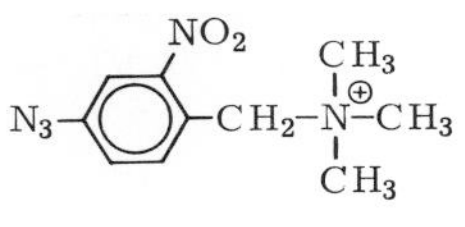

4-Azido-2-nitrobenzyl-trimethylammonium
photoaffinity reagent nicotinic cholinergic antagonist

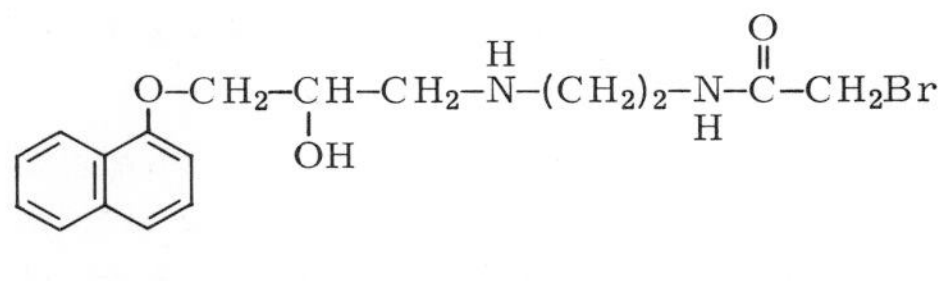

Propanolol derivative
irreversible β-adrenergic antagonist

Fig. 9.4. Some typical irreversible receptor ligands.

I. Radioactive labelling of a tyrosine residue

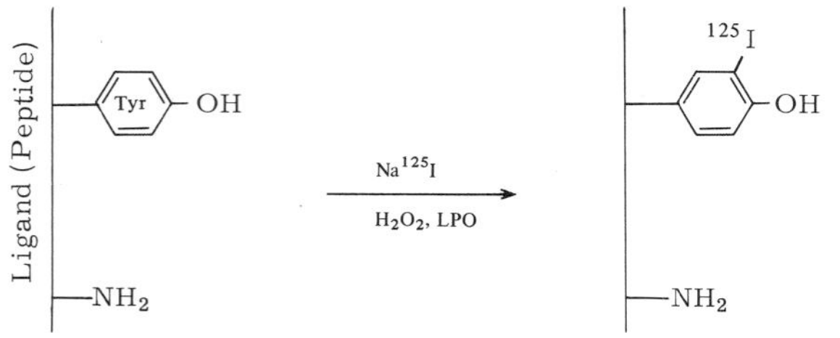

II. Introduction of the photoreactive nitrophenylazide (NAP) group

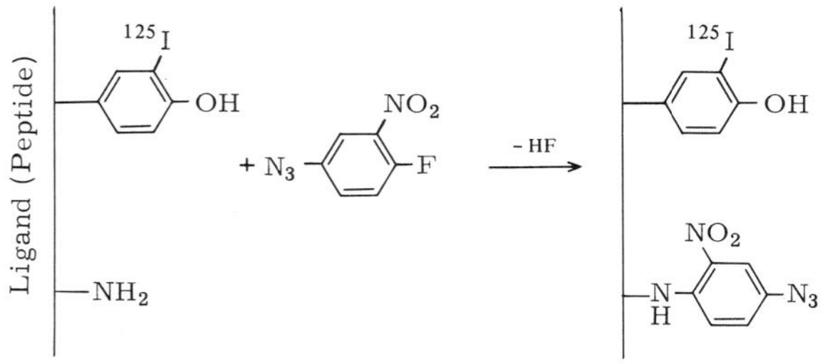

III. Irreversible *cross-linking* of peptide and receptor

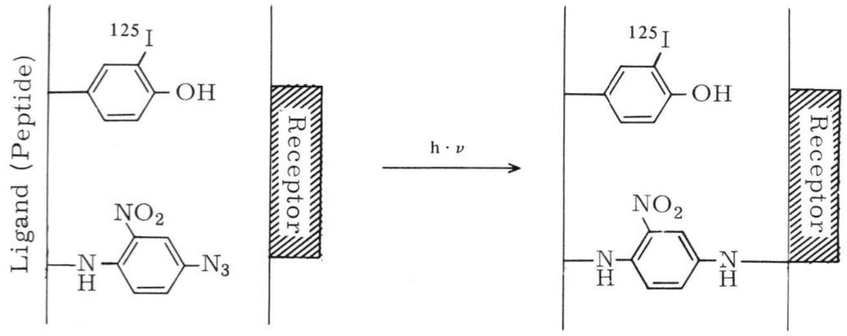

Fig. 9.5. Photoaffinity labelling of a receptor with a specific peptide ligand, a method which has been used to identify receptors for insulin, acetylcholine and other compounds (LPO, lactoperoxidase).

Mobile receptors: the "floating receptor" hypothesis

There are examples of ion regulator complexes in which the receptor and the ion channel appear to reside in different molecules. Thus some acetylcholine receptors found in the neurons of *Aplysia* after binding with acetylcholine increase the sodium conductivity. Other acetylcholine receptors in the same organism induce a rapid increase in the chloride conductivity while yet others a slow increase in the potassium permeability [6]. If it is assumed that the binding component is the same – which is in no way proven – then it would have to act in varying combinations with potassium, sodium and chloride channels [7]. While these combinations appear to be permanent the following observations led to the formulation of the *"floating" or "mobile" receptor hypothesis,* which states that receptors are not permanently bound into complexes but "floating" in the membrane and interacting with different active structures – transport systems, enzymes etc. (Fig. 9.6). There is, for example, only one type of receptor for insulin which, however, separately regulates a number of membrane functions: glucose transport, adenylate cyclase, phosphodiesterase, Na^+, K^+-ATPase, Ca^{2+}-ATPase, and amino acid transport. In contrast in the fat cells of the rat there are at least eight different receptors all of which regulate adenylate cyclase. The binding of the eight effectors is noncompetitive, yet their action is not additive. This

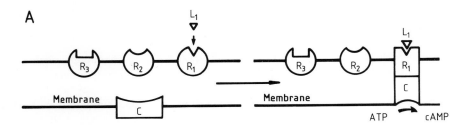

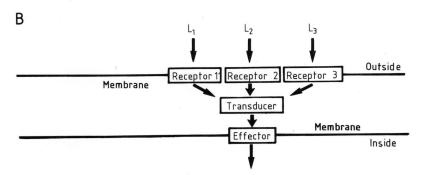

Fig. 9.6. Floating receptor hypothesis. (A) Of the various receptors, R_1, R_2, R_3, present in a membrane only the one activated by its ligand (L_1, an agonist or hormone) couples to the adenylate cyclase, C, and activates it. A third floating component, the G-proteins (coupling proteins) have been omitted for simplicity (see Fig. 9.14). In this way several signals, L_1, L_2, L_3, can converge via a variety of receptors on a single cyclase (B). The G-proteins function as signal transducers shuttling between R and C. A three-component complex receptor-transducer-adenylate cyclase according to the "collision-coupling hypothesis" postulated by Levitzki is not formed during the process of signal transduction through the membrane.

can only be explained by the assumption that the different ligand-receptor complexes compete for one cyclase. The mobile receptor hypothesis has been supported by experiments with cell cultures. Mutants from which either one or another receptor or the enzyme is missing have been derived from cells with β-adrenergic and prostaglandin receptors which regulate the activity of adenylate cyclase independently of each other. This has shown that receptor and cyclase are separate gene products (see further below). Analogous results have been obtained for the chemotaxis receptors of bacteria (Chapter 11), thus the principle that membrane functions are economically built up out of functional building blocks seems to have developed very early in evolution.

Besides the mobile receptors more stable complexes with transport systems exist. One example of these is the nicotinic acetylcholine receptor, but this time that of vertebrates (see below). This will be the first receptor we shall discuss in detail as its biochemical characterization is the most advanced.

Acetylcholine receptors

Electrophysiologically and pharmacologically two types of cholinergic synapses can be distinguished, the nicotinic and the muscarinic (see p. 162). Correspondingly there are two types of acetylcholine receptor of which only the nicotinic has so far been well characterized biochemically. Two favourable factors have contributed to this: the discovery of snake venom α-toxins (α-bungarotoxin, *Naja naja* α-toxin – polypeptides of relative molecular mass 7 800; see Chapter 8) which bind to the nicotinic receptors with high affinity and specificity, and the availability, in the electric organs of the electric ray (*Torpedo*) and eel (*Electrophorus*) of a rich source of the receptor. The snake toxins have provided an assay for the receptor *in vivo* and *in vitro* and the basis for its purification by affinity chromatography. The nicotinic acetylcholine receptor will be dealt with in some detail, firstly because in almost no other area has neurophysiology and molecular biology interacted so fruitfully, and secondly because the methods used have provided a model for research on other receptors.

The nicotinic acetylcholine receptor – First level: intact cells

The receptor in electric tissue has been investigated, on three levels [5] (Fig. 9.7). Schoffeniels and Nachmansohn showed that intact electric cells called electroplaques can be isolated from the electric organ of *Electrophorus* and investigated electrophysiologically. In *Torpedo* this is much more difficult to do as its cells are thinner, but there too the main properties of the acetylcholine receptor in the intact tissue have been investigated. For example, it has been shown that different agonists can cause a depolarization of the cell and that the dose-response curve (the dependence of the main part of the depolarization on the agonist concentration) is sigmoid in shape. Nachmansohn proposed that an analogy existed between receptor-transmitter and enzyme-substrate interaction [3]. Extending this analogy, the sigmoid dose-response curve could be interpreted as a cooperative phenomenon. Today we know that both the binding of the agonist to the receptor and the subsequent step of the opening of the ion channel are cooperative phenomena.

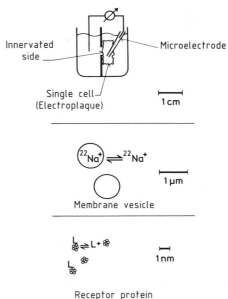

Fig. 9.7. Three levels of receptor investigation. As an example the nicotinic acetylcholine receptor from the electroplaques of the electric eel has been selected. Above: Electrophysiological studies with intact cells, a preparation introduced by Nachmansohn and Schoffeniels. Centre: Investigation of receptor function (ligand binding and the flux of radioactive sodium ions, developed by Kasai and Changeux) using vesiculated fragments of postsynaptic membrane. Below: Ligand (L) binding studies on purified receptor protein.

Second level: receptor membrane

At the second level of receptor research postsynaptic membrane fragments have been isolated from electric tissue by density gradient centrifugation of homogenized material. These contain the receptor in high purity and in its natural ligand environment, but without cytoplasm and the usual cell constituents (Fig. 9.8). As mentioned above binding studies have shown that the binding of acetylcholine to the membrane-bound receptor is cooperative. The Hill coefficient here as *in vivo* is 1.7. However the dissociation constant is two orders of magnitude smaller (1.8×10^{-8}) than the K_D derived from the dose-response curve. We shall return to this when we discuss the mechanism of pharmacological desensitization.

The postsynaptic membrane fragments have an important property which helps to bridge the gap between the physiology of the intact organism and its biochemistry. Closed vesicles are formed (not to be confused with presynaptic transmitter-containing synaptic vesicles) and these retain fundamental biological properties e.g. the ion flux through the membrane is stimulated by acetylcholine and other agonists and inhibited by α-neurotoxins and the

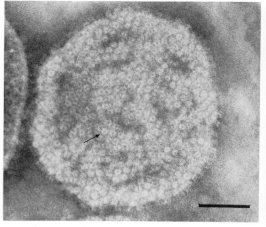
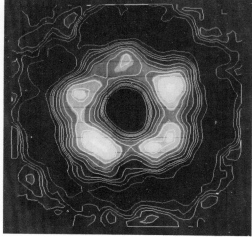

Fig. 9.8. Transmitter receptor in its membrane.
(A) Electron micrograph of the postsynaptic membrane of a nicotinic cholinergic synapse isolated from *Torpedo* electric tissue; bar 50 nm; doughnut-like structures are the receptors.
(B) Image processing aided by computers yields a more detailed picture of the receptor (M. Giersig and M. van Heel, Fritz-Haber-Institut, Berlin).
(C) A model summarizing the biochemical data. The nicotinic acetylcholine receptor is depicted as a pentamer of five glycoproteins spanning the membrane. MIR, main immunogenic region, a site located on the alpha-polypeptide chains against which an especially large number of monoclonal antibodies are directed. 43 K represents a protein of M_r 43 000 associated with the receptor on the inside of the membrane. The zig-zag lines extending from the upper (extracellular) part of the receptor represent the oligosaccharide chains. Reproduced, with kind permission, from Lindstrom et al., *Cold Spring Harbor Symp. Quant. Biol.* **48**, 89–99 (1983).

other antagonists. These vesicles have been filled with $^{22}Na^+$, the suspension diluted in a physiological buffer and an aliquot filtered off at known time intervals in order to measure the quantitative outflow of radioactivity (see Fig. 9.9). If the diluting buffer contains an agonist the ^{22}Na-outflow is increased. A dose-response curve is also obtained here very similar to that obtained *in vivo*.

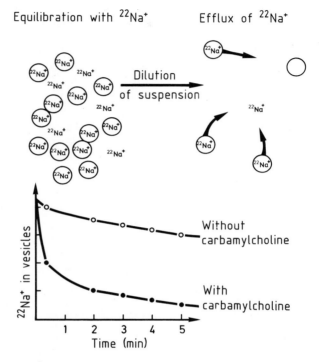

Fig. 9.9. Measurement of ion permeability in a biological membrane *in vitro*. The theory of this method is that the postsynaptic membrane forms closed vesicles in this preparation. These can be filled with radioactive ions, a radioactive gradient imposed by dilution and the outflow of ions determined by filtration of a sample of the suspension at various time intervals. The outflow is stimulated by agonists (here by carbamylcholine).

Receptor-rich membrane vesicles are a valuable model system; they contain few other proteins and with them both transmitter binding and the regulation of ion permeability can be tested. The latter is no longer possible with the purified receptor, since it has been deprived of its lipid environment. It must also be borne in mind that there are quantitative differences between the ion flux *in vivo* and that through the vesicle membrane. As with all model systems this one gives an aspect of the total picture but not one that is complete in every detail.

Third level: receptor molecules

Acetylcholine receptor protein can be isolated in quantities of 10 mg from electrical tissue particularly of the receptor-rich *Torpedo* [5,8]. It consists of glycoprotein with a relative molecular mass of 270 000. The receptor is a protein complex consisting of five polypeptide chains with the stoichiometry $\alpha_2\beta\gamma\delta$ (see also p. 171). It has been shown by means of affinity reagents that binding sites for low molecular mass effectors and for α-neurotoxins from snake venom are localized in a polypeptide chain of relative molecular mass 40 000, the α-chain. Little is known about the function of the other higher molecular mass polypeptide chains. They may perhaps take part directly in ion transport through the membrane and in the regulation of the response to agonists.

The polypeptide chains of relative molecular mass 40 000 contain a disulphide bond. If this is cleaved by reduction the cooperativity of the ligand binding is abolished, the Hill coefficient becoming 1.0; at the same time the affinity for acetylcholine is reduced. This effect is reversible; reoxidation e.g. with DTNB (5,5'-dithiobis-(2-nitrobenzoate) restores the original state.

The acetylcholine receptor regulates the ion permeability of the postsynaptic membrane, possibly by a conformational change in the receptor protein. Evidence has been obtained for conformational changes after ligand binding by measuring changes in the intrinsic (tryptophan) and extrinsic fluorescence, (for the latter the local anaesthetic quinacrine (Fig. 8.11) may be used as a fluorescent reporter group).

Pharmacological desensitization: model for synapse modulation

If the postsynaptic membrane is exposed to an increased concentration of acetylcholine (and if the acetylcholinesterase is simultaneously blocked) a slow decrease of postsynaptic response is observed. The membrane apparently becomes more insensitive to agonists. This is called desensitization. The phenomenon takes place at all three levels of organization, in the intact tissue, in the membrane vesicles and in the isolated receptor. The ion flux through the membrane is inhibited but not because the receptors bind the agonists more weakly, but because the ion channels are no longer opened. Pharmacological desensitization is not confined to the acetylcholine receptor; it is observed in many systems – in receptors for peptide hormones and in β-adrenergic receptors, for example.

It is not known whether the desensitization of acetylcholine receptors has a physiological function; however, it is thought to be one way in which synapse transmission is regulated. It has been shown that the affinity of agonists for the acetylcholine receptor is increased by up to two orders of magnitude, if they are in contact with it for a long enough period. The increase takes place in seconds to minutes and is accompanied by conformational changes in the protein. These, as mentioned before, are followed kinetically by measuring changes in the extrinsic or intrinsic fluorescence or by a reduction in the reactivity of SH groups of the protein. The desensitization is accelerated by Ca^{2+} and by local anaesthetics. Antagonists do not cause desensitization. From all these observations it has been concluded that the acetylcholine receptor exists in at least three states:

One inactive (I, channel closed, low affinity), one active (A, channel open) and one desensitized (D, channel closed, high affinity). The agonist activates the receptor and opens its ion channel by binding to A, whereas the antagonist binds to I, stabilizing the receptor in its inactive (channel closed) state. (For an alternative mechanism, the "induced fit", see p. 207).

Acetylcholine receptors from muscle tissue

As about 2% of the membrane protein of *Torpedo* electroplaque consists of acetylcholine receptor, this protein needs to be purified only fifty-fold to obtain pure receptor protein. By contrast, isolation from muscle requires more than 10 000-fold purification. Even this is attained only through the artificial device of using denervated muscle, since the denervation itself causes a twenty-fold increase in receptor concentration (see below).

Muscle receptors show cross-reactivity with antibodies to *Torpedo* receptor. This indicates chemical similarity. Recent determinations of the primary structures (by the cDNA technique) showed considerable sequence homologies.

Hypersensitization: another model for receptor modulation?

If the motor nerve is cut, the whole surface of the muscle fibre becomes excitable to acetylcholine within a few days (hypersensitization) [8]. The acetylcholine receptors are no longer limited to their characteristic area, the subsynaptic membrane, but are distributed equally over the whole fibre. The increase in extrasynaptic receptor density is not due to diffusion of synaptic receptors but to new synthesis. Some inhibitors of protein biosynthesis inhibit the hypersensitization or sensitivity of the extrasynaptic membrane to agonists. If [^{35}S]methionine is added to the preparation at the moment of denervation it is incorporated into the new receptors.

It is not yet certain whether the newly synthesised, extrasynaptic receptor is chemically identical with the subsynaptic. There are some points against this. Its isoelectric point is at 5.3 i.e. about 0.15 pH units (or five ion charges) higher than that of the subsynaptic (or of that isolated from innervated muscle). Also the kinetic data for the respective ion channels are different: after activation by agonists those of the denervated muscle remain open about three times longer than those of the innervated. However they have only about one third of the conductivity. On the other hand the affinity for ligands and the immunological properties are practically the same. The ion specificity of the receptor channel is unchanged (potassium and sodium ions permeate, chloride ions do not). Curare and α-bungarotoxin inhibit the activation. Possibly both receptor types are products of the same genes, but are modified in transit to the membrane (posttranslational modification). It is this aspect which makes the denervation and associated hypersensitivity so interesting. It prompts the questions. How is the distribution of receptors on the cell surface regulated [7]? What produces the disappearance of the extrasynaptic receptor during ontogenesis and the subsynaptic aggregation? It is thought to be associated with the molecular mechanism of synaptic coupling between cells during embryonic development and to be a model of the regulation of learning processes and behaviour patterns. We shall return to this later (Chapter 11).

The next important point is that hypersensitivity can be reversed by reinnervation. However, this is not produced as first thought by a trophic factor from the nerve, for direct electrical stimulation of the muscle fibre causes a disappearance of extrasynaptic receptors in the same way as nerve stimulation. Conversely a denervation need not be caused only by nerve section: blocking of synaptic transmission by botulinum toxin (presynaptic) or α-bungarotoxin (postsynaptic) will suffice.

The pattern of stimulation or impulse frequency is evidently a vital factor for the distribution of receptors on the cell surface and the frequency further influences gene expression, the differentiation of the cell by the induction of protein biosynthesis. Thus the enzyme pattern of a rapid muscle fibre can be induced in a slow one if the appropriate pattern of stimulation has been applied for sufficiently long [9]. Perhaps Ca^{2+} acts as an inducer, but this is only speculative. In any case the reduction in receptors in the special region of the surface membrane by a specific stimulation pattern can be conceived as a characteristic of differentiation and thus of chromosomal origin. There is still discussion about the theory that a stabilization of the synaptic receptors by a covalent posttranslational modification occurs and that this, in contrast to the extrasynaptic receptor, makes them resistant to proteolytic degradation.

Phosphorylation of acetylcholine receptors

Such a covalent modification could be the phosphorylation of the receptor in the presence of ATP, as has been observed in *Torpedo* [10] and *Electrophorus* [5]. The postsynaptic membrane contains a cAMP dependent kinase and a phosphatase. Carbamylcholine protects the receptor against phosphorylation. A functional difference has not yet been discovered between the two interconvertible forms – the phosphorylated and dephosphorylated receptor.

Myasthenia gravis, an autoimmune disease of the nicotinic cholinergic synapse

Myasthenia gravis is a disease which is characterized by weakness and fatigue of the striated musculature. It is caused by a defect of synaptic transmission at the neuromuscular endplate. Symptoms consist usually of progressive fatigue, drooping of the eyelids, slow chewing, and in severe cases rapid exhaustion on muscular movement or work. Its incidence is about three in a hundred thousand people, in women more commonly than in men (in the 40 year age group). There is evidence for a hereditary component, but there is not yet proof of this. The disease can be fatal, however, with new treatments, there is a life expectancy of over fifteen years.

Evidence for the localization of the defect in the synapse is that inhibitors of acetylcholinesterase, (neostygmine or edrophonium) cause alleviation of the symptoms, whereas there is an extreme sensitivity to curare. Marked morphological changes in the structure of the endplate have been observed. The distance between the pre- and postsynaptic membrane is significantly greater, the postsynaptic membrane is less folded and the subsynaptic surface seems to be diminished. Presynaptic changes make their appearance in an advanced stage of the disease. Synthesis, packaging of transmitter into synaptic vesicles and presynaptic release of transmitter are not affected.

In recent years great progress has been made in the analysis of myasthenia gravis at the molecular level. It is now known that the postsynaptic membrane responds more weakly to agonists, because the number of acetylcholine receptors is reduced by more than 70%. The properties of the remaining receptors and specifically the open time and the conductivity of the individual ion channels are unchanged.

Myasthenia gravis is classified as an autoimmune disease. Previous observations had already indicated the involvement of the immune system for myasthenia was frequently accompanied by tumours or enlargement of the thymus. Surgical removal of the thymus had alleviated the symptoms in a certain number of cases. More definite evidence for the autoimmune nature of the disease was obtained when antibodies against the acetylcholine receptor were detected in the serum of myasthenic patients by means of a sensitive radioimmune assay (RIA) based on the blocking by antibodies of the binding of ^{125}I-labelled α-bungarotoxin to the receptor. The autoimmune hypothesis was supported by the observation of Patrick and Lindstrom that an injection of purified acetylcholine receptor from electric fish (*Torpedo* or *Electrophorus*) into rabbits produced typical symptoms of myasthenia gravis. Thus an experimental model of the disease was discovered which will make possible an understanding of its mechanism and the development of an improved therapy. At present the exact role played by the antibodies is not yet clear. *In vitro* experiments have shown that they do not block synaptic transmission directly at the neuromuscular endplate. There is evidently no direct competition between transmitter and antiserum at the receptor; also *in vivo* the effect of the injected receptors occurs late, long after the formation of the antibodies. Work is now being done on this disease by various groups and perhaps there will soon be further information on the molecular mechanism of action of the anti-receptor antibodies and the production of myasthenic symptoms. It has been found that proteolytic degradation of the receptor is accelerated by the antibodies. Perhaps they thereby reduce the receptor density in the endplate by increasing receptor "turnover".

The muscarinic acetylcholine receptor

Acetylcholine receptors of striated skeletal muscle are nicotinic, those of the smooth muscle, e.g. the gut, are muscarinic. Central nervous system receptors and peripheral parasympathetic postganglionic receptors are predominantly muscarinic. (In the central nervous system of vertebrates the ratio of muscarinic to nicotinic acetylcholine receptors is about three to one). In the ileum and in the caudate nucleus receptor density (ca. 200 receptors per μm^2) is comparable to that of the nicotinic receptors in the electroplaques of *Electrophorus*. Their biochemical characterization is however less advanced than that of the nicotinic receptor, mainly because there is no tool as good as α-bungarotoxin [11]. The development of covalent bond-forming specific affinity reagents has opened up the possibility of isolation. Thus, using the alkylating reagent propylbenzilylcholine mustard (see Fig. 9.4), a polypeptide chain with relative molecular mass 87 000 was identified as part of the receptor. Its extraction from the membrane was made difficult by its sensitivity to mild, non-ionic detergents.

Antagonists (Fig. 8.4) are bound to the muscarinic receptors of brain and smooth muscle in accordance with the law of mass action with a Hill coefficient of unity at a single type of binding site. The agonist binding in contrast is complex and cannot be described by a simple dissociation constant. If the effectivity of the agonist is investigated by the displacement of tritiated antagonists, Hill coefficients are obtained which are mostly under but never above unity. The dose-response curves derived from the contraction induced by an agonist on a strip of ileum are correspondingly complex. The results indicate that there are two different populations of muscarinic acetylcholine receptors. Like the nicotinic receptors they are allosteric proteins which regulate the ion permeability of the membrane. However they perform different roles: muscarinic excitation is accompanied by a decrease in potassium permeability, muscarinic inhibition in other cells by an increase. In each case the response of the muscarinic membrane to acetylcholine is very much slower than that of the nicotinic. In both, desensitization after long incubation with agonists is observed. However, unlike nicotinic, muscarinic activation seems to be accompanied by a decrease of affinity for agonists.

The role of the muscarinic acetylcholine receptor is not limited to the regulation of channels for alkali ions, but – as we have already shown in Chapter 2 – it affects phosphorylation and dephosphorylation of phosphatidylinositol and stimulates the formation of cGMP. Receptor activity causes an increase of free intracellular Ca^{2+}. The molecular and functional relationship of these observations is not yet clear.

The catecholamine receptors

The catecholamines adrenaline, noradrenaline and dopamine act through a variety of receptors. We shall first consider the adrenoceptors, the targets of adrenaline and noradrenaline and then the dopamine receptors. In Chapter 8 we have already discussed the principle of "multiple receptors" which means that there appear to be several subtypes of receptors interacting with a given transmitter. We suggested that the "purpose" of this multiplicity may be to increase the number of effects that a relatively small number of neurotransmitters can produce. There are at least four types of adrenoceptors, called α_1-, α_2-, and β_1-, β_2-adrenoceptors (adrenergic receptors) and possibly two dopamine receptors, called D_1 and D_2. This subtype classification should not be carried too far; it is largely based on binding studies and pharmacological properties, not on biochemical differences in the isolated receptor proteins.

β-adrenergic receptors: interaction of receptor, cyclase and regulator (R, C and N*)

In Chapter 8 we have already introduced Ahlquist's concept of two different adrenergic effects. A smooth muscle cell for example, has two types of receptor for adrenaline, the α-adrenergic and the β-adrenergic, which are pharmacologically distinguishable; *α-adrenergic receptors* cause contraction of the cell as they interact with adrenaline; somewhat weaker agonists being noradrenaline, phenylephrine and on a smaller scale isoprenaline. Typical α-blockers, i.e. antagonists, are ergot alkaloids, phenoxybenzamine and phentolamine (Fig. 8.19). *β-adrenergic receptors* do not differ in principle, but mainly in the different potency of the effectors. Adrenaline and noradrenaline are again agonists (at β-receptors almost equipotent). Isoprenaline however has a stronger action (the

Table 9.2. Classification of adrenergic receptors.

Type	Potency	Effect	Mechanism
α-Adrenergic		Vasoconstriction, uterine contraction, inhibition of intestinal peristalsis, pupillary dilation	
α_1	Antagonist prazosin > yohimbine	Postsynaptic	Ca^{2+}-concentration is increased by agonists
α_2	Antagonist yohimbine > prazosin	Presynaptic and post-synaptic	cAMP is decreased by agonists
β-Adrenergic		Vasodilation, inhibition of uterine contraction, myocardial stimulation	cAMP is increased by agonists
β_1	Isoprenaline > adrenaline = noradrenaline	Adipose tissue (fatty acid mobilization), cardiac stimulation	Subtype-differentiation relates to ligand recognition; both β_1 and β_2 stimulate adenylate cyclase
β_2	Adrenaline >> noradrenaline	Bronchodilation, vasodepression	

* Recently for the GTP-binding regulatory protein the letter G has been agreed upon by most laboratories.

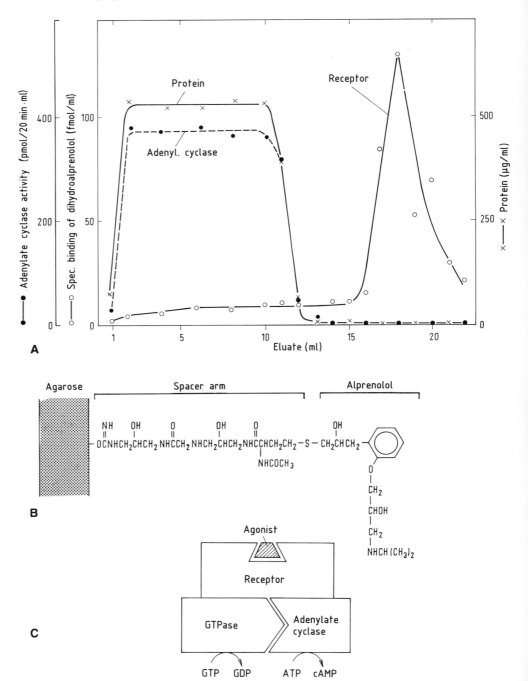

response in this case is relaxation of the muscle cell). Propanolol and dichlorisoprenaline are typical β-adrenergic antagonists (Table 9.2).

β-receptors activate adenylate cyclase. We will now deal with them in detail in order to compare them as "enzyme regulators" with the "ion channel regulator" of the acetylcholine receptor.

The biochemical characterization [12] of the β-adrenergic receptor (β_1) was, like that of the nicotinic acetylcholine receptor, made possible by two fortunate discoveries. It was found that avian erythrocytes were a rich starting material and that the detergent digitonin would dissolve the receptor without inactivating it. Turkey erythrocytes contain ca. 0.2 pmol β-receptor/mg of membrane protein, and it is not only possible to purify it 12 000 times by affinity chromatography but, as the elution profile of Fig. 9.10A clearly shows, adenylate cyclase is not an integral component of the receptor, but a separable subunit. Previously a GTP binding component had been similarly isolated by affinity chromatography.

Using these three building blocks – the receptor, the cyclase and the GTP-binding protein – a model of the mechanism of action of the β-adrenergic receptor could be devised in which the binding protein is assigned a regulating or transducing role.

In this model (Fig. 9.10C) the β-adrenergic receptor activates adenylate cyclase not directly but via the GTP-binding regulator protein. We shall return later to this GTP-binding protein, usually called G, N or G/F, which we have already met as "transducin" when we discussed amplification in the visual process (Chapter 1). Binding of the agonist to the receptor causes a conformational change which is transmitted to the regulator subunit (N) and then, after complex formation with GTP, to the adenylate cyclase. As a modification of the mobile receptor hypothesis (p. 212) this transient interaction between receptor, N-protein and cyclase has been summarized in the "collision-coupling-hypothesis". The GTP effect is only temporary, as the regulator has GTPase activity and thus hydrolyses its own effector. The cAMP level in the cell is thereby raised by adrenaline or noradrenaline under the regulatory control of guanidylnucleotides.

The question of the mechanism of action of the β-adrenergic system is however not answered by this description, but only transferred one step further into the cell. It now becomes: what role is played by cAMP, and how does it reproduce the physiological action of β-adrenergic agonists? Since the basic work of Sutherland, cAMP has been known as the *second messenger* in non-neural, hormone-stimulated tissue, and when more neurotrans-

---

← Fig. 9.10. Biochemistry of the β_1-adrenergic receptor [12].
(A) Affinity chromatography of the detergent-solubilized receptor indicating that adenylate cyclase and receptor are indeed physically separable proteins eluting differently from the affinity column.
(B) Chemical structure of the affinity resin: a ligand of the β-adrenergic receptor is attached (over a spacer arm) to Agarose beads.
(C) Simplified model of the receptor, depicting it as a ternary complex of three proteins, the receptor (binding protein, R), the enzyme (adenylate cyclase, C) and a regulatory protein with intrinsic GTPase activity (called variously G, G/F or N). The three proteins never really come together, but stimulation of C is a sequential process (see below). Reproduced, with kind permission, by the author and by Elsevier Science Publishers.

mitters with an adenylate cyclase regulating action became known, it was tempting to formulate a common hypothesis for hormone and transmitter. One such was proposed some years ago by Greengard and has since been increasingly supported by experimental data [13,14].

Greengard's hypothesis: cAMP regulates via protein phosphorylation

In 1968 Edwin Krebs and his coworkers showed that a cAMP-dependent protein kinase was involved in the stimulation of glycogenolysis. Later, Greengard's group found protein kinase activity in almost all animal tissues, and it was postulated that cAMP produces its many sided physiological effects via this new class of enzymes (Fig. 9.11A, 9.12). cAMP-dependent protein kinases are tetrameric enzymes, each consisting of two regulatory and two catalytic subunits (Fig. 9.11B). The binding of cAMP to the regulatory subunits causes their dissociation from the catalytic subunits, which are thereby activated. The protein kinase found predominantly in nervous tissue, the so-called Type II, actually phosphorylates its own regulatory subunits; this autophosphorylation is not observed in the other Type I protein kinase. However their activating mechanism is the same.

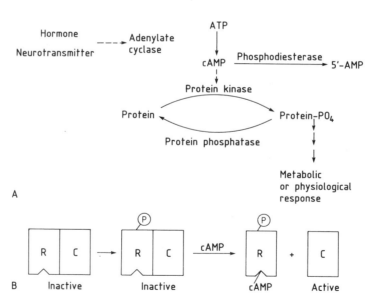

Fig. 9.11. cAMP and protein phosphorylation. (A) Scheme of the reaction; (B) activation of the cAMP-dependent protein kinase. R, regulatory, C, catalytic subunit of the enzyme.

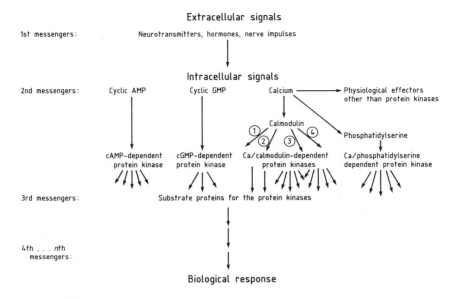

Fig. 9.12. Signals in the brain. Extracellular signals (first messengers) produce specific biological responses in target neurons via a series of intracellular signals (second, third, etc. messengers). Second messengers in the brain include cyclic AMP, cyclic GMP and calcium. Cyclic AMP and cyclic GMP produce most, and possibly all, of their second messenger actions through the activation of virtually one type of cyclic AMP-dependent protein kinase and one type of cyclic GMP-dependent protein kinase, respectively. The former enzyme exhibits a broad substrate specificity and the latter a more restricted specificity. Calcium exerts many of its second messenger actions through the activation of calcium-dependent protein kinases, as well as through a variety of physiological effectors other than protein kinases. Calcium activates protein kinases in conjunction with calmodulin or phosphatidylserine. There are at least four types of calcium/calmodulin-dependent protein kinases in brain: (1) phosphorylase kinase which phosphorylates only phosphorylase (and possible glycogen synthase); (2) myosin light chain kinase, which phosphorylates only myosin light chain; (3) calcium/calmodulin-dependent protein kinase I; and (4) calcium/calmodulin-dependent protein kinase II. The substrate specificities of the latter two enzymes have not been established, but both appear to phosphorylate numerous proteins. The activation of individual protein kinases causes the phosphorylation of specific proteins in target neurons. In some cases, these proteins, or third messengers, appear to be the immediate effector for the biological response. In other cases they seem to produce the biological response indirectly through fourth, fifth, sixth etc. messengers (Figure and legend, with kind permission, from P. Greengard, [14]).

Fig. 9.13 summarizes schematically the possible sites of action of the cAMP/protein kinase system in a synapse. Presynaptically it could, for example, regulate transmitter synthesis. In Chapter 8 we have described the activation by cAMP-dependent phosphorylation of tyrosine hydroxylase, the enzyme which catalyses the first, rate-determining step of catecholamine synthesis. In addition the system could affect presynaptically the axonal transport of transmitter, its exocytosis and other important constituents of the neuron. A cAMP-dependent phosphorylation of a protein component of the microtubular system has been reported.

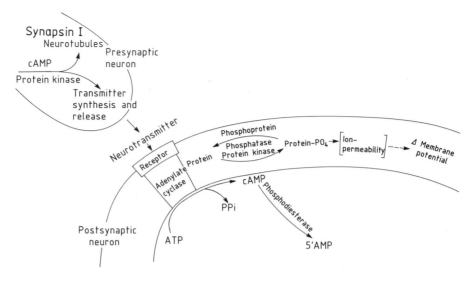

Fig. 9.13. cAMP and synapse function [13]. Reproduced, with kind permission, from P. Greengard and Macmillan Journals Ltd.

Synapsin I, substrate of various protein kinases

One of the most abundant substrates of protein kinase is a protein termed synapsin I. It appears to be a peripheral protein of synaptic vesicles in many nerve terminals. It contains multiple phosphorylation sites, one of which (site 1) is phosphorylated both by cAMP-dependent protein kinase and by calcium/calmodulin-dependent protein kinase I. Two other sites (sites 2 and 3) are phosphorylated specifically by calcium/calmodulin-dependent protein kinase II. The phosphorylation has been shown to be regulated by serotonin, dopamine, and noradrenaline. It is thought to be involved in the regulation of transmitter release from the nerve terminal. On SDS-polyacrylamide gels synapsin I appears as a doublet corresponding to apparent relative molecular masses of 80 000 and 86 000.

Cyclic nucleotides and Ca^{2+}, two classical "second messengers"

With some hormones and neurotransmitters cGMP replaces cAMP as the cyclic nucleotide which is regulated (Fig. 9.12). Thus while dopamine, serotonin, adrenaline (β_1, β_2, α_2 receptors), histamine (H_2 receptor), octopamine and the peptide transmitters regulate the cAMP system, acetylcholine (muscarinic) and histamine (H_1 receptor) act through the cGMP system. With the latter group, the cyclic nucleotide may be replaced by yet another, rapidly acting second messenger. For this Ca^{2+} is a possible candidate, but there may be others, e.g. IP_3, see chapter 2. However, in spite of the large number of neurotransmitters, hormones and other effectors with their multiplicity of action, there seems to be only a limited number of these chemical messengers. The specificity of action of cAMP, for example, in a particular cell depends on the substrate specificity of the protein kinase and on that of the receptors contained in its membrane.

Physiological regulation does not occur solely by the synthesis and circulation of active substances and chemical messengers. It must also be possible for the circulated signals to be inactivated. At the cyclic nucleotide level cAMP is hydrolysed to 5'AMP by phosphodiesterase. At the phosphorylated protein level, phosphoprotein phosphatases ensure the reversibility of protein phosphorylation.

Recently the regulation of neural function by cAMP has been complicated by the discovery of additional factors. We have already mentioned the important role of Ca^{2+} in transmitter release (Chapter 8). Calcium ions also regulate adenylate cyclase and phosphodiesterase indirectly via a Ca^{2+}-binding protein called calmodulin (see Chapter 10). The same protein also regulates the phosphorylation of proteins of the synaptosomal membrane and is here inhibited by another protein, calcineurin. These new regulators are somewhat complex in their action and clarification of this must await further investigation. In Chapters 10 and 11 we shall present further evidence for the important role that phosphorylation regulated by second messengers plays in neuronal activity.

α-adrenergic receptors (α-adrenoceptors)

Both types of β-receptors stimulate adenylate cyclase. They differ in their ligand recognition portion (R). We meet a different situation when we turn to the α-adrenergic receptors. Here $α_1$ appears instead to regulate primarily the intracellular level of the other second messenger Ca^{2+}; whereas $α_2$ not only does not stimulate adenylate cyclase but appears to inhibit it. Our present picture is that $α_2$-receptors interact with the adenylate cyclase (C) via an inhibitory regulatory protein (N, G). There are two different types of regulatory proteins, one stimulatory, termed N_s, and one inhibitory, called N_i. Both have been purified (from liver, brain, and erythrocytes) and their quaternary structure has been elucidated. They are composed of three different polypeptides two of which (β, γ) are identical in both proteins. The N-proteins are also the site of exogenous effects as, for example, the action of F^- or the bacterial cholera and pertussis toxins (see Chapter 2 for the structure and function of cholera toxin). A summary of the present knowledge of the structure and regulation of signal transduction via adrenoceptors is given in Fig. 9.14, A and B. Fig. 9.14B also gives some details of the mechanism of the sequential interaction between R, N, and C, indicating that the transmitter or hormone first activates N through its interaction with its receptor. Activation of N is based on a GDP/GTP exchange. Activated N then interacts with C. This is only transient because N inactivates itself by cleaving bound GTP by its intrinsic GTPase activity. Once again it is interesting to note the similarity with the rhodopsin/transducin/phosphodiesterase-interaction discovered in visual transduction (Chapter 1). This similarity is more than just an analogy.

The model (Fig. 9.14) is applicable to other receptor systems too. Dopamine receptor D_1 stimulates adenylate cyclase in the same way whereas D_2 mediates the inhibitory effects of dopamine (see below). The model can be further generalized by including various hormone receptor systems. Glucagon, ACTH, some prostaglandins and histamine (H_2 receptor) are examples of stimulatory effectors, opiates including the endogenous opioid peptides (see below) act as inhibitory effectors of adenylate cyclase.

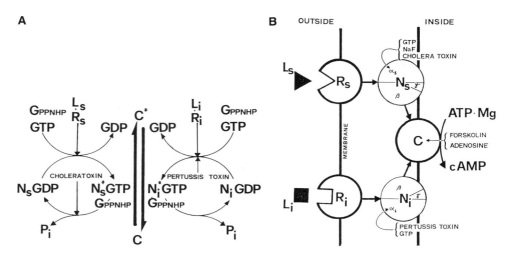

Fig. 9.14. Regulation of membrane bound adenylate cyclase.
(A) Stimulatory, L_s, or inhibitory, L_i, transmitters or hormones are bound to their respective receptors, R_s, R_i. The L·R-complex interacts with high affinity with the N-protein* causing a GDP/GTP exchange. The N·GTP-complex dissociates from R and activates (left side of the scheme) or inhibits (right) adenylate cyclase. The activating or inhibiting effect is terminated by a GTPase inherent to the N-proteins cleaving GTP to GDP + P_i. According to this model the N-proteins serve as shuttles carrying the stimulatory or inhibitory signal from the respective receptor on the outside to the adenylate cyclase on the inside of the membrane thereby undergoing themselves a cycle of activation/inactivation. The active state of the regulatory proteins, N_s^*, N_i^*, can be prolonged artificially with G_{ppNHp} or GTBγS, GTP analogues which are not cleaved by the GTPase. Cholera toxin has the same effect but only with N_s. Pertussis toxin on the other hand inhibits the inhibitory action of N_i by a still unknown mechanism. C*, activated adenylate cyclase.
(B) Scheme of the components of this signal transducing system. The site of action of the regulatory effectors GTP, cholera toxin, F⁻ (N_s) and GTP, pertussis toxin (N_i) are the respective α-subunits of the N-proteins. The diterpen forskolin activates adenylate cyclase directly. Adenosine also interacts with the enzyme directly on the inside of the membrane, binding to an inhibitory binding site called the "P-site" (not be confused with the activating adenosine receptor located on the outer membrane surface).

* see footnote p. 223

Toxins as tools: cholera toxin and pertussis toxin cause ADP-ribosylation of N-proteins*

We have already encountered several instances where natural toxins have served as valuable tools in neurochemistry for elucidating key mechanisms or isolating important molecules of the nervous system (see p. 121). Here we shall give briefly another example of this "technology": The regulatory N-proteins are the site of action of certain bacterial exotoxins; as mentioned on p. 40 and in Fig. 9.14 B cholera toxin activates adenylate cyclase permanently by activating N_s. The mechanism of this effect is based on an ADP-ribosylation, i.e. a transfer of ADP-ribose from NAD to the α-subunit of N_s. The consequence of this covalent modification is the dissociation of N_s into its subunits. α_s appears to be the subunit interacting with the adenylate cyclase in the activation step of the enzyme. The β-subunit prevents α_s from doing this in the intact N_s complex. Dissociation of α_s from N_s therefore causes activation of adenylate cyclase.

Pertussis toxin, a protein excreted by *Bordetella pertussis*, has no effect on N_s but it inhibits the inhibitory action of N_i. It has been shown to block, by ADP-ribosylation of α_i, both the interaction between R_i and N_i and that between N_i and C (see Fig. 9.14).

These two proteins will give valuable information on the mechanism of coupling between the components of cyclase-regulating receptor systems. By their ADP-ribosylating capacity they have served to identify N_s and N_i in many tissues, thus with ^{32}P-labelled NAD as substrate the radioactively labelled N proteins can be detected by SDS polyacrylamide gel electrophoresis and autoradiography.

Both toxins are similar in structure. They both consist of a binding part (B) and an enzymatically active polypeptide chain (A) (see Chapter 2).

Dopamine receptors

Dopamine receptors of the central nervous system can be accurately characterized by means of a range of radioactive ligands [15,16]. They are of particular pharmacological interest because their *in vitro* affinity for a group of neuroleptic drugs (i.e. those which alleviate or remove the symptoms of schizophrenia) can be correlated with the neuroleptic activity of these drugs *in vivo* [16]. Parkinson's disease (see below) is also associated with damage to dopaminergic systems (although probably not on the receptor level). Dopaminergic pathways of the CNS are involved in the basic human activities of conscious perception, emotions and memory, reasons enough to deal with dopamine and its receptors more extensively.

Anatomically there are three main dopaminergic systems in the brain of higher vertebrates (Fig. 9.15):

---

* see footnote p. 223

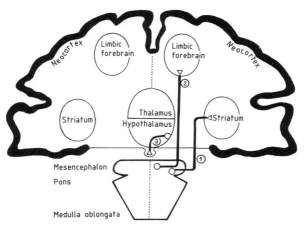

Fig. 9.15. Schematic representation of the three main dopaminergic pathways of the CNS (for explanation see text).

1. **The nigrostriatal system.** This is the part of the motor system concerned with posture and movement and is known as the extrapyramidal system. It connects the basal ganglia with structures in the midbrain. The cell bodies are located in the *substantia nigra* and its axons project primarily to the *putamen* and the *nucleus caudatus* which form the *neostriatum*. It is involved in Parkinson's disease.

2. **The mesocortical system.** The cell bodies are located medial and superior to the *substantia nigra* and the axons project to the limbic system. Since this part of the brain is involved in memory and emotions it is being investigated in connection with mental and emotional disorders.

3. **The tuberoinfundibular system.** This is important for neuroendocrine regulation, i.e. for the interaction between the hypothalamus and the pituitary. The cell bodies are located in the arcuate nucleus of the median eminence and the axons project to the pituitary stalk.

Pharmacologically there are five major classes of drugs affecting the dopaminergic systems of the CNS, the catecholamines, the phenothiazines, the butyrophenones, the ergot – alkaloids and the thioxanthenes, and a variety of other compounds as for example butaclamol, sulpiride and pimozide, all of which are summarized in Fig. 9.16.

Physiologically there are two, types of dopamine receptors [16,17]. One type, termed D_1, was discovered in 1972 by Greengard and was defined by him and by Iversen as the binding site through which dopamine stimulates adenylate cyclase in the brain. D_1-receptors occur in the striatum, in the superior cervical ganglion and in the retina. The function of D_1 in these tissues is unknown, but recently substrates for cAMP – dependent protein kinase regulated specifically by drug – D_1 interactions, have been discovered. D_1 is the only type of dopamine receptor found in the parathyroid, and there it is known to regulate parathyroid hormone release. Another type of dopamine receptor, termed D_2, appears to inhibit adenylate cyclase, perhaps by a mechanism similar to the one shown in Fig. 9.14. It is the only type of dopamine receptor found in the anterior pituitary (and in the *pars intermedia*), where it mediates dopaminergic inhibition of prolactin release. It also occurs in the striatum where its function is unknown. D_1 and D_2 can be distinguished by binding assays with selective ligands: butyrophenones, e.g. spiroperidol, bind to D_2 of bovine pituitary with an affinity (K_D) of 0.3 nmol/l whereas its affinity for D_1 is three orders of magnitude less. A third type of dopamine receptor, called D_3, is possibly not a separate receptor protein, e.g. an autoreceptor on presynaptic membranes, but another affinity state of D_1.

Biochemically very little is known about all of these dopamine receptors. Purification of receptor protein will prove difficult because of the absence of a rich source of the receptors, for example the number of butyrophenone binding sites (D_2) in the bovine striatum is only about 450 fmol/mg protein or 50 pmol/g tissue.

Dopamine receptors occur in two interconvertible states. This will be illustrated with D_2: the butyrophenone [^{3}H]spiroperidol binds to D_2 from bovine pituitary with a K_D of 0.3 nmol/l to a single class of binding sites (Hill coefficient of unity). Displacement of this antagonist by other antagonists shows similar binding characteristics. Displacement by agonists exhibits negative cooperativity (Hill coefficient, 0.58) and two types of

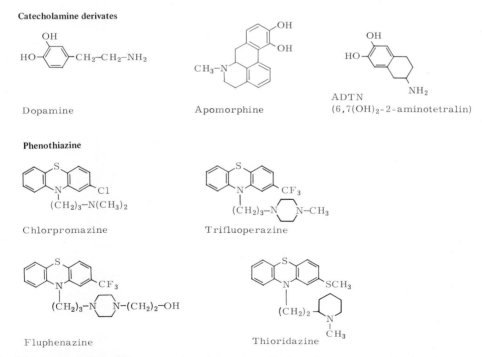

Fig. 9.16. Some of the most important drugs interacting with dopamine receptors. (Continued next page)

Butyrophenones

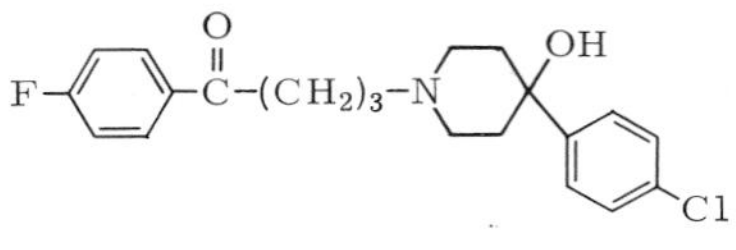

Haloperidol

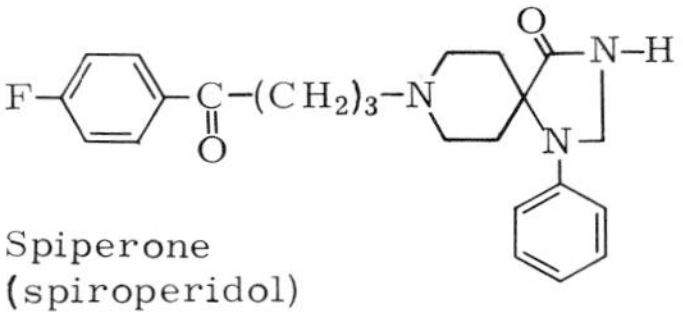

Spiperone
(spiroperidol)

Ergot derivatives

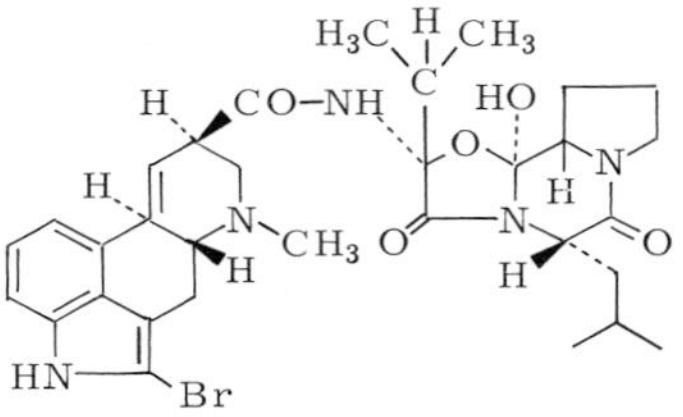

Bromocryptine

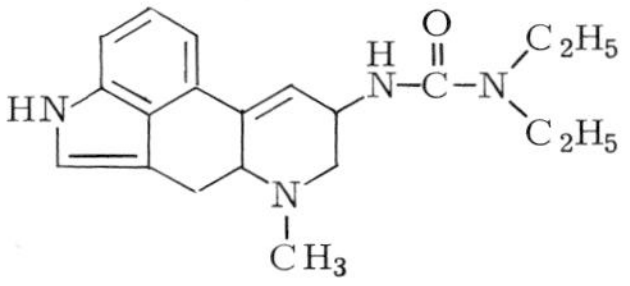

Lisuride

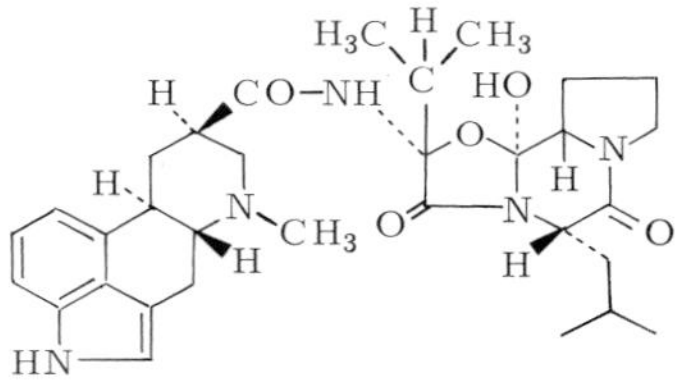

DHEC
(dihydroergocryptine)

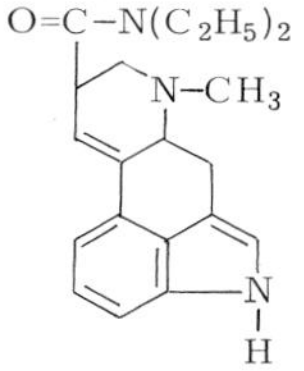

LSD
(lysergic acid diethylamide)

Thioxanthene and others

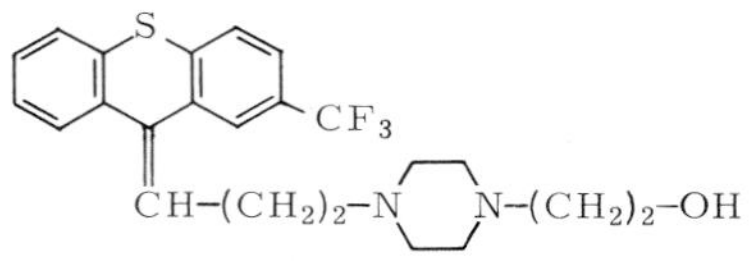

Flupenthixol

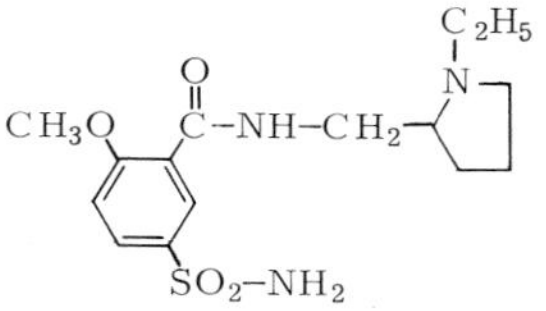

Sulpiride

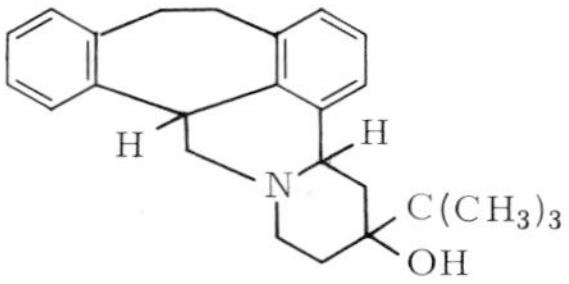

Butaclamol

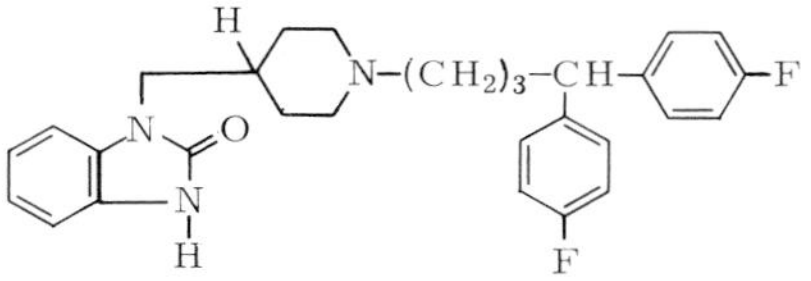

Pimozide

Fig. 9.16. (continued)

binding sites, one with a low affinity in the μmol/l range (R_L), the other with a much higher affinity in the nmol/l range (R_H). GTP and even more the non – cleavable GTP analogue GPPNHP has the effect that agonists only bind to one class of binding sites (Hill coefficient of unity), with a K_D of 360 nmol/l (R_L).

GTP, probably via a GTP – binding protein similar to N_i of Fig. 9.14, mediates an agonist – induced interconversion according to the scheme

$$\text{(high antagonist, high agonist affinity)} \quad R_H \underset{}{\overset{GTP}{\rightleftharpoons}} R_L \quad \text{(high antagonist, low agonist affinity)}$$

The affinities are often determined in displacement experiments of this kind and are expressed as K_I or IC_{50}, the inhibitor concentration necessary for 50% displacement of the radioactive ligand from its binding site. The relation between K_I, IC_{50} and K_D, the dissociation constant of the antagonist to be displaced is given by the formula

$$K_I = \frac{IC_{50}}{1 + \frac{[I]}{K_D}}$$

where [I] is the concentration of the antagonist to be displaced and K_D its dissociation constant.

Such displacement tests are very convenient and widely used since they allow affinity determinations for a variety of ligands when only one ligand of sufficiently high specific radioactivity, in our case the antagonist I, is available. The displacement test shows the possible existence of two states, the agonist binding (A in Fig. 9.2) and the antagonist binding state (I in Fig. 9.2); these are detected if different values are obtained for the affinity (expressed as the inhibitor constant K_I in the displacement test) for a given ligand, according to whether it has to displace a (radioactive) agonist or antagonist. Such a discrepancy of K_I values can only be explained by assuming that the ligand is bound to the receptor with different affinities in the two states. If for example the agonist [^{3}H]dopamine is displaced with unlabelled dopamine, an inhibitor constant K_I of 17.5 nmol/l is obtained. If on the other hand, the antagonist [^{3}H]haloperidol is displaced the K_I for dopamine is 670 nmol/l. The unlabelled dopamine has thus generated a receptor form with a ca. 40 times higher affinity for the radioactive agonist than that formed via the antagonist. The antagonist haloperidol has a 600 times higher affinity for the I-state than for the A-state (K_I = 1.4 nmol/l against K_I = 920 nmol/l). (Conversely the displacement test can also determine if an unknown substance is prospectively an agonist or an antagonist – a valuable biochemical test for the pharmaceutical laboratory!). The method is only applicable when the transformation from A to I occurs slowly.

Dopamine and schizophrenia

Neuroleptic drugs, especially the phenothiazines and the butyrophenones (Fig. 9.16) have an interesting property which may lead to understanding one of the most widespread mental diseases: schizophrenia. Neuroleptics are psychopharmacological drugs which alleviate some of the major symptoms of schizophrenia, and their potency in doing so parallels their binding affinity to D_2-receptors. Since dopamine-releasing substances as, for example, the amphetamines (Table 8.4) may induce in normal individuals psychotic symptoms resembling paranoid schizophrenia, it is reasonable to assume that some disturbance in the dopaminergic system is involved in this disease. At present this "dopamine hypothesis" of schizophrenia is favoured by many investigators, and there is evidence that an increase occurs in the number of D_2-receptors in the neostriatum of schizophrenic brains, even in brains of patients who have apparently never been treated with neuroleptics. But one has to be very careful. The observed correlation between antipsychotic potency and D_2-affinity for neuroleptics may be artifactual: antipsychotic potency is usually assessed after continued neuroleptic administration, which may alter the system under investigation. Similarly changes in receptor density may be the consequence and not the cause of the mental disturbance, and it is very difficult to be certain that there has been no treatment with psychopharmacological drugs before the *post mortem* brain samples have been obtained. In other words one has to be open to other theories on the causes of disturbances of the conscious human brain, for which no animal model is available. Some investigators believe that noradrenaline may regulate the sensitivity of the dopamine system and that this regulation is altered in the patient's brain.

At present it can be considered to be a major breakthrough that neuroleptics are of real help both to patient and clinician and that the pharmacologist can predict their potency from the results of binding assays *in vitro*. This makes the screening of new drugs easier. The situation is much clearer with another disease involving central dopaminergic tracts, Parkinson's disease.

Parkinson's disease

Parkinson's disease is a progressive, frequently fatal disorder of the central nervous system, which is characterized by stiffness of the muscle, difficulty in moving, and tremor. Patients can be recognized by their staring facial expression, slightly stooping stiff posture and laborious bodily movements. In other patients the jerking and trembling of the limbs are the predominant symptoms. The disease is traceable to a degeneration of a region of the central nervous system responsible for motor control. It occurs in the second half of life, and the causes and triggering factors are largely unknown. The breakthrough in the area of chemotherapy was a classic example of the application of methods of neurochemistry to neurology.

The disease affects primarily the *substantia nigra* and *corpus striatum*, regions which are rich in the transmitter substance dopamine. The most striking biochemical findings are the reduced concentration of dopamine, its degradation products (e.g. homovanillic acid) and of enzymes like tyrosine hydroxylase which are involved in its synthesis. The recognition that the function of the dopaminergic tract was damaged led to a very effective therapy: a dramatic improvement of symptoms was experienced by patients with Parkinsonism when L-DOPA (L-dihydroxyphenylalanine), the precursor of dopamine was taken orally. Dopamine itself is without action as it cannot penetrate the blood-brain barrier.

Two conclusions can be drawn about the therapeutic effect of L-DOPA: firstly, the process for the conversion of DOPA to dopamine must still function in Parkinson's disease; secondly, dopamine receptors must still be available so the degeneration must be limited to the presynaptic region. As the DOPA decarboxylase activity is very much reduced, it has been concluded that the decarboxylation of DOPA is taken over by neighbouring noradrenergic and serotonergic neurons. From there the dopamine produced probably diffuses to its site of action. The measured increases in dopamine concentration after application of L-DOPA are nevertheless slight so it is assumed that there is a supersensitivity of the dopamine receptors in the postsynaptic membrane of the neurons concerned. This would also explain the remarkable selectivity of L-DOPA, whereby the dopamine produced from exogenous DOPA has its primary action here and not in other dopaminergic regions of the central nervous system.

Only a small percentage of the administered L-DOPA reaches its site of action. The daily dose therefore varies between 2 and 15 g. It may be decreased considerably if it is given together with a decarboxylase inhibitor which inhibits the peripheral decomposition of DOPA. This is particularly desirable as high doses of DOPA can produce toxic side effects.

These side effects have only partially reduced the success of DOPA therapy. Initial nausea disappears after a few days. The release of involuntary, uncontrolled movements which occurs in more than 70% of patients is more distressing. There can also be emotional and behavioural changes which can vary from restlessness and sleep disturbance to psychosis. Finally the most dangerous effects are those on the heart and blood vessels. In unfortunate cases where there is prolonged treatment with L-DOPA such severe disturbance of heart rhythm can be caused that the drug has to be discontinued.

Opiate receptors

When the God of dreams was named Morpheus by the ancient Greeks, they could not have imagined the significance his name would later acquire in molecular neurobiology and pharmacology! Morphine was named after him. This alkaloid isolated from the milky fluid of unripe poppies together with a class of derivatives, the opiates, have served to characterize the receptors of the central nervous system concerned with sensitivity to pain. Since receptors are normally the sites of action of endogenous, physiological substances, the discovery of morphine receptors strongly suggested that there is an endogenous substance which resembles the exogenous drug in action but not necessarily in structure. This was the basis for the isolation of the opiate-like peptides endorphine, dynorphine, and the enkephalins; these peptides may well be transmitter substances in the brain (see Chapter 8). The molecular structure of the binding site for opiates and the enkephalins, the opiate receptor (of which in fact there now appear to be several; see p. 239), is being increasingly elucidated [18]. It is stereoselective, i.e. it only binds the (-)-isomers of opiates (Fig. 9.17) and it distinguishes between agonists and antagonists. The affinity of the agonists corresponds to their biological activity, and is measured as the inhibition of contraction of the guinea pig ileum induced by electrical stimulation. Oddly enough the digestive tract and the mouse *vas deferens* are the only tissues beside the central nervous system which contain opiate receptors; their function there is unknown.

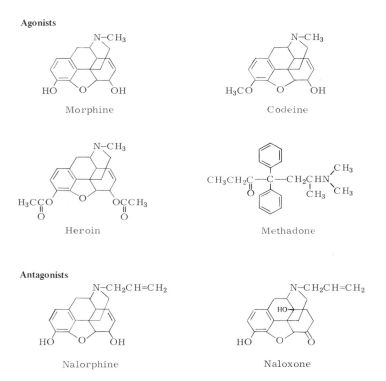

Fig. 9.17. Opiates: agonists and antagonists.

When 50% of the binding sites are occupied 50% of the maximal action is also achieved; there are evidently no "spare receptors" of lower affinity here which are brought into action at higher agonist concentrations as has been postulated for other receptors. The opiate receptor has an essential SH group, the first evidence that it is also a protein.

Opiate receptors exist in two interconvertible forms, one of which has a high affinity for agonists, the other one for antagonists (see Fig. 9.18). The interconvertibility, probably a conformational change, is released by Na⁺ ions. This follows from the observation that Na⁺ (1 mmol/l) reduces the affinity for agonists and raises it for antagonists. K⁺, Rb⁺ and Cs⁺ do not show this action, Li⁺ only weakly. Na⁺ seems to fix the receptor in the antagonist-binding conformation. This also explains the higher effectivity of opiate antagonists *in vivo* as the Na⁺ concentration is high in the brain, and the opiate receptor exists predominantly in the "antagonist state".

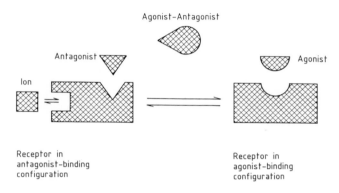

Fig. 9.18. Opiate receptor: two state model. The change from the agonist-binding to the antagonist-binding configuration is caused by Na⁺ ions.

As indicated by the title of this section there are several opiate receptors, another example of the concept of "multiple receptors" mentioned in Chapter 8. The three types of opiate receptors now relatively well characterized are called $\mu, \varkappa$, and δ. A fourth type identified by its preferential binding of the ligand N-allylnormetazocine (SKF 10047) and called by some authors the σ-receptor is considered by others not to be truly an opiate receptor. It is difficult to discriminate between them because of the lack of specific ligands. The highest selectivity is shown by the $\varkappa$-receptor which binds only dynorphine (and leu-enkephalin with an extended C-terminal sequence). $\varkappa$-receptors are found mainly in the spinal cord where they appear to be involved in the regulation of pain transmission. The enkephalins bind to μ and δ, with a slight preference for the μ-receptor.

At present it is not clear whether the multiple opiate receptors represent different receptor proteins or just different conformational states of one molecule. In the striatum an interconversion (Na⁺-,GTP-dependent) between μ and δ has been claimed. But the biochemistry of this whole exciting field is in its infancy.

Presynaptic inhibition by opiates

Opiates have both an analgesic and euphoric action. The distribution of opiate receptors in the brain has been investigated with radioactive ligands or fluorescent antibodies against enkephalins. The greatest density is found in the limbic system – the developmentally old region in which emotional excitation occurs, and where the euphoric and emotional component of the pain-relieving action of the opiates is localized [19]. The direct effect on pain conduction is found in the spinal cord. This part of the CNS provides direct evidence of the physiological role of the opiate receptors. They are probably presynaptic and so to be found in nerve endings. It has been reported that opiates inhibit the release of substance P [20] – the substance thought to be the transmitter in the pain tract. In this model, an enkephalin, as an endogenous opiate, would act on the nerve ending of an excitatory neuron (Fig. 9.19). There it would cause inhibition of transmitter release by a still obscure mechanism, perhaps by inhibiting influx of calcium necessary for exocytosis of substance P into the synaptic cleft.

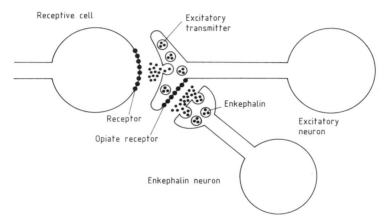

Fig. 9.19. Presynaptic inhibition of pain conduction by enkephalin: the excitatory transmitter is probably substance P. Hypothetical model of action after [19]. Reproduced, with kind permission, from the authors and Macmillan Journals Ltd.

Addiction and tolerance – a molecular model

The medical use of opiates is limited since the pain-relieving action diminishes with time making increasingly high doses necessary. This phenomenon called *tolerance*, like that of addiction cannot be averted by the continued development of new opiates. Addiction, the "I-cannot-live-without-drugs" syndrome, is not a psychological problem but is a true dependence of the body in the physiological sense. Investigations with *neuroblastoma x hybridoma*-cells (see Chapter 12 and Fig. 9.20) have given an insight into a possible

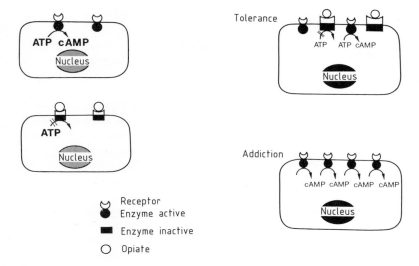

Fig. 9.20. Tolerance and addiction: molecular models of the opiate effect. Left: Opiate blocks adenylate cyclase. Right: The blocked receptor cyclase complexes are replaced by new ones (tolerance, habituation). After removal of the opiate the total cyclase activity is increased; the excess of cAMP causes withdrawal symptoms (addiction). After [21].

molecular mechanism; it was found that, as *in vivo*, opiates including the endogenous opiates inhibited the adenylate cyclase of the cell membrane and thus decreased the cAMP level [21]. This effect however was only temporary, as the cell compensated for the inhibited enzyme by the synthesis of additional cyclase.

This higher total enzyme concentration demands a higher opiate concentration to produce the same inhibitory effect. This corresponds exactly to the medical condition of tolerance. If the inhibitor is now removed by washing it out, the cAMP concentration increases to far more than the normal level; for there is more than the normal amount of cyclase present. Due to its multiple action in cell metabolism this excess disturbs the metabolic balance and damage is caused as in human drug withdrawal. The cell now needs the opiates as an inhibitor and is addicted.

It has been shown in this series of experiments that the number and affinity of receptors are unchanged, and thus it is primarily the cyclase which hypertrophies in response to the inhibition. (This can be seen as further support for the "floating receptor" model). A theory could be put forward that the analogue of "hypertrophy" as a response to increased transmitter release are the processes of plasticity (see Chapter 11) involved in learning and memory. It is not yet quite clear why endogenous opiates do not give rise to tolerance and addiction. A feed-back control of opiate release has been postulated which keeps the opiate in the synaptic cleft constant.

GABA receptors, inhibitory and allosteric protein complexes

As mentioned in Chapter 8 GABA is, together with glycine, the major inhibitory transmitter in the CNS. It inhibits by opening Cl^--channels thereby hyperpolarizing the postsynaptic membrane. It is of considerable interest, since a variety of drugs – tranquillizers such as the diazepines and the barbiturates and convulsants such as bicuculline and picrotoxin – interact with the GABA/chloride-channel system. A GABA receptor from various vertebrate brains has been isolated recently by Barnard and his coworkers. They have identified a single protein species of relative molecular mass 220 000 containing the binding sites for GABA agonists, GABA antagonists, benzodiazepines (agonists and antagonists see p. 198), barbiturates and chloride dependent convulsants. An $\alpha_2\beta_2$ quaternary structure was found, and the mRNA coding for this receptor was obtained from brain and translated in the *Xenopus* oocyte.

Most investigations so far have been concerned with binding studies and the pharmacology of the system *in vivo*. The results of these investigations can be summarized as follows.

GABA exerts its effect through a GABA receptor which may in fact be a group of distinct receptor subtypes or a set of multiple states of one type of receptor. $GABA_1$-receptors with low and $GABA_2$-receptors with high agonist affinity have been identified. The latter is distributed fairly evenly over various brain areas; the former is much more abundant in the cerebellum, cortex and hypothalamus than in other areas such as the *substantia nigra* or the striatum. The agonists GABA and muscimol have been shown to bind to the GABA recognition site of the receptor. Among the antagonists, bicucullin competes with GABA for this site while picrotoxin appears to interact with the Cl^--channel.

Benzodiazepines and barbiturates enhance GABA effects allosterically

In addition to the GABA and picrotoxin binding sites there are several further effector binding sites which interact with the GABA-receptor/Cl^--channel complex. Both benzodiazepines and barbiturates enhance the inhibitory action of GABA. Separate benzodiazepine [22] and barbiturate receptors have been postulated, though according to the definition of a receptor given at the beginning of this chapter we should simply call them "binding sites" since no endogenous ligands have been found for them. But as soon as such a ligand has been identified unequivocally such binding sites can be promoted to true receptors. An intensive search for an endogenous ligand corresponding to the benzodiazepines has been made during the last few years [23]. Such a compound would be extremely interesting. Benzodiazepines (e.g. Valium[R], Librium[R] and two dozen similar tranquillizers of this profitable market) have a fourfold effect. They are sedative, anxiolytic, anticonvulsant and muscle relaxant. They are taken for stress-symptoms and anxiety and prescribed to epileptics. Pre-anaesthesia almost always starts with the administration of a benzodiazepine. If all these effects are normally mediated by an endogenous analogue of the benzodiazepine it would be of the utmost importance to identify it, since it might well be responsible for many basic features of our mental state as well as for distressing pathological disturbances.

The search for the endogenous ligand of the benzodiazepine receptor has resulted in a variety of candidates, none of which actually appears to be the suspected compound [23]. Inosine and hypoxanthine, nicotinamide, thromboxane, harmane and ethyl β-carboline-3-carboxylate were favoured at various times mainly due to their high affinity for diazepine binding sites. Some of these ligands were not even true endogenous compounds but mere preparative artifacts. But their investigation turned out to be especially useful and led to the discovery of potent benzodiazepine antagonists (e.g. the β-carboline carboxylates) and even so-called "inverse agonists", compounds having the opposite effect to diazepines.

The search goes on. One apparently significant result was the discovery of a new CNS-neuropeptide, the so-called GABA-modulin. This is a peptide of relative molecular mass 16 500 which non-competitively reduces the number of $GABA_2$-receptors. It appears to be an allosteric regulator which is in turn influenced by diazepine binding. Stimulation of its binding by GABA is prevented by GABA-modulin. Its physiological role remains to be determined.

To summarize: the inhibitory neurotransmitter GABA appears to act through a complicated protein complex consisting of a GABA-recognition site and a chloride channel; it is probably an allosteric protein existing in several affinity states and regulated by a variety of allosteric (and steric) effectors [24]. Again we should be careful not to call every newly discovered binding site a "receptor" as many of the most important drugs that act by modulating the inhibitory effect of GABA (e.g. the barbiturates and the benzodiazepines) may simply interact with different allosteric binding sites on the GABA receptor complex. Studies with purified GABA receptors confirm this concept.

Glycine receptors are also inhibitory; the first central receptor isolated

While GABA is the most important inhibitory transmitter in the brain, glycine largely replaces it in the spinal cord and brain stem. The putative transmitters taurine and β-alanine are further inhibitory agonists of the glycine receptor; the most potent exogenous antagonist is strychnine, an alkaloid from *Strychnos nux vomica*. Strychnine does not block inhibitory effects mediated through GABA receptors while the GABA antagonist bicuculline does not block glycine receptors. The glycine receptor is linked to a Cl^--channel which again is not the same Cl^--channel as that linked to the GABA-receptor.

The natural toxin strychnine (Fig. 9.21) turned out to be a useful tool for the isolation and biochemical characterization of the glycine receptor. Its tritiated derivative has an affinity high enough to be used in binding assays of the receptor protein (K_D = 11.3 nmol/l) and immobilization of strychnine resulted in an affinity column which enabled Betz and his coworkers to make a 1000-fold one-step purification of the Triton-solubilized receptor. Furthermore strychnine, a photosensitive compound, can be applied without any chemical modification as a photoaffinity label. UV-irradiation of the strychnine receptor complex results in covalent labelling of a single polypeptide (relative molecular mass 48 000).

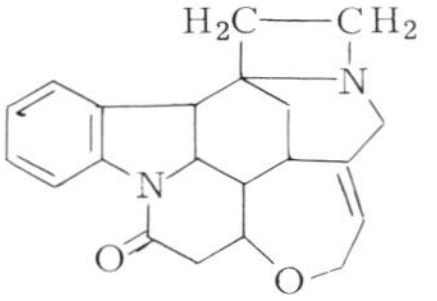

Fig. 9.21. Strychnine.

These are some of the biochemical characteristics of this first neurotransmitter receptor isolated from the CNS: the glycine receptor is a glycoprotein composed of several polypeptide chains (relative molecular mass as determined by SDS polyacrylamide gel electrophoresis 48 000, 59 000 and 92 000); the strychnine binding site appears to be located on the smallest polypeptide chain, the function of the other chains being unknown at present; the receptor protein contains a single class of independent binding sites for strychnine.

Displacement of strychnine from its binding site by glycine shows a sigmoidal concentration dependence curve. This implies a two-state model for the glycine receptor similar to the other receptors described so far. The Hill coefficient for the displacement of strychnine by glycine is 1.7. Glycine and strychnine appear not to compete but to act cooperatively by interacting at two different sites on the receptor. The interaction between agonist and antagonist is removed by NEM. Strychnine is thought to bind to the Cl^--channel. In solution the glycine receptor requires phospholipids for stability; its sedimentation coefficient in detergent is 8.3. GABA shows little affinity for the glycine-receptor.

Glutamate receptors, serotonin receptors and the many other receptors of general interest have not yet been isolated

The counterparts to the inhibitory transmitters GABA and glycine are the excitatory amino acid neurotransmitters glutamate and aspartate. They are especially abundant in the hippocampus; since glutamate has been implicated in long-term synaptic modifications related to learning and memory this has been the most thoroughly investigated brain region. However other regions, e.g. cerebellum and cortex, also contain significant concentrations of these excitatory transmitters.

In the hippocampus and elsewhere there appear to be at least two classes of glutamate-receptors, one being Na^+-dependent, the other not. The hippocampal receptors increase in number during stimulation. This effect is long lasting and correlated with "postsynaptic potentiation" (see Chapter 11); interestingly it also appears to be related to Ca^{2+}-dependent proteolysis of a peripheral membrane protein, called *fodrin*, by the thiol protease calpain I [25].

There are several structural analogues of the excitatory transmitter glutamate which are convulsants and at higher concentrations neurotoxic agents. One of the most potent agonists is kainic acid (see Fig. 8.26); this is used to create specific lesions (p. 195) but it does not compete with glutamate for a common receptor.

The pattern of "multiple receptors" is also found with serotonin receptors (5-hydroxytryptamine or 5-HT-receptors). HT_1 has a high affinity for serotonin (nanomolar K_D) and appears to be linked to a stimulatory N-protein (GTP-dependent) and an adenylate cyclase. By contrast, HT_2 has a low affinity for serotonin. Both HT_1 and HT_2 bind the serotonergic antagonist LSD with high affinity. Serotonin receptors appear to be involved in regulating sleep, mood and pain perception and in disorders such as depression and infantile autism. HT_1 has been solubilized from bovine cortical membranes with detergents and tentatively identified as a protein of relative molecular mass 58 000 occurring in interconvertible affinity states.

The serotonin receptor is of particular interest as one of the probable binding sites for hallucinogenic substances like lysergic acid diethylamide (LSD). It binds serotonin and LSD is displaced from it with negative cooperativity (Hill coefficient, 0.5) i.e. each compound is bound to a different conformation of the receptor. The same considerations apply as before: there are pure agonists (tryptamine derivatives) and antagonists, but LSD is a mixed agonist-antagonist. Agonists block the binding of serotonin more effectively by one or two orders of magnitude than LSD. Antagonists are in contrast better blockers of the LSD binding. Thus, as with the receptors discussed earlier the picture in general is consistent with a two state model though there is little direct evidence for it.

Autoreceptors and some remarks concerning the fine tuning of nervous activity

Until now receptors have been presented almost exclusively as postsynaptic structures. However evidence has existed for some time that monoamine transmitters, for example, also affect their own release presynaptically. This introduces a further interesting possibility for the regulation of synaptic transmission: the transmitter concentration in the synaptic cleft regulates the transmitter release by a feed-back signal conveyed by presynaptic receptors.

The local distribution of the receptor can also influence synaptic function in a second way. We have presented the structure of the cholinergic endplate (Fig. 8.2) as a prototype of the synapse. The monoaminergic nerve endings of sympathetic neurons, by contrast, form "varicosities" (Fig. 9.20) whose geometry is quite different from that of the neuromuscular synapse. The receptors do not seem to be concentrated here in specialized structures (postsynaptic densities) to the same extent. The path of the transmitter to the receptor can be thus lengthened and the information transfer modulated by interaction with other transmitter release sites.

In addition there are autoreceptors in cholinergic and in other synapses. Recently in synaptosomes from the electric organ of *Torpedo* which is a model system for nerve – muscle transmission discovered and developed by Whittaker, acetylcholine receptors have been found in which occupation by agonists blocked the release of acetylcholine. In contrast to the nicotinic receptor of the postsynaptic membrane these presynaptic receptors are muscarinic. Here too, the autoreceptor seems to be the receiver of a feed-back signal which regulates transmitter release.

In recent years it has become evident that such modulatory effects play an important role in neuronal activity. We have already mentioned in Chapter 8 the immunohistochemical evidence that a variety of neuropeptides are packaged together with classical neurotransmitters in nerve endings. Enkephalins for example have been found to coexist in certain pathways with adrenaline and noradrenaline and also with serotonin and acetylcholine. Substance P, itself a proven transmitter in certain pathways, has been found together with serotonin or acetylcholine in others. In these cases the peptide appears to modulate the transmitter action. This modulatory effect is usually much longer lasting than the transmitter effect itself.

Even in nerve impulse propagation we must be aware that the simple "all-or-nothing" response may be subject to considerable modulation. We shall see in Chapter 11, that "neural plasticity" is based on modulatory interactions of various effectors; for example the sensitization of the gill withdrawal reflex in *Aplysia*, a behaviour taken as a learning model, is based on a modulation of axonal potassium channels. This modulation takes place via serotonin which through the second messenger cAMP stimulates phosphorylation of a specific class of potassium channels thereby inhibiting their closure and prolonging the action potential [26]. This is probably only one extreme case of a large group of modulatory effects varying considerably in extent and mechanism.

Receptors are under regulatory control

Receptors as primary receivers of extracellular signals are not rigid, but react flexibly to the intensity of this signal. They are regulatory proteins, their activity is affected by a variety of factors as several specific examples have already shown. To summarize the regulation mechanisms: the activity of receptors – including hormone and neurotransmitter receptors – is regulated by

a) the number of receptors
b) the receptor affinity.

The density of receptors can vary absolutely i.e. the number per total cell surface may change, or locally, at definite sites on the cell surface. An example of the former is the hypersensitivity to acetylcholine after denervation of a muscle fibre and of the latter the phenomena of *patching* and *capping*, i.e. the formation of receptor aggregates and their accumulation at one pole of the cell. Local increases of receptor concentration can be evoked by the ligand but also via polymerizing agents like lectins and antibodies (note that the monovalent fab fragment unlike its bivalent precursor does not cause patching). The polymerization alone can be sufficient for regulation as is shown by the insulin-like action of lectins at insulin receptors. The absolute number of receptors per cell decreases if the hormone or transmitter concentration is raised for a considerable time. This response is not evoked by all ligands, but only by the agonist itself, when present in excess. Thus β-adrenergic agonists reduce the number of β-adrenergic receptors in frog erythrocytes; the antagonist propanolol inhibits this effect but cannot itself evoke it.

The decrease in the number of receptors is usually brought about by the internalization of the receptor-ligand complex. Examples are insulin and the nerve growth factor (NGF) receptors. Subsequently this complex is degraded by lysosomes. The purpose of this mechanism of regulation is undoubtedly to compensate for the excess of ligand by decreasing the responsiveness of the cell. The regulation of receptor activity by a change in affinity e.g. the desensitization of the acetylcholine receptor by continuous incubation with acetylcholine (see p. 218) serves a similar purpose. Desensitization has been reported in many receptors. In the β-adrenergic receptor it is based on a synergistic decrease in affinity for the transmitter and an increase in affinity for adenylate cyclase related to receptor phosphorylation.

We have already encountered the reverse process: the opiate receptor compensates for the receptors occupied by ligands by an increase in the number of receptor-cyclase complexes. In sum this results in hypersensitization.

Summary

Receptors are proteins which, as the sites of binding and action of physiological effectors (hormones, neurotransmitters), transfer extracellular signals to the interior of the cell. They consist of recognition and binding proteins, which receive the signal, and an effector portion which transforms the signal into action. The latter can be an ion channel, a transport system or an enzyme. Various models for the mechanism of coupling between ligand (hormone, transmitter) binding and effect have been discussed, of which the most plausible is that involving an allosteric modification of the receptor protein. The functions of binding and effect are possibly carried out by different subunits of a receptor complex. An example of this is the hormone-sensitive adenylate cyclase which, as an effector, can be separated from its binding portion and biochemically purified. According to the *floating receptor* hypothesis this enzyme diffuses laterally in the cell membrane and can be regulated by a variety of receptors. The extracellular signal is transferred to the enzyme through a third component, a group of coupling proteins called N-proteins. They can have stimulatory (N_s) or inhibitory (N_i) actions. The N-proteins in turn are activated by GTP, and the function of the receptor appears to be to trigger an exchange of bound GDP for GTP. The signal is terminated by hydrolysis of the bound GTP by a GTPase activity intrinsic to the N-protein.

There are relatively few neurotransmitters but variability is enhanced by the multiplicity of receptors, i.e. there appear to be in general more than one type of receptor interacting with a given transmitter. Examples of this concept are the adrenoceptors (α_1, α_2, β_1, β_2), the dopamine-receptors (D_1, D_2), the cholinergic receptors (muscarinic and nicotinic), and the opiate receptors ($\mu, \varkappa, \delta$).

The nicotinic acetylcholine receptor complex is a binding protein and ion channel (for Na^+ and K^+), and was the first receptor to be isolated and characterized biochemically. The receptors of the inhibitory transmitter GABA and glycine which regulate Cl^--channels are of the same type; receptors acting by means of a "second messenger" – the dopamine and opiate receptors – are of the β-adrenergic type.

The physiological function of receptors is that of regulatory proteins. Their number, affinity and activity are under the control of various regulation mechanisms. They are also the site of action of numerous exogenous effectors, namely drugs or toxins. Some diseases of the nervous system are receptor diseases (myasthenia gravis, and perhaps schizophrenia). Some so-called receptors, especially drug-binding sites, may just be regulatory binding sites or subunits of true neurotransmitter-receptor complexes. Thus the minor tranquillizers, the benzodiazepines and the barbiturates which enhance the inhibitory action of GABAergic neurons appear to do so by stimulating the binding of GABA to its receptor.

References

Cited:

[1] Yamamura, H.I., Enna, S.J., and Kuhar, M.J. (eds.): *Neurotransmitter Receptor Binding*, 2nd edition. Raven Press, New York 1985.
[2] Karlin, A., "On the application of a 'plausible model' of allosteric proteins to the receptor for acetylcholine", *J. Theoret. Biol.* **16**, 306-320 (1967).
[3] Nachmansohn, D., *Harvey Lect.* **49**, 57-99 (1955).
[4] Belleau, B., "A molecular theory of drug action based on induced conformational perturbations of receptors", *J. med. Chem.* **7**, 776-784 (1964).
[5] Changeux, J.-P., "The acetylcholine receptor: an 'allosteric' membrane protein", *Harvey Lect.* **75**, 85-254 (1981).
[6] Kehoe, J., "Three acetylcholine receptors in Aplysia neurons", *J. Physiol.* **255**, 115-146 (1972).
[7] Cuatrecasas, P., "Membrane receptors", *Ann. Rev. Biochem.* **43**, 169-214 (1974).
[8] Fambrough, D.M., "Control of acetylcholine receptors in skeletal muscle.", *Phys. Rev.* **59**, 165-227 (1979).
[9] Pette, D., Müller, W., Leisner, E., and Vrbova, G., "Time dependent effects on contractile properties, fibre population, myosin light chains and enzymes of energy metabolism in intermittently and continously stimulated fast twitch muscles of the rabbit", *Pflüger's Arch.* **364**, 103-112 (1976).
[10] Gordon, A.S., Davis, C.G., Milfay, D., and Diamond, I., "Phosphorylation of acetylcholine receptor by endogenous membrane protein kinase in receptor-enriched membranes of *Torpedo californica*", *Nature* **267**, 539-540 (1977).
[11] Birdsall, N.J.M., and Hulme, E.C., "Biochemical studies on muscarinic acetylcholine receptors", *J. Neurochem.* **27**, 7-16 (1976).
[12] Strosberg, A.D., Vauquelin, D., Durieu-Trautmann, O., Delavier-Klutchki, C., Bottari, S., and Andre, C., "Towards the chemical and functional characterization of the β-adrenergic receptor", *TIBS* **5**, 11-15 (1980).
[13] Greengard, P., "Possible role for cyclic nucleotides and phosphorylated membrane proteins in postsynaptic actions of neurotransmitters", *Nature* **260**, 101-108 (1976).
[14] Nestler, E.J., and Greengard, P., "Protein phosphorylation in the brain", *Nature* **305**, 583-588 (1983).
[15] Creese, I., Burt, D.R., and Snyder, S.H., "Biochemical actions of neuroleptic drugs: focus on the dopamine receptor", *Handb. Psychopharmacol.* **10**, 37-89 (1978).
[16] Creese, I., Sibley, D.R., Hamblin, M.W., and Leff, S.E., "The classification of dopamine receptors: relationship to radioligand binding", *Ann. Rev. Neurosci.* **6**, 43-71 (1983).

[17] Stoof, J.C., and Kebabian, J.W., "Two dopamine receptors: Biochemistry, physiology and pharmacology", *Life Sci.* **35**, 2281–2296 (1984).
[18] Robson, L.E., Paterson, S.J., and Kosterlitz, H.W., "Opiate receptors", In: *Handbook of Psychopharmacology.* Iversen, L.L., Iversen, S.D., and Snyder, S.H. (eds.), Vol. 17, p. 13-80. Plenum Press, New York and London 1983.
[19] Khachaturian, H., Lewis, M.E., Schäfer, M.K.-H., and Watson, S.J., "Anatomy of the CNS opioid systems", *Trends Neurosci.* **8**, 111–119 (1985).
[20] Jessel, T.M., and Iversen, L.L., "Opiate analgesics inhibit substance P release from rat trigeminal nucleus", *Nature* **268**, 549-551 (1977).
[21] Lambert, A., Nirenberg, M., and Klee, W.A., "Tolerance and dependence evoked by an endogenous opiate peptide. *Proc. Natl. Acad. Sci. USA* **73**, 3165-3167 (1976).
[22] Möhler, H., and Okada, T., "Biochemical identification of the site of action of benzodiazepines in human brain by ^{3}H-diazepam binding", *Life Sci.* **22**, 985-995 (1978).
[23] Tallman, J.F., and Gallager, D.W., "The GABA ergic system: a locus of benzodiazepine action", *Ann. Rev. Neurosci* **8**, 21-44 (1985).
[24] Braestrup, C., and Nielsen, M.: "Benzodiazepine receptors". In: *Handbook of Psychopharmacology.* Iversen, L.L., Iversen, S.D., and Snyder, S.H. (eds.), Vol. 17, p. 285-384. Plenum Press, New York and London 1983.
[25] Lynch, G., and Baudry, M., "The biochemistry of memory: a new and specific hypothesis", *Science* **224**, 1057–1063 (1984).
[26] Shuster, M.J., Camardo, J.S., Siegelbaum, S.A., and Kandel, E.R., "Cyclic AMP-dependent protein kinase closes the serotonin-sensitive K^+ channels of *Aplysia* sensory neurons in cell-free membrane patches", *Nature* **313**, 292–295 (1985).

Further reading:

Lamble J.W. (ed.): *Towards Understanding Receptors,* 1981; *More about Receptors,* 1982; *Receptors Again,* 1984 (edited together with Abbot, A.C.); Elsevier Science Publishers, Amsterdam, New York, Oxford.
Yamamura, H.I., Enna, S.J., and Kuhar M.J., (eds.): *Neurotransmitter Receptor Binding,* 2nd edition. Raven Press, New York 1985.
Lajtha, A. (ed.): *Handbook of Neurochemistry,* 2nd edition, Vol. 6. Plenum Press, New York and London, 1984.
Lajtha, A., "Molecular neurobiology", *Cold Spring Harbor Symp. Quant. Biol.* **48**, 1983.
Strange, P. (ed.): *Cell Surface Receptors.* Ellis Horwood Publishers, Chichester 1983.
Drachman, D.B., "The biology of myasthenia gravis", *Ann. Rev. Neurosci.* **4**, 195–226 (1981).
See also ref. 5, 8, 13, 16, 18, 19, 23, 24.

Chapter 10

Neuronal Cytoplasm and Neuron-Specific Proteins

In the first chapter of this book it has been shown that almost all typical neuronal functions are more or less direct membrane functions. In particular nerve impulse propagation, the cell-to-cell electrical and chemical transmission of nerve impulses, active ion transport, cell recognition and synapse development, and interactions with neuromodulators, neuropharmacological substances and neurotoxins are all basically membrane phenomena. We will now correct this one-sided view by a consideration of the cytoplasm. Although it is essentially similar to the cytoplasm of other cells in terms of its organelles (except synaptic vesicles) and its enzymes (apart from those involved in the metabolism of transmitters) the neuronal cytoplasm is specifically adapted to the functions of the neuron.

Axonal transport, the intracellular communication system

We have defined the nervous system as the organ of communication and the neuron as its elementary unit. The neurons are the connecting links between parts of the organism which are often very far apart, sometimes many centimetres (or even several metres in the case of the giraffe and whale!). This means that cell bodies and nerve terminals can be separated by distances which cannot be travelled in a reasonable time by molecules moving by passive diffusion. How can the nucleus in the cell body regulate the action at, for example, the tip of the growing axon? Paul Weiss, the founder of the developmental biology of the nervous system, discovered the answer about 30 years ago: in the axon there is a transport system for macromolecules and smaller molecules. This "axonal transport" [1] occurs in both directions, i.e. *anterograde,* from the cell body to nerve endings, and *retrograde* i.e. from the axonal terminal back to the cell body. (There is also transsynaptic transport from neuron to neuron, from glial cell to neuron and vice versa; little is yet known about these processes). This transport property of the neural cytoplasm is not a unique property of the neuron but is analogous to intracellular transport in other cells. It can however be particularly well investigated in nerve cells due to their characteristic shape. Two basic methods have been developed for this: axonal ligation and the use of radioactive markers. The first method was devised by Paul Weiss who demonstrated transport by *ligating* or tying-off axons (Fig. 10.1). At the ligature everything transported in an anterograde direction is dammed up on the proximal side (the side nearest the cell body); the products of retrograde transport accumulate on the distal side (the side nearest the synapse); thus the axon is depleted of the products of anterograde transport in its distal region and of those of retrograde transport in its proximal (central) region. In a development of this technique local cooling of the axon is substituted for the ligature; this procedure is less damaging and has the advantage of being reversible.

The second method utilizes radioactive precursors; these are injected into experimental animals in the region of the cell bodies of defined tracts and traced to their projections; often injection into the eye is used and the label is followed along the optic tract. More complete information can be obtained by investigating the distribution of radioactivity at various time points by cutting the nerve into equal segments and measuring their radioactivity; this procedure can be combined with ligation or with an investigation of the terminal cytoplasm using the technique of isolating the nerve terminals as synaptosomes by subcellular fractionation.

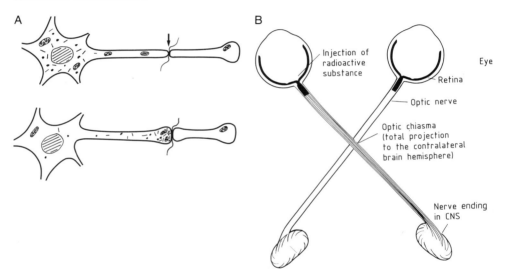

Fig. 10.1. Axonal transport. (A) Evidence from ligature of axon, (B) from analysis of target area of the nerve (here the optic system of the bird).

When the distribution of radioactivity in the nerve measured at different time points is compared, the kinetics of the transport can be worked out (Fig. 10.2). If the individual segments are then subjected to SDS-polyacrylamide gel electrophoresis the transported radioactive components can be identified.

For investigations utilizing the optic tract, fish and birds are particularly suitable (Fig. 10.1B) as their whole optic nerve projects to the opposite half of the brain (and is not, as in mammals, separated into right and left visual fields). The olfactory nerves of certain long-snouted fish (garfish, pike) provide excellent systems, and for mammals, the dorsal root ganglion-ischial nerve of the cat has been particularly useful (see Fig. 10.2).

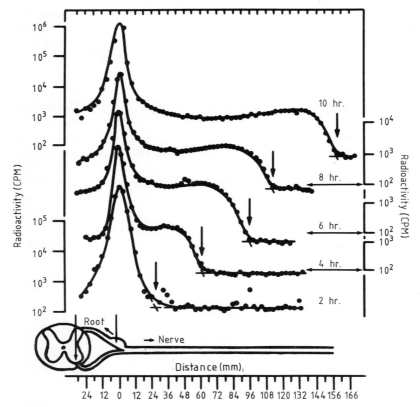

Fig. 10.2. Axonal transport. Evidence from measurement of the distribution of radioactivity in counts per minute (cpm) at different points in time after the injection of a radioactively labelled amino acid into the dorsal root of the spinal cord. After 10 h the radioactivity has travelled about 160 mm along the ischial nerve (fast transport). Reproduced, with kind permission, from Ochs and the American Association for the Advancement of Science, *Science* **176**, 253, Fig. 1 (1972).

Everything is transported: proteins, lipids, transmitters, mitochondria etc.

Protein synthesis takes place almost exclusively in the cell body; proteins of the axonal and presynaptic membranes and especially the enzymes of metabolism and transmitter synthesis are therefore all transported anterogradely. Indeed transmitter synthesis occurs not only in the nerve ending but also in the cell body, so that acetylcholine, catecholamines and GABA are also transported. The question as to whether ribosomes are transported and therefore whether protein synthesis takes place in the nerve ending was disputed for a long time; it is, however, now clear that whole mitochondria travel along the axon and that the protein synthesis observed was mitochondrial. Among the transported molecules are also phospholipids, glycolipids and glycoproteins. If radioactive fucose, a constituent of the carbohydrate side chains of many glycoproteins, is injected into the cell body, it is found to be incorporated into protein before leaving it, and not at the nerve terminal. N-acetylneuraminic acid (NANA) behaves differently: it is probably added to the glycoprotein or glycolipid only at the nerve ending.

Kinetics of transport: different rates for different components

There are marked variations in the rate of transport. Originally it was divided into slow axonal transport (or "axonal flow") at a rate of 1-4 mm/day, and fast axonal transport at 200-400 mm/day. Since then at least one intermediate speed of 15-50 mm/day has been added, and some authors believe that as many as 5 speeds can be distinguished [2]. It is important to note however that the molecules of one species are not transported at several different rates but that all molecules of one substance are transported at the same rate. The following proteins, some of which we will discuss in further detail, travel by axonal flow (slow axonal transport): tubulin, neurofilament subunits, actin, myosin and myosin-type proteins, and in addition, the soluble enzymes of intermediary metabolism. If the axon is separated from the cell body, the slow flow ceases. Retrograde slow axonal transport has not been observed. The mitochondria travel at an intermediate speed; glycoproteins, glycolipids and also enzymes of transmitter metabolism such as dopamine-β-hydroxylase and acetylcholinesterase at a fast speed. Acetylcholinesterase is also transported retrogradely with approximately the same fast speed.

Mechanism: energy dependent, membrane-bound and carried by filamentous structures

The axoplasm is a gel-like mass in which macromolecules especially could not diffuse at the rates observed. A further point against a passive transport system is that uncouplers of oxidative phosphorylation block the transport. 2,4-Dinitrophenol, cyanide and azide ions inhibit it as does fluoride, an inhibitor of glycolysis. Oxygen and ATP must be present. Fast axonal transport is independent of the cell body and can be observed in isolated axons in Ringer's solution and also in salt-free sucrose solutions. Electrical excitability is not a requirement and blocking of action potentials with tetrodotoxin has no effect. Blockers of protein synthesis do not inhibit fast transport; low temperatures however do damage the transport mechanism.

Colchicine, an alkaloid from the meadow saffron (*Colchicum autumnale*) (Fig. 10.3) is a potent inhibitor of axonal transport and this poison provides evidence for the mechanism of transport. Colchicine is a classical inhibitor of mitosis, and its effect is due to its ability to dissociate the filaments of the spindle apparatus. The axoplasm is also interlaced longitudinally with numerous filamentous structures, which make up most of the protein and produce the gel-like consistency. Three types of filaments can be distinguished: the *microtubules*, the *microfilaments*, both also present in non-neural cells, and *neurofilaments*

Mechanism: energy dependent, membrane-bound and carried by filamentous structures 255

which are characteristic for neurons and glial cells. Microtubules are hollow tubes of 24 nm diameter, neurofilaments are fibres of 10 nm, microfilaments of 8 nm diameter. The microtubules which consist of the protein tubulin are depolymerized into their subunits by the alkaloids colchicine and vinblastine. In numerous cases it can be shown that this depolymerization occurs in parallel with inhibition of axonal transport, but a direct causal relationship is still in doubt; thus cases have been described where colchicine inhibits transport without damaging the microtubules. The action of colchicine may not be limited to affecting the state of aggregation of the tubulin.

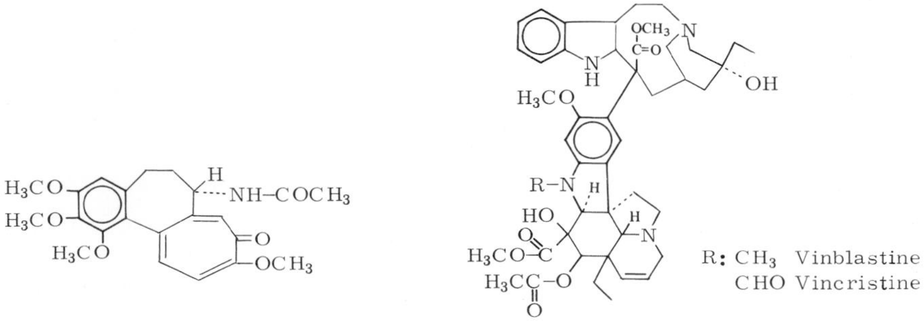

Colchicine   Vinblastine

Fig. 10.3. Inhibitors of axonal transport. The alkaloids, colchicine (*Colchicum*) and vinblastine (*Vinca*) which due to their dissociating action on the microtubules of the spindle apparatus also cause inhibition of mitosis.

There are many hypotheses about the mechanisms involved in axonal transport, and as has so often been mentioned in this book, we must await future results for clarification, contenting ourselves with just a few questions and provisional answers. Are transported components packed into vesicles, or do they flow in solution through the axon? Some evidence supports the vesicle hypothesis. In particular the proteins involved in fast transport move in association with membrane components, and it can be seen electron-microscopically that membranous organelles are present and transported. A somewhat similar mechanism must be involved in secretion. There too, the substance must be packaged and transported to the site of secretion, a process which, incidentally, can also be blocked by colchicine.

A further important question is whether there is any selectivity or specificity with respect to the substances transported or whether every molecule present is despatched down the axon. There is considerable evidence for a certain degree of specificity. No acetylcholine is transported in adrenergic neurons when it is injected into the cell body and there is no transport of serotonin in cholinergic neurons. This selectivity however is more likely to be due to the specificity of the transmitter storage vesicles than to the transport mechanism and is a further argument for vesicular transport. The same applies to the transport of enzymes; dopamine-β-hydroxylase is transported specifically by catecholaminergic neurons, because it is incorporated into catecholaminergic storage vesicles. Apart from these observations hardly any differences can be seen between transport in differing neurons.

The same applies to retrograde transport; here also the selectivity of the uptake mechanism of the presynaptic membrane determines the specificity of the transport. Thus, adrenergic neurons transport nerve growth factor (NGF) retrogradely but not other hormones, and neurotropic viruses (*poliomyelitis, herpes*) are transported retrogradely because there is a binding site and uptake system for them at the nerve ending.

A third question is linked with the previous two and concerns the contribution of the endoplasmic reticulum to the transport. This probably forms a system of continuous channels along the whole length of the axon, and radioactive molecules are concentrated here particularly rapidly during transport. These channels might constitute an ideal transport system, but even if there were evidence for them their existence would in itself provide no indication as to the nature of the transport mechanism. What is their interaction with microtubules, micro- and neurofilaments and actin? Space does not permit a discussion of the various unproven mechanistic models. Most assert that the mechanistic work of transport is calcium-dependent and is carried out by an actin/myosin system with a mechanism similar to that of muscle contraction. Perhaps the most plausible is the model proposed by Droz, a protagonist of the reticular hypothesis (Fig. 10.4), as it seems to account for most of the known facts [3].

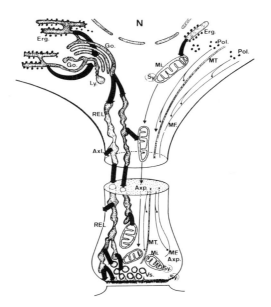

Fig. 10.4. Model of axonal transport (from Droz [3]). N, nucleus; Mi, mitochondria; RER, rough endoplasmic reticulum; SER, smooth endoplasmic reticulum; Go, Golgi apparatus; Ly, lysosome; Axl, axolemma; Axp, axoplasm; Pol, polyribosome; MT, microtubules; MF, microfilaments; Sy, site of synthesis for hydrophobic polypeptides; Vs, synaptic vesicle. For explanation, see text [3].

The diagram (Fig. 10.4) shows the route taken by a protein synthesized in the perikaryon from the rough endoplasmic reticulum to the Golgi apparatus, and from there via the smooth endoplasmic reticulum to the nerve ending. On the right hand side of the diagram slow axonal flow is depicted i.e. the synthesis of the components of the tubular and filamentous structures and their assembly and transport.

Tubulin and associated proteins

Biochemists feel much safer describing structures rather than mechanisms. Some axoplasmic proteins have been well investigated and will be briefly presented here.

Tubulin [4] can be isolated from the brain as a dimeric protein with a relative molecular mass of 110 000 and a sedimentation coefficient ($S_{20,w}$) of 6 S. It consists of two non-identical subunits, one (α) of 53 000 and the other (β) of 57 000. The amino acid sequence shows considerable homologies, evidence that during evolution they resulted from gene duplication. Microtubules are poylmers of tubulin. Thirteen linear "protofilaments" of tubulin subunits are arranged together in a tube-shaped structure, so that the cross section shows a 13-fold axial symmetry; seen from the side, the subunits are stacked helically (Fig. 10.5). Individual tubulin molecules are in an association-dissociation equilibrium with the microtubules, which is affected by various parameters (temperature, Ca^{2+}, GTP, phosphorylation). Very pure tubulin forms microtubules only under conditions of very high protein and magnesium concentrations. In the cell other factors seem to contribute to the association. In particular it is facilitated under physiological conditions by

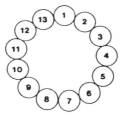
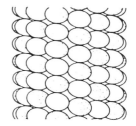

Fig. 10.5. Microtubules. The cross section (left) shows the 13-fold axial symmetry. In the electron micrograph the α and β subunits of tubulin cannot be distinguished from one other; it is assumed that they alternate. The side view (right) shows the protofilaments i.e. vertical rows of tubulin units and the helical progression of one subunit to the next (from [4]).

specific *microtubule-associated proteins* (abbreviated to MAP) [5] and a *tubulin assembly protein* (TAP or τ-factor) which are difficult to separate in the purification of tubulin. In addition tubulin contains two GTP binding sites per α, β-dimer. Association to form the tubular structure results in the hydrolysis of GTP to GDP by a GTPase activity of tubulin. The significance of this reaction is shown by the easier depolymerization of the GDP-tubulin-complex; GTP and GDP appear to induce different conformations of tubulin in which the GTP form associates more easily and the GDP form dissociates more readily. Less clear is the significance of a cAMP-dependent kinase and of a Ca^{2+} calmodulin-dependent kinase (see below) which phosphorylate the α and β subunits of the tubulin on serine residues. These covalent modifications are presumed to alter the association-dissociation equilibrium.

The large number of effectors is evidence of a complex regulation of microtubule formation in the cell. The association evidently starts at one point in the cell and proceeds unidirectionally by addition of monomers to the growing or "head" end of the polymer while at the tail end depolymerization predominates. The sum of these vectorial processes might account for the observed slow axonal flow of tubulin.

Regulation of tubulin association plays an important role in the regeneration of damaged neurons and in neurite growth in the early stages of development of the neuron. Microtubules are at first formed very slowly, although in the early stages of differentiation there is already plenty of tubulin available in the cell. Further differentiation is characterized by intensive neurite growth and an increasing rate of association of tubulin which is accompanied by characteristic changes in the MAPs.

Neurofilaments

We have already mentioned the second class of filamentous structures in the axoplasm, the neurofilaments. With a diameter of 10 nm they lie between the 24 nm neurotubuli and the 6 nm actin filaments. They are therefore considered to be a class of *intermediate filaments* [6] which have been found in various cells and to which the keratin filaments of the epithelial cells, glial filaments and desmin filaments of muscle cells belong. A role in the mechanical ordering and maintenance of cellular space is ascribed to them. In the electron microscope they show fibrous ramifications. Neurofilaments from rabbit nerve consist of three proteins with relative molecular masses of 68 000, 150 000 and 200 000. Only two neurofilament proteins with relative molecular masses of 200 000 and 60 000 have been isolated as yet from the squid giant axon [7]. They are sensitive to a Ca^{2+}-dependent protease and therefore not easy to obtain intact. All neurofilament proteins are phosphorylated by a cAMP-dependent kinase.

Actin, myosin: a role in mechanical work?

The contractile proteins actin and myosin are found in all neurons. Actin exists in such high concentrations that it suggests a role in axoplasmic transport; however no definitive evidence for this plausible hypothesis has yet materialized.

Calmodulin, a mediator protein for calcium regulation

Axoplasmic transport is a calcium-dependent process and in neurons, as in probably all other cell types, a protein is found which is involved in many Ca^{2+}-regulated processes; this is calmodulin [8]. It is a very stable protein of relative molecular mass 17 000 with four Ca^{2+}-binding sites per polypeptide chain (Fig. 10.6). The amino acid sequence is notable for the presence of the rare amino acid trimethyllysine.

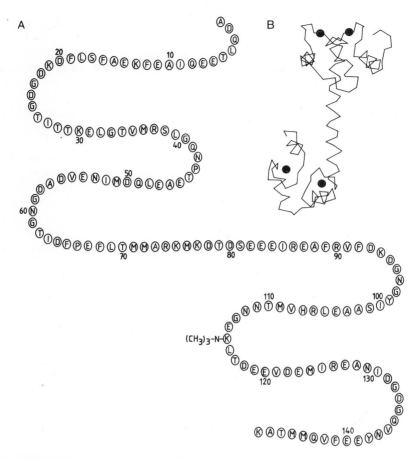

Fig. 10.6. Calmodulin. The molecule contains four Ca^{2+}-binding domains. (A) Primary structure, (B) Tertiary structure, according to Babu et al., *Nature* **315**, 37–40 (1985), (●) Ca^{2+}.

Ca^{2+}-calmodulin exerts its regulatory roles here and in general either directly as an allosteric regulator of certain proteins or indirectly through the action of Ca^{2+}-calmodulin-dependent protein kinases which appear to be especially abundant in the brain and comprise as much as 0.3% of the total brain protein. Participation of calmodulin in a process can be easily ascertained in biochemical assays with neuroleptic drugs as e.g. chlorpromazine which inhibits Ca^{2+}-effects mediated by calmodulin. A variety of drugs including some local anaesthetics have a similar effect. When such substances are used medicinally it is not clear whether their main therapeutic action is accompanied by side-effects on other processes in the nervous system and elsewhere where Ca^{2+} calmodulin plays a major role *in vivo*.

Calmodulin in turn may be regulated by other proteins and here a protein termed *calcineurin* appears to be of interest. Calcineurin interacts with high affinity with calmodulin and thus inhibits its regulatory function. It contains four Ca^{2+}-binding sites per molecule and consists of two subunits, calcineurin A (relative molecular mass 61 000) calcium-dependent and interacting with calmodulin, and calcineurin B (relative molecular mass 15 000) also having Ca^{2+}-binding sites. Recently it was discovered that calcineurin possesses phosphoprotein phosphatase activity [9]. It appears to be identical with the so-called phosphoprotein phosphatase 2B which is most abundant in nervous tissue but is also found in other organs.

Tubulin association is inhibited by Ca^{2+} and this effect is mediated by calmodulin. Furthermore calmodulin plays a specific role in the regulation of neurotransmitter release from nerve endings – also a Ca^{2+}-dependent process (see Chapter 8). The very high calmodulin concentration in the brain (10 µmol/l) as well as the unusual conservation of the amino acid sequence during evolution gives an indication of the significance of this protein. In addition to its action in neurons calmodulin has been identified as a mediator of the calcium regulation of adenylate cyclase, phosphodiesterase, phosphorylase kinase and above all the phosphorylation of numerous membrane proteins (see Chapter 9).

Neuron-specific proteins

There is a range of proteins of as yet unknown function which, as they are found exclusively in neural cells, some in the cytoplasm, others in membranes, must be ascribed a special role there. Neuron-specific proteins are the receptors, ion channel components and neurofilaments already described in this and earlier chapters. In addition there are certain enzymes of neurotransmitter metabolism, myelin proteins, synaptic proteins (e.g. postsynaptic density protein), and binding proteins for neurotoxins like β-bungarotoxin, botulinum- and tetanus toxins. Synapsin also deserves mention (see Chapter 9).

The *S-100 protein* is a protein specific for the nervous system [10]; it is widely present in both neurons and glial cells. It has been described as both cytoplasmic and membrane-bound. Its relative molecular mass is 20 000, and it consists of two identical Ca^{2+}-binding polypeptide chains. Actually it appears to belong to a group of very similar proteins, of which one, the PAP Ib-protein (abbreviation for *phenylalanine-rich acidic protein*) was recently sequenced [11]. S-100 has considerable sequence homology with the Ca^{2+}-binding muscle protein troponin C. Its function is unknown; we shall return in Chapter 11 to its possible involvement in neural development and plasticity.

The *14-3-2 protein* seems to be present only in neurons of the central nervous system [10]. It is a dimer of two polypeptide chains of relative molecular mass 39 000, and is demonstrable by means of immunofluorescence in all species of mammals and birds so far tested. Its function is unknown: recently however enolase activity has been found in highly purified 14-3-2 preparations.

P-400 protein has been found in the cerebellum but not the cerebral cortex of mice and is associated with motor control. P-400 has a relative molecular mass of 400 000 and is not present in the *staggerer* and *nervous* mouse behavioural mutants (see Chapter 12).

The search for neuron-specific proteins linked with genetically caused nervous diseases is tedious but has great promise. An example is the protein PcI Duarte, a variant of a brain specific protein, whose appearance is correlated with schizophrenia [12]. An ever increasing number of nerve specific components are being discovered by immunological methods. As their function is unknown they are denoted by symbols such as NS1, NS2, L1, etc., (NS stands for *nervous system antigen*). The application of the technique of monoclonal antibodies is leading to the discovery of cell-surface antigens specific for various types of cells of the nervous system. One such neuron-specific surface is the **n**euronal **c**ell **a**dhesion **m**olecule (N-CAM) which appears to have an important function in the development of the nervous system [13]. We shall come back to this in the next chapter. It has been estimated that the brain expresses some 30 000 genes the products of which are mostly still awaiting identification.

Summary

The cytoplasm of the neuron is in constant motion. This motion, called axonal transport, is the functional link between the cell body and its nucleus on the one hand and the nerve ending, often as far as a metre or more distant, on the other. It is involved in the growth and maintenance of the axon, its regeneration following lesions and adaptations to synaptic activity. Axonal transport occurs in both anterograde and retrograde directions so that components and the signals they embody can travel not only from cell body to synapse but also in the reverse direction. There is a slow anterograde axonal flow (1-4 mm per day) and intermediate (15-50 mm per day) and fast (200-400 mm per day) rates of axonal transport. Each molecular species is transported at a characteristic rate. Tubulin, neurofilament subunits, actin and myosin are transported slowly; mitochondria at the intermediate rate; membrane proteins, glycoproteins, glycolipids, enzymes of transmitter synthesis and transmitters at the fast rate. DNA, RNA and gangliosides are not transported. Retrograde transport removes synaptic degradation products, enzymes and also substances taken up by the presynaptic membrane like NGF, tetanus toxin and neurotropic viruses. Evidence for axonal transport has been derived from axonal ligature, from analysis of the target area (e.g. optic tectum for transport from retinal cells) or from tracing the nerve distribution of a radioactive protein precursor injected into a region rich in neuronal perikarya, for example the dorsal root ganglion. Transport requires energy, calcium and microtubules. These consist of tubulin, whose association is regulated in a complex fashion by MAP proteins, TAP, GTP, Ca^{2+}, calmodulin, phosphorylation and other factors. Colchicine inhibits microtubule assembly and axonal transport. Microtubules, microfilaments and neurofilaments interlace the actin/myosin complex of the axoplasm and give it a gelatinous consistency. Beside these, the neuronal cytoplasm contains, among other components, the enzymes of intermediary and transmitter metabolism, and special effector proteins like calmodulin, which mediates the regulatory effect of Ca^{2+} in metabolism, transmitter release etc.; in addition calcineurin and a range of neuron-specific proteins – 14-3-2 protein, S-100 protein (also in glia), and P-400 protein (cerebellum) – are present. Also neuron-specific are a series of surface membrane antigens (NS1, NS2, L1) whose function is not yet known, and N-CAM, the cell adhesion molecules important in the development of the nervous system. The brain expresses some 30 000 genes. Many of these gene products may be "brain specific" but have not yet been identified [14].

References

Cited:

[1] Schwartz, J.H., "Axonal transport: components, mechanisms, and specificity", *Ann. Rev. Neurosci.* **2**, 467–504 (1980).
[2] Lasek, R., "Axonal transport: a dynamic view of neuronal structures", *TINS* **3**, 87–91 (1980).
[3] Droz, B.: "Synthetic machinery and axoplasmic transport: maintenance of neuronal connectivity", In: *The Nervous System*. D.B. Tower (ed.), Vol. 1, p. 111–127. Raven Press, New York 1975.
[4] Snyder, J.A., and McIntosh, J.R., "Biochemistry and physiology of microtubules", *Ann. Rev. Biochem.* **45**, 699–720 (1976).
[5] Mareck, A., Fellous, A., Francon, J., and Nunez, J., "Changes in composition and activity of microtubule-associated proteins during brain development", *Nature* **284**, 353–355 (1980).
[6] Lazarides, E., "Intermediate filaments as mechanical integrators of cellular space", *Nature* **283**, 249–256 (1980).
[7] Roslansky, P.F., Cornell-Bell, A., Rice, R.V., and Adelman, W.J., "Polypeptide composition of squid neurofilaments", *Proc. Natl. Acad. Sci. USA* **77**, 404–408 (1980).
[8] Klee, C.B., Crouch, T.H., and Richmann, P.G., "Calmodulin", *Ann. Rev. Biochem.* **49**, 489–515 (1980).
[9] Stewart, A.A., Ingebritsen, T.S., Manalan, A., Klee, C.B., and Cohen, P., "Calcineurin", *FEBS Lett.* **137**, 80–84 (1982).
[10] Bock, E., "Nervous system specific proteins", *J. Neurochem.* **30**, 7–14 (1978).
[11] Isobe, T., and Okuyama, T., "The amino-acid sequence of S-100 protein (PAP I-b protein) and its relation to calcium-binding proteins", *Eur. J. Biochem.* **89**, 379-388 (1978).
[12] Comings, D.E., "PC I Duarte, a common polymorphism of a human brain protein, and its relationship to depressive disease and multiple sclerosis", *Nature* **77**, 28–32 (1979).
[13] Edelman, G., "Cell adhesion molecules", *Science* **219**, 452-457 (1983).
[14] Sutcliff, J.G., and Milner, R.J., "Brain specific gene expression", *TIBS* **9**, 95–99 (1984).

Further reading:

Weiss, D.G. (ed.): *Axoplasmic Transport*. Springer-Verlag, Berlin, Heidelberg, New York 1982.
Forman, D.S., "New approaches to the study of the mechanism of fast axonal transport", *TINS* **7**, 112–116 (1984).
Bray, D., and Gilbert, D., "Cytoskeletal elements in neurons", *Ann. Rev. Neurosci.* **4**, 505-523 (1981).
See also ref. 2, 6, 8, 13, 14.

Chapter 11

Development, Stabilization and Plasticity of the Nervous System

The mature human nervous system consists of a network of more than 10^{10} nerve cells, each of which forms synaptic contact with up to 10^4 other nerve cells. In the introductory chapter we have defined the function of the nervous system as the gathering, processing, storing and transmission of information, not through coding it in a pattern of impulses in the individual cells, but through the synaptic connections of cells with one another, i.e. the specificity of the network. One of the central questions of neurobiology is the origin of the mechanism of this specificity in the ontogeny of the nervous system.

This must obviously be considered in the context of the more general question of the mechanism of cell differentiation. The fertilized ovum, like every cell of the mature organism except erythrocytes, has a complete set of chromosomes containing the DNA responsible for conveying all the hereditary information. Differentiation of a cell means the suppression of the greater part of this information and the promotion of the special part of it necessary for the particular composition and function of the specialized cell. Molecular biology today is still far from understanding the molecular mechanism of differentiation.

In vertebrates, embryonic development of the nervous system begins at the gastrula stage with the formation of a cell aggregate called the *neural plate*. The subsequent development of the neuron cannot be dealt with here in detail. It can be summarized by the two terms: *migration* and *growth*. The developing neuron migrates, mainly through numerous cell layers of the developing organism towards its target site, where it becomes integrated as part of the mature organ.

At present there is no indication of a specific recognition label by which each migrating cell finds its destination in the maturing tissue. On the contrary there appears to exist only a limited number of cell adhesion molecules (CAMs) which mediate cell-to-cell adhesion and fix migrating cells in the right place. The best characterized of these, discovered by Edelman, is the neuronal adhesion molecule (N-CAM). Another one is the neuroglia adhesion molecule (Ng-CAM) mediating adhesion of neuronal cells to glial cells. N-CAM is a large glycoprotein (M_r 200 000) with a high content of sialic acid (N-acetylneuraminic acid). It undergoes dynamic changes during development in amount, distribution and carbohydrate content. The latter, especially the sialic acid content appears to determine the affinity towards the cell surface.

The neuron also *grows* by sending out extensions, the dendrites and the axon. The axon must frequently extend over great distances on the way to *its* target cell.

The maturation of the neuron to become a differentiated cell depends on:
(1) the portion of the DNA that is expressed;
(2) the time at which this expression occurs;
(3) the site at which it takes place.

Since the "when" (point 2) could be a molecular automatism of DNA transcription and translation, the "what" combined with the "where" (points 1 and 3) is of particular interest to us. For if it happens that the site (i.e. the cellular environment which the developing cell encounters on its way from the neural plate to its specialized target organ) has an influence on gene expression, then this implies a restriction of the genetic determination of an organism. In fact there is much evidence that cells mutually influence one another during development. This occurs either by direct contact for which the molecular mechanism is not yet clear, or through the despatch of chemical signals, called growth factors. The latter we shall discuss under the term "trophism", the former will be illustrated by the formation and stabilization of synapses. It should be noted that it is not the genetic programme alone which determines the final structure of the neural network, but the position of the individual cells in space and time. That saves us from a dilemma which is shown by the following comparison: the human genome contains about 10^6 genes, but the number of synapses is more than 10^{14} (10^{10} neurons each with 10^4 synapses, see above), so it is unlikely (though not impossible as demonstrated by the tremendous variability of antibodies created by a limited number of genes) for the specificity of each individual synapse to be programmed by a separate gene site. We will return to this important point when we deal with synaptogenesis, i.e. the formation and stabilization of specific synapses. The development of the nervous system seems to be controlled by several factors, genetic and trophic and by a decisive third factor as we shall see, the activity of the neuron.

In the course of maturation (see Fig. 11.1) the cell encounters many alternative paths and many decisions must be made in the choice of route. The cells of the matrix, for example, can develop either into glioblasts or neuroblasts. The glioblast is further differ-

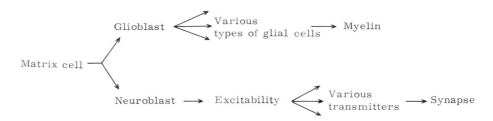

Fig. 11.1. Scheme of differentiation of nerve cells.

entiated into various glial cell types of which some respond to another signal of differentiation to form myelin. The neuroblast differentiates in the direction of nerve cell function, electrical excitability, synthesis of a transmitter and synapse formation. The biochemical mechanisms underlying these stages of differentiation are largely unknown but there are already a range of descriptive investigations. The autonomic nervous system has proved to be valuable for this research due to its relative simplicity and homogeneity [1].

We will deal initially with a review of this and then with the recent biochemical investigations of nerve cell differentiation in cell culture [2] and finally with some results of research in other areas of the nervous system.

Differentiation of transmitters

Most nerve cells of the mature nervous system utilize only one transmitter and only have the tools for synthesis, packaging and releasing of this one transmitter, but recently the presence of more than one amine transmitter including the enzymes needed for their synthesis has been discovered in single identified neurons of *Aplysia*, and neuropeptides have been detected in adrenergic neurons of mammals. The fact remains however that a neuron must at some stage of maturation decide for one transmitter and against another. The time and trigger for this decision has been observed in sympathetic nerve cells of rats during embryonic development and in cell culture.

Cells of the sympathetic nervous system are adrenergic or cholinergic. Sympathetic ganglia contain both cell types, and also non-neuronal cells. If ganglion cells of newborn rats are cultured in the absence of non-neuronal cells, they produce exclusively noradrenaline and form the characteristic synaptic vesicles of adrenergic neurons. In contrast, if non-neuronal cells are present the transmitter acetylcholine is produced [1]. Further it was shown that the non-neuronal cells themselves do not need to be present but that the culture medium in which they grew, the so-called conditioned medium, causes this switch from adrenergic to cholinergic. A protein of M_r 45 000 was identified as the signal which acted between the two cell types. This operated in fact as the "switch" in the neuron. It was even possible to identify cells in process of switching: they formed simultaneously cholinergic and adrenergic synapses.

Transsynaptic regulation: orthograde and retrograde

The above example has shown that "trophic factors", chemical signals from surrounding cells, can influence events in the neuron. These can, as here, be sent out from non-neuronal cells without direct contact with the nerve cell, but they can also originate from another nerve cell and act via the connecting synapse. One example of such a transsynaptic effect is the induction of tyrosine hydroxylase (and dopamine-β-hydroxylase) by preganglionic cholinergic stimulation (Fig. 11.2) [3]. As has been described in Chapter 8 tyrosine hydroxylase is the "pace-maker enzyme" of catecholamine synthesis. Both its synthesis and its activity are regulated by acetylcholine via cAMP as *second messenger*, the latter probably by means of cAMP-dependent protein kinase. This process has been well investigated using as a model the superior cervical ganglion of newborn mice and rats. In this it has been shown that the activity of tyrosine hydroxylase increases about eightfold following birth. If however, the preganglionic cholinergic pathways are cut, this activation does not occur. Blockers of nicotinic cholinergic receptors can cause the same effect (but not α-bungarotoxin, the typical blocker of peripheral acetylcholine receptors which does not act on the ganglionic receptors).

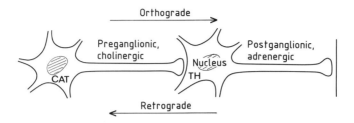

Fig. 11.2. Transsynaptic regulation. Induction of tyrosine hydroxylase (TH) by preganglionic cholinergic innervation (orthograde). In reverse the activity of the postganglionic adrenergic system acts on the choline acetyltransferase (CAT) in the cholinergic cells (retrograde).

Thus in the "physiological direction" (orthograde or anterograde transsynaptic) the development of the postsynaptic adrenergic neuron seems to be regulated by the nicotinic cholinergic nerve ending. In the opposite direction (retrograde transsynaptic) there is also a regulatory effect, for if the postsynaptic cell is selectively damaged there is an inhibition of the postnatal increase in activity of choline acetyltransferase.

These however are modulating effects on transmitter synthesis occuring after the definitive stage of differentiation has already taken place. The molecular mechanism of this modulation is not yet known in detail; the transmitter and probably the ion milieu and trophic factors play a part. Tyrosine hydroxylase is also induced by the nerve growth factor (NGF, see below) which is taken up by the nerve ending by pinocytosis, and from there conveyed by retrograde axonal transport to the cell nucleus.

Differentiation of ion channels and excitability

Nerve cells are characterized not only by their transmitters but also by their excitability. We will discuss the origin of *chemical* excitability together with synaptogenesis. *Electrical* excitability, or the ability to produce and transmit action potentials, originates very early in embryonic development [4]. In many species the ova produce action potentials, but they are different in many respects from those of the differentiated cell. The ion channels underlying the electrical excitability go through fundamental changes during development. In many cases the inward flux is not initially conveyed by sodium but by calcium ions. Where the ion channels also transport sodium ions, the sodium current is insensitive to tetrodotoxin (TTX). The change in ion dependence is accompanied by a decrease in the duration of the action potential. The mechanism and biological significance of these changes are unknown. The chemo-sensitive ion channels, like those regulated by acetylcholine, do not go through these changes in ion specificity.

There are various differentiation programmes

It could be imagined that a common "regulatory gene" activated by the corresponding differentiation signal initiates the biosynthesis of all the cell components characteristic for a specific step of differentiation. This does not seem to be the case. Thus for example the steps necessary for morphological differentiation are regulated independently of those necessary for transmitter synthesis. This area of research is however still very much in a state of flux so we must be content with only a brief mention of it in order not to be submerged in a flood of preliminary observations. One method of investigation is to stimulate a suitable cell line by a biological (e.g. nerve growth factor, NGF. Fig. 11.3) or an artificial stimulus (e.g. dibutyryl-cAMP, dimethylsulphoxide) until neurites are extended and then find out, for instance, how the pattern of enzymes or membrane proteins changes in parallel. It is interesting to note that only quantitative changes have been observed in the protein composition of such cell lines during the *in vitro* differentiation. In spite of the sensitivity of the method of two-dimensional electrophoresis which is used, no new protein has appeared nor has there been a loss of one previously present.

Trophic factors

Differentiation is the result of the interaction of the genetic programme and the environment. Diffusible factors which effectively stimulate differentiation and growth of a cell are called trophic factors. For a given neuron they can be produced by its target organ, by the surrounding glial cells or by one of its other innervating neurons. If we recall the example of the sympathetic ganglionic nerve cells we have seen that the effect from the non-neuronal cells is conveyed both orthogradely (or anterogradely) and retrogradely. In addition to this transsynaptic regulation, trophic factors play a role in cell survival, cell migration, the homing of the developing neurites (axons or dendrites) on to their target and the formation and stabilization of specific synapses. In addition to their activity during development trophic factors also fulfil these functions in the mature organism.

Biochemically speaking, the term "trophic factor" is not as yet well defined. Only in a few cases have such factors been isolated and chemically characterized; in most cases their existence has been merely inferred from observations or simply postulated. A trophic factor can be a protein, as is shown in the NGF (nerve growth factor, see below), a low molecular substance (transmitter, metabolite) or an ion (Ca^{2+}, K^+). The latter might be the case when the "trophic factor" is constituted by functional interaction between neuron and target cell. We shall return to this in the section on synapse stabilization and plasticity. The mechanisms of action of trophic factors are as many-sided as the molecular structures. They can either be taken up by the cell and act directly on the cell nucleus or they can exert their effect on the cell membrane via a *second messenger*.

Nerve growth factor (NGF)

The best researched trophic factor is the nerve growth factor (NGF) discovered by Levi-Montalcini and Hamburger in 1950 [5]. It stimulates the growth of peripheral sensory and sympathetic neurons and promotes the survival of the mature sympathetic neuron. It also stimulates the sprouting of axon-type extensions of embryonic ganglion cells in culture (Fig. 11.3B), a reaction which is utilized as a biological assay for the presence and isolation of NGF. NGF is found in many neural and non-neural tissues, but not in blood.

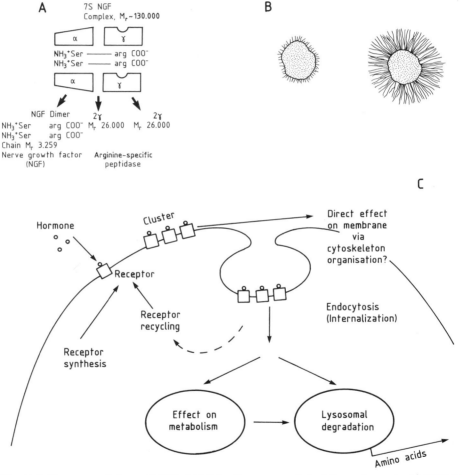

Fig. 11.3. Nerve growth factor (NGF). (A) Subunit structure of the 7 S complex and of β-NGF. From Server and Shooter, *Adv. Protein Chem.* **31**, 339-409 (1977). (B) Sensory ganglion of a chick embryo. Left, without NGF; right, with NGF. (C) Model of NGF interaction with a target cell which may also apply to other peptide hormones (kindly supplied by P. Layer, Tübingen).

It has been shown in cell culture that NGF is synthesized and secreted by a number of cell types. It occurs in high concentrations in the submaxillary salivary gland of the male mouse and can be isolated from it in milligram quantities. This high concentration does not occur in the female and immature male mouse; the physiological significance of this is not clear. In any case the submaxillary salivary gland does not seem to be the only site of synthesis and as most of the NGF is secreted in the saliva, it must have an exocrine rather than an endocrine function there. If the gland is removed, the concentration of NGF decreases in the various tissues of the animal, though only temporarily, and it increases again to its normal value without the gland having regenerated. This indicates a further, unknown site of synthesis.

The molecule NGF

NGF is a protein free from carbohydrate and lipid. Up to now structural studies have been mainly carried out with mouse NGF, but it has been also found in similarly high concentrations in snake venoms (*Crotalidae, Viperidae, Elapidae*). The active subunit of the mouse NGF at a concentration of a few ng/ml is a polypeptide of 118 amino acids with a relative molecular mass of 13 250. Its secondary structure contains few α-helical regions; its tertiary structure is stabilized by three disulphide bonds. The amino acid sequence has been elucidated.

The protein isolated from the submaxillary salivary gland has a much more complex structure (Fig. 11.3A); it is considered to be a dimer of two non-covalently linked identical chains, called β-NGF. This dimer in turn is part of a protein complex, called 7S NGF because of its sedimentation coefficient and contains in addition two α- and two γ-chains each with a relative molecular mass of 26 000. The relative molecular mass of the whole complex is thus ca. 130 000. The α- and γ-chains are not necessary for biological activity; there is no evidence for the function of the α-chain, except that it inhibits β-NGF. By contrast, the γ-subunit has arginine-specific protease activity and a sequence homology with trypsin. It takes part in the proteolytic conversion of a high molecular mass precursor of β-NGF. Such a pro-β-NGF with a relative molecular mass of 22 000 has already been discovered.

β-NGF has properties in common with other hormones in addition to its formation from a pro-hormone: it shows extensive sequence homologies with insulin, relaxin and non-suppressible-insulin-like-activity (NSILA). It evidently has an evolutionary connection with these peptide hormones.

The mechanism of NGF action is unknown

Neuronal plasma membranes (and also membranes of some non-neuronal cells) contain specific receptors (NGF receptors) which bind NGF at first with low then with high affinity. The high affinity receptors have been shown to cluster and, together with the bound NGF, are internalized by endocytosis and transported within the cell partly to lysosomes (for degradation) and partly to the nucleus. When taken up at nerve endings, receptor and NGF are transported by retrograde axonal transport. Similar processes may be involved in other types of hormonal regulation and NGF therefore serves as a model for other hormones and growth factors.

The mechanism of action of NGF within the cell is not known. Protein phosphorylation in response to NGF has been observed and the involvement of a cAMP-dependent protein kinase has been postulated. Several substrates for this NGF-stimulated phosphorylation have been identified (among them tyrosine hydroxylase, the ribosomal protein S6, histone H1 and H3, and nonhistone nuclear proteins) but no causal relationship between these effects and the physiological function of NGF has been demonstrated.

NGF has other actions: it stimulates the uptake of uridine, the formation of polysomes, protein-, lipid-, and RNA synthesis, and the utilization of glucose. Such actions enable it to promote the growth and survival of sympathetic neurons and some sensory ones. Its ability to stimulate the growth of axons and dendrites is thought to be effected via its control of microtubule assembly. If a mouse is injected with antibodies against NGF, the sympathetic nervous system degenerates. Its role as a trophic factor is exemplified by its property of induction of tyrosine hydroxylase, the key enzyme of catecholamine synthesis.

NGF regulates differentiation, survival and target-oriented growth of nerve cells

NGF is not only a "survival factor" but also a "differentiation factor". This is clearly shown with pheochromocytoma (PC 12) cells. PC 12 is a tumor cell line derived from the (non-neuronal) chromaffin cells of the adrenal medulla. Unlike nerve cells they divide in cell culture and like the chromaffin cells they are capable of synthesizing, storing, and releasing catecholamines. In response to NGF they differentiate further in the direction of nerve cells: they stop proliferating and extend neuritic processes; they become electrically excitable and sensitive to acetylcholine; and they develop the capacity to form synaptic connections.

Besides survival and differentiation the third main field in which NGF is involved appears to be directed neuronal growth. *In vitro* it can be shown that neurites grow up an artificial NGF concentration gradient, a phenomenon resembling chemotaxis (see Chapter 12). A much favoured hypothesis for the physiological role of NGF is therefore at present the following: NGF is produced by peripheral target tissues and chemotactically directs growing axons to their proper target. The limited supply of the compound may even restrict the number of axons reaching this target.

Are there more nerve growth factors?

It is very unlikely that NGF is the only protein of its kind, and, in fact, there is plenty of evidence for further trophic factors [6]. Sensory nerve fibres for example have been shown to grow into their target tissue *in vitro* even when all NGF has been removed by NGF-antibodies. Furthermore a protein of M_r 12 300 has been isolated from pig brain, which supports the survival of cultured chick sensory neurons and the development of fibre outgrowth from them but which is clearly distinguished from NGF by its antigenic and functional properties. As a final example there is considerable indirect evidence for the existence of a motor neuron growth factor (MNGF) supporting somatic motor neurons. This may become the next important trophic factor to be characterized biochemically. However because of their very low concentrations in the respective tissues it will require recombinant DNA techniques to make them accessible for further investigations.

Synaptogenesis

The development of a synapse involves the formation of a postsynaptic specialization in the membrane of the target cell, the appearance of specific transmitter receptors and the invasion of the postsynaptic area by the ending ("growth cone") of the innervating axon. Undoubtedly these processes are coordinated. Certainly, as mentioned before, there are many separate differentiation programmes and not simply one "synapse gene" which controls the synthesis of all components of a synapse.

We shall now return to the postsynaptic side and so to the following question: when do the specific transmitter receptors appear during the development of an excitable cell? This has been particularly well investigated in primary cultures of muscle cells [2]. The development of polynuclear myotubes from myoblasts can be followed *in vitro* morphologically and biochemically. The process is calcium-dependent and proceeds in parallel with the appearance of acetylcholine receptors. However this parallelism does not imply a causal relationship: it is true that in the absence of Ca^{2+} fusion does not take place, but acetylcholine receptors are nevertheless formed and also attain the same concentration as in the presence of Ca^{2+}. This was established by experiments with a muscle tumor cell-line in which the cells had lost the ability to fuse and form myotubes. In these cells too acetylcholine receptors were synthesized after the termination of the logarithmic growth phase. Acetylcholine receptors and acetylcholinesterase seem to be regulated by a common mechanism. Changes in the activity of one always occur simultaneously with those in the other.

In the mature innervated muscle cell acetylcholine receptors are not distributed equally over the whole cell membrane, but are concentrated in the subsynaptic area of the endplate about a hundred-fold relative to the extrasynaptic area. In Chapter 9 we saw that this concentration is associated with the presence of nerves or electrical activity. Acetylcholine receptors are also not uniformly distributed in non-innervated myotubes, but form *clusters*. These can be seen optically by means of α-bungarotoxin labelled with a fluorescent dye or autoradiographically with radioactively labelled toxin. The receptors are practically immobilized in the clusters while the free receptors in the membrane are able to diffuse and move around. It would be an attractive hypothesis to suppose that the clusters form the target areas of the muscle cell surface for the ingrowing neurites but the reverse seems to be the case: the site where the nerve fibre first comes into contact with the muscle cell becomes an area of high receptor concentration. This has been shown for example in mixed cultures of spinal cord neurons and muscle cells from the toad *Xenopus*. Recently involvement of an *aggregation-promoting protein factor* has also been shown.

The question therefore arises of the role of the transmitter in synaptogenesis, or in the wider sense, the significance of the activity of the participating cell. If a muscle cell line is grown in the presence of agonists e.g. carbamylcholine, contrary to expectations the receptor concentration decreases (up to 50%). On the other hand the presence of agonists and antagonists does not affect the formation of functional synapses. If the acetylcholine receptors are blocked with α-bungarotoxin, synapses are still formed. While tetrodotoxin increases the concentration of receptors it has, like the electrical activity of the muscle cell, no influence on synaptogenesis. Acetylcholinesterase nevertheless does not appear in curare-blocked synapses and in electrically inactive muscle cells.

The investigation of the biochemistry of synaptogenesis is made more difficult because it has not yet been possible to demonstrate functional synaptogenesis between cloned cell-lines of nerve cells. One is dependent on the few homogeneous cell populations from primary cultures and thus from cells which have been obtained by dissociation from differentiated organs like various ganglia or the retina. Definite progress has resulted from the discovery that hybrid cells, obtained from the fusion of neuroblastoma with glioma cells are able to form chemical synapses [7]. This result also throws an interesting light on the possible helper function of glial cells for nerve function.

Up to now most of our statements on synaptogenesis have been negative: receptor synthesis is not dependent on the formation of myotubes, or on the presence of transmitter or Ca^{2+}; synaptic contact is not conditional on the presence of transmitter, electrical activity or functional receptors. None of the descriptive research done so far has provided a comprehensive solution either to the problem of the mechanism of formation, specificity and stabilization of synapses or to that of the subsequent formation of the neuronal network responsible for the higher functions of the nervous system. At the beginning of this chapter we have highlighted this question as one of those central to neurobiology. We shall therefore follow it a little further.

One of the most dramatic indications of the significance of interaction between cells in the formation of synapses is "neuronal cell death". At a specific point of time in embryonic development 50-80% of neurons die. This dying off corresponds to the moment when the axons arrive at their target area. This suggests that all neurons which do not form permanent synaptic contacts die off, and leads further to the hypothesis that there is a trophic factor which is transported retrogradely from the synapse to the soma and is necessary for the stabilization of the cell.

In mature mammalian skeletal muscle each muscle fibre is innervated by only one neuron. In contrast in newborn rats and cats there are many axons for one muscle fibre. During development therefore, the excess axons must have been removed while one "survived".

"Selective stabilization" of synapses, a plausible hypothesis

The above observations together with others have led to the hypothesis of "selective stabilization" as the mechanism for the formation of functional synapses and the neuronal network [8]. It postulates that first of all there is a certain "redundancy" in the formation of labile contacts only a few of which will be converted into stable synapses. The alternative hypothesis originating from Sperry assumes a specific chemical affinity between the axon endings and the postsynaptic cell which is genetically preprogrammed in advance of its utilization to result inevitably in a stable synapse.

Selective stabilization has the advantage over the preprogramming theory that far fewer genes would be required for the formation of 10^{14} synapses. Whole groups of neurons could be specified by one gene, and many such groups in their growth to their target organ could be controlled by one common regulatory mechanism. The fine detail of the connections of the neuronal network would be created by its activity which would thus constitute an epigenetic mechanism.

These alternative hypotheses may not be mutually exclusive. There is much evidence that a chemical affinity exists between pre- and postsynaptic cells [9].

As previously mentioned an axon can grow a considerable distance to its target, passing countless potential target-cells which remain unaffected by it. There are also two hypotheses for this directed growth which again are not mutually exclusive: either the axon is "led" on its way by microfilaments (but it is not clear how these would mark out such a specific course); or, according to Sperry, it grows against a chemical gradient sent out from the target, a specific signal comparable perhaps to that of chemotaxis. In any case the axon finds and recognizes its target, a process only imaginable as yet as being analogous with the receptor-ligand or antigen-antibody interaction. The interaction is not however permanent. Films of cell cultures have shown that growing neurites are in constant motion, growing out and retracting again, seemingly testing and probing the target cell surface before forming a permanent contact. The specificity of the interaction is also not absolute; if the target cells are damaged, synapses can be formed with other cell types. An example of this is provided by experiments with the cerebellum: afferent cerebellar mossy fibres normally form synapses with the dendrites of the granule cells; by selective damage of the latter they form functional synapses with the spines of the Purkinje cells (see also Chapter 12). The genetically determined chemical specificity is thus not rigid but flexible or plastic. This plasticity implies the existence of re-enforcement mechanisms, which fill the gap from the genetic side. The activity or utilization of the synapse could be such a mechanism. The importance of sensory input from nerve tracts for the construction of a functional nervous system has been shown, for instance, by the outstanding experiments of Hubel and Wiesel on the optic system of the cat.

The hypothesis of "selective stabilization" is attractive because not only does it solve through its gene-saving mechanism the dilemma mentioned at the beginning of this chapter – the discrepancy between the number of genes and the number of specific synapses – but it is at the same time a model for the learning process and for memory, for the conversion of a non-functional or labile synapse into a stable one is already a process of information storage. With the formulation of this hypothesis the application of neurochemistry to the problem of synapse formation really begins. A number of questions can be asked about the mechanism and molecular process involved: What is the molecular nature of the postulated gradient and of the cell surface molecules proposed for recognition and the specificity of the contact? What kinds of molecular changes transform a labile synapse into a stable one? By what physical or chemical processes are these triggered? The neuromuscular synapse serves as the best experimental model to answer the last two questions. We have already encountered a number of examples of the mutual interaction of nerve and muscle fibres and have also discussed briefly the molecular mechanisms involved (Chapter 9). We should also remember the evidence for the changes in the postsynaptic membrane following denervation i.e. the interruption of synaptic activity (hypersensitization, p. 219). Such experiments however do not really answer our questions. They are only evidence for a capacity for change (plasticity) of synaptic components. Synaptic plasticity, however, is not only interesting for the development of the nervous system but as outlined above also for the higher functions like learning and memory. We will deal with these themes in the following sections.

Plasticity

The term plasticity subsumes the ability of a cellular system to modify the rigid determinism of the gene. A system is plastic when its configuration is determined both by internal regulation and by external influences. It thus possesses sufficient flexibility to adapt to changing environmental conditions. The relationship of genetic and environmental factors is particularly significant in the development and function of the nervous system. For teachers and psychologists it is the basis of much philosophical discussion.

Protein structure and activity are dependent on the chemical environment

There is little support nowadays for the extreme position that everything is predetermined by the chromosome. Such a position could claim that DNA determines precisely the primary amino-acid sequence of a protein, that this primary structure determines the folding of the polypeptide chain thus characterizing the secondary and tertiary structure of the protein and finally its quaternary structure and its association to form higher structures (self-assembling systems). The activity of the protein would thus be genetically determined, and with it the form and function of the organism. Here however the vital element of plasticity comes in. Secondary, tertiary and quaternary structures are not only dependent on the primary structure, but on the surrounding "milieu" of pH, ionic strength, concentration of effector molecules etc. This milieu is directly affected by external influences (often mediated through nervous activity) and by the metabolism which in turn is influenced by the intake of nutrients and by hormonal and neuronal regulation.

There is a lot of evidence that genetic control over the development of the nervous system is limited. For example Levinthal showed that in genetically identical water fleas with a similar gross nervous system structure, the number and site of synaptic contacts and dendritic fine structure are different. The number and size of dendritic spines can be varied in experimental animals by behavioural training. The thickness of the rat cortex depends on the amount of input received from the animal's environment (we will return to this topic below). But the decisive evidence for the flexibility of our genetic programme lies in our own ability to learn, our potential to store things in our CNS which, since they are unanticipated in the course of evolution, cannot have been laid down in the chromosome.

Biochemical basis of learning

All the evidence indicates that learning does not consist of the storage of information in single molecules like the storage of genetic information in DNA molecules. The search for "memory molecules" has long been popular, as for example when a peptide named scotophobin (the word means a substance concerned with fear of the dark) was isolated from the brain of rats to which a fear of the dark was imparted by training. Although experiments of this kind were a false trail, they do not exclude the possibility that learning is associated with very specific molecular changes in the nervous system and can be ultimately explained biochemically.

Biochemical research into memory can be distinguished as progressing along two different paths: one is the search for a biochemical correlate of learning i.e. molecular changes which are associated with the learning process; the other is that of intervention research i.e. the method of selectively blocking or accelerating a molecular process and then investigating the effect of this on learning. Correlation and intervention methods are however limited in their descriptive power. Both give results averaged over many single chemical reactions and cells of the nervous system; they describe at best the activity "learning", but not the mechanism of storage of items of information. Also they can only distinguish, in a limited number of cases, if the observed process is really the storage of information or its mechanism of recall, since memory consists of at least three stages, the acquisition of information by the storage mechanism, the storage mechanism itself and the recall process.

Learning depends on the plasticity of the nervous system

Memory is investigated in the laboratory by using animals which are taught specific tasks. It is indicated by changes in their behaviour following teaching, and therefore we define memory as the storage of information which is demonstrated as learned behaviour. This biological behavioural definition assumes that there is no difference in principle between the complex memory performance of our brains and the simple conditioning reflexes of a lower organism [10-12] the term "in principle" referring to the unit events and the elementary molecular changes accompanying them in the participating nerve cells.

Thus we have already shown a bias for a certain theory of learning which has the advantage that it can be investigated with neurochemical methods [13]: namely that memory is a property of nerve cells which depends on the plasticity of the neuronal network. This hypothesis was originated by the Italians Tanzi and Lugaro at the end of the 19th century and was further developed by Ramón y Cajal. It is in contrast to the "dynamic" hypothesis of learning of Alexander Forbes and especially Lorente de Nó (1938). The latter postulated neuronal circuits in which nerve impulses circulate for an indefinite period (see Fig. 11.4). The plasticity hypothesis requires chemical changes which can lead to morphological modifications like the sprouting of neurites and the formation of synapses whereas the dynamic hypothesis demands electrical changes. The trace left behind by stored information, the so-called *engram* is therefore either chemical or electrical in nature.

Certain experiments have given results incompatible with the electrical nature of the engram. These have shown that when the nervous system is submitted to electric shocks or to cooling to a temperature at which there is no demonstrable nervous activity, learned information is not lost. These procedures however prevent learning i.e. the storage process itself. Thus it is now thought that there are at least two stages in memory-formation – a short-term memory and long-term memory – of which only the first is based on the temporary conservation of nerve impulse in neuronal circuits as shown in Fig. 11.4B. According to one of several theories advanced by Hebb, information if persistent or augmented, is transferred from the oscillatory electrical circuits which temporarily encode it into a chemical storage device involving molecular changes in the neuron.

Learning paradigms

There is a hierarchy of complexity in learning and memory which suggests both suitable experimental models and ways of explaining the processes involved in neurochemical terms.

1. Every *process of adaptation* is in the widest sense a learning process. In Chapter 12, the chemotaxis of bacteria is described as a model for perception and behaviour. Among other "intelligent" properties of these unicellular organisms is their ability to learn: by induction of a suitable enzyme a bacteria can "learn" to grow in a specific nutrient; by induction of the corresponding receptor a chemotactic bacterium can "learn" to swim into an area of high ribose concentration instead of, as previously, high galactose concentration. Although in this case there is no nervous system involved, the elucidation by Adler and by Koshland of signal perception and processing in chemotactic bacteria can almost be said to have provided a successful chapter in neurobiology!

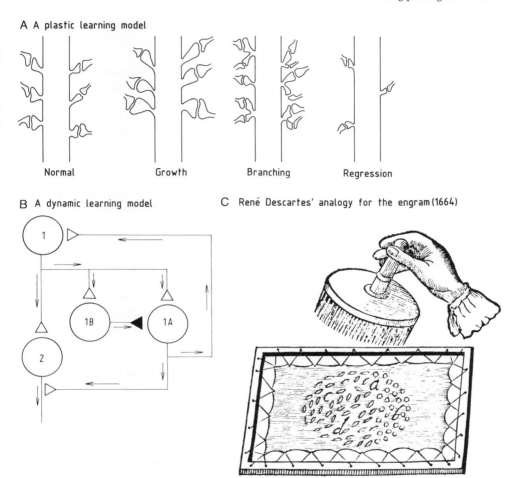

Fig. 11.4. Alternative learning models. (A) Plasticity of synaptic contacts occurring in growth and regression in response to intensity of use. The example is of dendritic spine synapses (after Eccles [14a]). (B) Storage of information by circulation of a nerve impulse in a closed nerve circuit. Cell 1 excites cell 2, 1B and 1A. Cell 1A excites in return cell 1 and thereby maintains the circulation of impulses (after Kandel [10]). (C) Descartes' model of learning: in the same way as an array of needles pierces holes in material, so information causes holes to open up in the membrane of the third cerebral ventricle (after Descartes, *Traité de l'Homme*, 1664).

2. A closer paradigm of the learning process is *habituation*. If a hammer is hit on a hard object next to a mouse it exhibits a flight reflex. If the stimulus occurs twice a day every 5 seconds for an hour, after 20 days the number of flight reflexes is reduced by 90% – the animal becomes accustomed to the noise. The reverse phenomenon, an increase in the number of reflex responses to a repeated stimulus is called *sensitization*. Both

phenomena can be demonstrated electrophysiologically in single nerve cells and are linked with *depression* and *potentiation*. Molecular mechanisms which could be associated with potentiation are discussed in detail below.

3. The third learning paradigm is *conditioning*. An ineffective stimulus provoking no response can produce one if it is combined with an unconditioned stimulus. In a dog the unconditioned stimulus can be food, the reflex response, the secretion of gastric juice. If the giving of food is associated for an extended period with the rattling of the food bowl, this acoustic stimulus, originally producing no response, can now stimulate secretion in the absence of the unconditioned stimulus. The dog has become conditioned and the flow of gastric juice in response to the acoustic stimulus is referred to as a "conditioned reflex". Here there must already be the cooperation of many nerve pathways, but the basic feature of the conditioned reflex can also be demonstrated in simple model systems and explained in terms of the properties of nerve cells and their impulse traffic. This has been done – particularly by E. Kandel and his group, using identified neurons in the ganglia of *Aplysia* [10] (see below and Chapter 12).

That higher memory function, for example the acquisition of language or abstract concepts, are only a higher level of complexity of the learning paradigms mentioned is as yet an unproven hypothesis. It is based among other things on the fact that in both lower and higher organisms, in invertebrates and vertebrates, a fundamental similarity is found in the properties of nerve cells, the mechanism of nerve conduction and transmission, modulation and plasticity. But it is also clear that the paradigms already described are inadequate for a complete description. In investigations of post-tetanic facilitation in the hippocampus (see below) all neuronal plasticity so far detected occurs in a period of seconds and minutes, occasionally of hours, whereas long-term memory persists for days, months and years, sometimes for the whole human life span.

The site of learning

From what has already been said it emerges that learning is localized in the nervous system and complex learning in the central nervous system. In the brain of higher organisms there seems however to be no special brain region in which information is stored, thus no memory organ in a narrow sense. Rather it seems that specialized information (visual, acoustic, sensory, motor etc.) is stored in regions of the cortex which also subserve the corresponding function. On the other hand memory in general seems to involve the cooperative interaction of relatively large areas of the cortex and other parts of the brain. This concept is supported by the work of Lashley in the thirties; he showed that the loss of memory resulting from ablation of the cortex was roughly proportional to the amount of tissue excised, but independent of the site of the operation. In contrast there are very many clinical observations which show that following damage to a part of the cortex through accident, tumors etc. another part of the brain can, after some training, take over the function of the affected part (*rehabilitation*). Thus if memory is limited to particular areas of the cerebral cortex, this localization is plastic, not rigid.

In addition to the neocortex, the hippocampus seems to have a special role. Lesions of this part of the limbic system do not cause the loss of previously learned information, but damage the storage of new information. Thus, the hippocampus does not seem to be a memory store as such, but rather an organ responsible for the consolidation of memory or the transfer of information from short-term to long-term memory.

In the search for the anatomical structures and mechanisms involved in memory one should bear in mind that it is almost impossible to distinguish the storage from the transfer and recall mechanisms. Loss of memory usually involves all three aspects and only in favourable cases can they be separated. The difference between *retrograde* and *anterograde amnesia* is shown by the following case. The ablation of the hippocampus which was performed on the now famous patient "H.M." as the treatment of last resort for his severe epileptic attacks, resulted in anterograde amnesia. By constant repetition, H.M. could retain for minutes specific facts in his short-term memory; however if he was distracted, he immediately forgot what he had learned, as the transfer into the long-term store did not function. Before the operation he could learn well; afterwards all he learned was immediately forgotten.

The reverse case is shown by the Korsakoff syndrome. Patients with this disease, triggered by severe poisoning, typhoid fever or alcoholism, suffer from retrograde amnesia. They progressively forget increasingly distant events but can still learn new facts. This suggests that in this type of amnesia it is not so much the storage of information that is affected, but rather the recall mechanism or the store itself.

Stored information must leave behind traces (engrams) in the cellular and molecular structure of the nervous system. The most likely site of such a trace is the synapse: as mentioned in Chapter 1, the quality of a piece of information is not encoded in the pattern of action potentials but mainly involves the network of synaptic couplings. Post-tetanic facilitation may therefore provide us with a correlate of the engram at the level of the synapse.

Post-tetanic facilitation, a synaptic "memory"?

If a nerve pathway is stimulated presynaptically for some time at a frequency of, for example, 40 Hz it is observed that sometimes minutes after this stimulation there is an increased postsynaptic response to a presynaptic impulse (Fig. 11.5). This phenomenon which was discovered by Larrabee and Bronk in 1947 is termed post-tetanic facilitation or *potentiation* (recently the term *long term potentiation* – LTP – is most commonly used) and provides a classical example of the plasticity of the nervous system at the cellular level. If the high frequency (tetanic) stimulation is applied postsynaptically, facilitation is not observed, thus it is not the action potential mechanism *per se* that is preconditioned by repeated activation.

If a cell has several afferents facilitation only occurs in those fibres which have been tetanically stimulated; the response of the cell to an impulse via the other afferent pathways remains unchanged. The basis of the facilitation is therefore to be found only

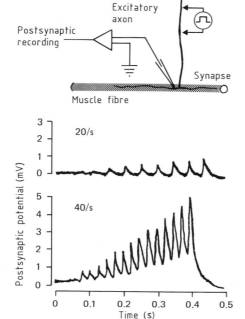

Fig. 11.5. Sensitization at a crustacean endplate by stimulation with 20 and 40 stimuli per second respectively. The height of the postsynaptic potential released steadily increases (after Dudel and Kuffler). Reproduced, with kind permission, from Dudel and the *Journal of Physiology* [18].

in those synapses which have been subjected to tetanic stimulation. Electrophysiologists refer to this as "homosynaptic facilitation". A more complete analysis has shown that the tetanic stimulation causes an increase in the number of quanta of transmitter liberated presynaptically in a given time i.e. in the frequency, but not the amplitude, of miniature endplate potentials. Transmitter applied from outside the synapse causes no facilitation, but on the contrary produces a desensitization of the postsynaptic receptors (see Chapters 8, 9). Facilitation thus apparently depends on a promotion of presynaptic transmitter release, but also occurs when the sodium channels are blocked by tetrodotoxin and the cells depolarized via electrodes. It is therefore not due to sodium ions penetrating into the nerve ending during the action potential, but to the raised intraterminal calcium concentration caused by Ca^{2+} flowing into the nerve ending during depolarization. Normally Ca^{2+} is immediately removed by the mechanisms discussed in Chapter 7, but if the impulses follow each other too rapidly these mechanisms cannot handle the quantity of inflowing calcium, and the resulting rise of calcium concentration favours the release of transmitter. Facilitation can thus be given a molecular basis; protein phosphorylation of components of the cytoskeleton, perhaps by protein kinase C which is Ca^{2+}/phospholipid dependent, may be involved.

One interesting hypothesis places the molecular events underlying LTP on the postsynaptic side: Lynch and Baudry observed an increase of glutamate receptors after tetanic stimulation of the hippocampus. The increase was not due to newly synthesized receptors but to the exposure of already existing ones "hidden" in the postsynaptic membrane. As mentioned in Chapter 9 these glutamate receptors become exposed through proteolysis of the brain spectrin *fodrin* which as a consequence of tetanic stimulation becomes hydrolysed by the Ca^{2+}-dependent thiol-protease *calpain* [14].

The opposite effect of a decrease in the postsynaptic response to a tetanic stimulus has been observed in certain nervous systems. Both, post-tetanic depression and facilitation show that the transmission of the impulse is not stereotyped but is affected by the previous history of synaptic activity. The synapse shows here a memory and many authors, above all Eccles [14a] consider this is the basis of the ability to learn in our central nervous system.

We shall now return again to the methods by which the engram or memory trace has been investigated.

Intervention experiments for the localization of memory

Inhibitors of DNA synthesis like arabinosylcytosine (see Fig. 11.6) do not prevent the storage of information. This is not as surprising as might be thought, for differentiated nerve cells no longer divide and have practically no DNA metabolism. Learning thus probably consists neither of the storage of learned information in DNA (which anyhow would be directly opposed to the central dogma of molecular biology) nor in the formation of new nerve cells. DNA synthesis and division of glial cells are also excluded by such intervention experiments. (In such experiments if an effect is observed the cytotoxicity of the intervening reagent must be taken into account.)

Inhibition of transcription and translation considerably affect learning in animal experiments. Hydén, in a series of classical experiments, has shown that the RNA content of nerve tissue increases markedly during learning and with the help of radioactive RNA precursors demonstrated that *de novo* synthesis of RNA accompanies training. Blockers of transcription produce amnesia, but here the cytotoxic effect is difficult to separate from a true amnesic effect. The increase in RNA during learning has been confirmed by other observations. Thus for example, the RNA content of the visual cortex of rats which have been reared in darkness is clearly less than that of others reared in the light. This indicates at least an increase in cell metabolism and in the size of active neurons, but may not have anything to do with actual information storage.

Without protein synthesis there is no long-term memory

Translation-blocking antibiotics (see Fig. 11.6) like puromycin, cycloheximide and acetoxycycloheximide inhibit learning ability. Most authors agree on this, but not on the interpretation of the finding. Inhibition of protein biosynthesis affects so many different enzymes, transport proteins etc. that amnesia could be a secondary result of the total metabolic inhibition. Cycloheximide inhibits learning at a concentration in the region of its

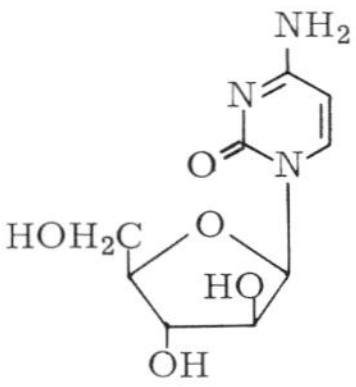

Fig. 11.6. Antibiotics. Blockers of DNA synthesis (arabinosylcytosin) have no effect, translation blockers (puromycin, cycloheximide) have a strong effect on learning and memory.

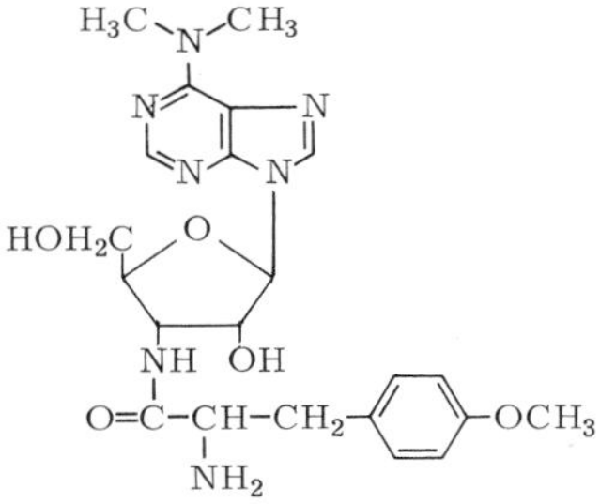

Arabinosyl-cytosine

Puromycin

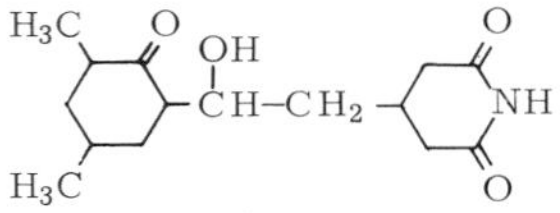

Cycloheximide

LD_{50} and for other antibiotics, for example puromycin, side effects are reported which could provide an equally valid explanation for the inhibition. Apart from the inhibition of protein synthesis puromycin increases the electrical activity of the hippocampus and it has been suggested that perhaps not the puromycin itself but a peptidyl-puromycin derivative interacts with adrenergic synapses. Puromycin causes mitochondrial swelling, stimulates glycogen degradation and resembles structurally (as an adenosine derivative) the second messenger cAMP. It is therefore not a very suitable reagent for investigations of the mechanism of learning.

Antibiotic research has however been useful in a quite different application, that of differentiating between short- and long-term memory [15]. Barondes has shown in a classic experiment that acetoxycycloheximide, in doses which block 90% of protein biosynthesis, does not fundamentally impair the learning ability of mice. However the learned material was forgotten again in a few hours (Fig. 11.7). Evidently the antibiotic prevented the consolidation of the learning into long-term memory, but not the intake and temporary storage into short-term memory. The latter as other authors have shown later, are blocked by ouabain, the inhibitor of Na^+,K^+-ATPase. Barondes showed that the time of the injection of the antibiotic in relation to the time of the training is critical. Inhibition of protein synthesis during or after training has little or no effect, but before training it prevents the extensive transfer of learned information into the long-term memory (Fig. 11.8).

We can thus conclude, with the reservations mentioned in the first section, that there are different stages and so different mechanisms of memory – a long-term memory dependent on protein biosynthesis and a short-term memory which has something to do with the ouabain-sensitive ion pump. Many authors postulate an additional stage, an electrically defined ultra-short term-memory.

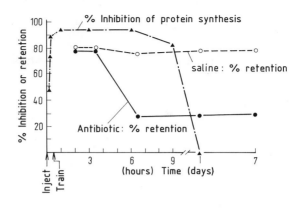

Fig. 11.7. Protein synthesis and learning. Immediately after the injection of the antibiotic acetoxycycloheximide into the brain of mice, protein synthesis was blocked by more than 90% (▲–▲–▲). The mice nevertheless learned their task, but forgot it after three hours (●--●). Control (○--○): With an injection of a salt solution instead of the antibiotic they retained the learning for days. Reproduced, with kind permission, from Barondes [15].

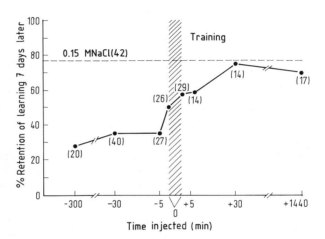

Fig. 11.8. Protein synthesis and learning. If protein synthesis is blocked during or after training the mice the effect on the memory is minimal. If it is blocked before training, almost everything is forgotten within seven days. Reproduced, with kind permission, from Barondes [15].

The search for memory-specific proteins: S-100 and others

The involvement of protein synthesis in long-term memory suggests a mechanism based either on the growth and remodelling of synapses or the synthesis of proteins which are specific for memory.

Hydén directed attention to the protein S-100, whose concentration was clearly increased during training in the specific area of the brain involved. S-100 has been known for a long time as a neuron-specific protein, but has meanwhile been found in even higher concentrations in glial cells [16]. Its name refers to its solubility in a 100% saturated solution of ammonium sulphate. It is a small (relative molecular mass 24 000), very acidic, weakly antigenic protein, which occurs relative to other tissues in about a 100 000-fold higher concentration in brain and is mainly present in the white matter. Particularly high concentrations are found in the cerebellum. It is also present in the peripheral nervous system. If a rat antiserum against S-100 is injected into the brain of an animal its learning ability decreases significantly (Fig. 11.9) [17].

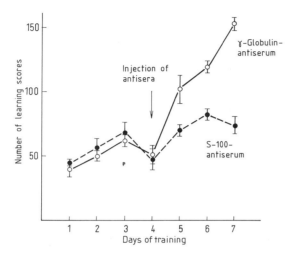

Fig. 11.9. Inhibition of learning by antiserum against S-100 protein. Two groups of rats during days of training learned increasingly frequently to grab the food with a particular foot. If one group was injected on the 4th day with an antiserum against S-100 protein these animals did not learn any more. The other group was injected with a control antiserum against γ-globulin of blood serum. This had no effect on the learning. Reproduced, with kind permission, from Hydén [17].

Analogous effects on long-term memory have also been observed with antisera against two further proteins (β, M_r 32 000; γ, M_r 26 000) [20]. During training these incorporate increased amounts of radioactive valine which implies that they have been more rapidly synthesized relative to other proteins than they would have been without the training. Antisera against synaptic membrane fragments and ganglioside G_{M1} block long-term memory, a further indication of the importance of the synapse for information storage. Of course the same reservations apply to these intervention experiments as those mentioned for the experiments with protein synthesis inhibitors.

General or specific effect?

It has been shown in a large range of experiments with rats that the environment has a clear influence on the growth of the brain [19]. In Fig. 11.10 a number of parameters are presented which significantly increase when rats are not kept isolated in laboratory cages in a stimulus-poor environment but are placed in a stimulating environment with "toys" and enriched surroundings. Brain mass and thickness of cortex, diameter of neuron cell bodies, number of glial cells, total protein, RNA and a range of enzyme activities were higher in the stimulatory environment. This should be considered when single parameters are being observed in relation to a learning process. Similarly morphological changes have been

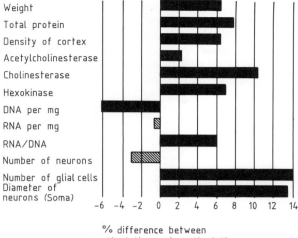

Fig. 11.10. Influence of environmental stimulus on some parameters of the rat brain. The table shows the percentage changes in rats in a stimulating environment compared with a control group in an unstimulating environment. Both striped bars represent data which were not statistically significant (after [19]).

described, like an increase in the number and size of dendritic spines and synaptic contacts, which only in occasional cases can be related specifically to the location in the brain and the training task. On the other hand there is no doubt that the efficiency of synaptic nerve impulse transmission is a quite fundamental factor in the storage of information. The most "specific" arguments for the synapse as the site of the engram are derived from experiments with primitive model animals like *Aplysia* where Kandel and his coworkers have investigated the changes in identified neurons during learning paradigms like habituation and sensitization [10].

Aplysia sensitization – a learning model completely described from behaviour to the molecular events

There is one model system where the mechanism of "learning" has been elucidated on all relevant levels, the whole animal (behaviour), the neuronal network (electrophysiology), and the molecular events (biochemistry): the mollusc *Aplysia californica*. [21]. This animal has a relatively simple central nervous system consisting of four paired and one single ganglion. It has also a relatively simple set of behavioural reflexes. One of these is the so-called "gill-withdrawal reflex", a defensive withdrawal of the animal's respiratory organ, the gill, in response to a tactile stimulus. This behaviour undergoes two simple forms of learning, habituation and sensitization, of which sensitization is probably based on what we have described above as facilitation; this will be discussed below in more detail. The neurons involved in the gill-withdrawal reflex are localized in one of the nine *Aplysia* ganglia, the abdominal ganglion. The *Aplysia* nervous system has, in addition to its simplicity, one further advantage: the unusual size of many of its neurons. They can be identified easily and reproducibly in every single specimen. As will be mentioned in Chapter 12 these *identified neurons* have been given code letters and numbers, and investigations by electrophysiological methods have yielded "wiring" diagrams and full descriptions of the circuitry of simple behaviour as, for example, the gill-withdrawal reflex.

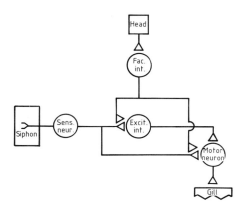

Fig. 11.11. Scheme of a neuronal circuitry in *Aplysia* mediating the "gill-withdrawal reflex". Excit. int., excitatory interneuron; fac. int., facilitating interneuron; sens. neur., sensory neuron. Some neurons in the fig. actually represent groups of equivalent neurons. For further explanations see text (according to Paris, C.G., Castelluci, V.F., Kandel, E.R., and Schwartz, J.H., in: Rosen, O.M., and Krebs, E.G. (eds.): *Cold Spring Harbor Conferences on Cell Proliferation.* Vol. 8, 1361–1375, 1983. Reproduced with kind permission).

The elegant set of experiments concerning this stems largely from the laboratory of E. Kandel and we shall follow his results and conclusions (Fig. 11.11). There are 6 motor neurons innervating the gill. They receive their excitatory input (via several interneurons) from 24 sensory neurons connected to sensory cells in the siphon (which is also part of the respiratory organ) receiving and transmitting the tactile stimulus. This circuit undergoes habituation upon prolonged repetitive stimulation. But here we shall focus on another circuit: when the tactile stimulus is applied not to the siphon but to the head of the organism, sensitization instead of habituation is observed. This sensitization is mediated by facilitating interneurons using serotonin as their transmitter.

As mentioned in Chapter 9 it is known that serotonin receptors are linked to an adenylate cyclase and serotonin stimulates cAMP production. According to the general scheme presented in Fig. 9.12 probably all intracellular effects of cAMP are mediated by cAMP-dependent protein kinases and it appeared logical to assume that the observed sensitization had something to do with phosphorylation of a key protein. Previous electrophysiological experiments had shown that facilitation was based on prolonged action potentials and increased transmitter secretion by the sensory neuron which was innervated by the facilitating interneuron. Exogenously applied serotonin or the K^+-channel blocker TEA had the same effect. Interestingly, injection of the catalytic subunit of the cAMP-dependent protein kinase into the sensory neuron stimulated facilitation and injection of protein kinase inhibitor abolished this effect. Kandel and his coworkers finally showed that serotonin as well as injected protein kinase caused, due to a decrease in K^+-conductance, a considerable prolongation of action potentials and this in its turn led to prolonged depolarization of the nerve ending and increased Ca^{2+}-inflow. A higher Ca^{2+}-concentration in the nerve ending produced an increase in transmitter release, i.e. a more efficient chemical transmission of nerve impulses from the sensory to the motor neurons. The molecular picture of *Aplysia* behaviour and learning was completed by the finding that phosphorylation of a single membrane protein of M_r 137 000, probably a potassium channel, is caused by serotonin via cAMP and the cAMP-dependent protein kinase. This molecular model of facilitation of the *Aplysia* gill-withdrawal reflex is shown in Fig. 11.12. Similar effects mediated by K^+-channel phosphorylation have been shown in a variety of cells; the specific nature of the K^+-channel involved may vary and in some cases the channels are activated, in others they are inhibited, but it appears to be a general scheme that protein phosphorylation can regulate the electrical activity of nerve (and muscle) cells and thereby modify behaviour.

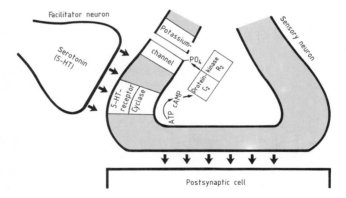

Fig. 11.12. Model of the molecular mechanism underlying presynaptic facilitation of the "gill-withdrawal reflex" in *Aplysia* (modified scheme from Paris, C.G., Castelluci, V. F., Kandel, E.R., and Schwartz, J.H., in: Rosen, O.M., and Krebs, E.G. (eds.): *Cold Spring Harbor Conferences on Cell Proliferation*. Vol. 8, p. 1373, 1983). For explanations see text.

The model (Fig. 11.12) is only a hypothesis (though a plausible one) and the observed parallelism between phosphorylation and change of behaviour is perhaps no more than a correlation, it may not represent a causal connection.

Do catecholamines, acetylcholine and pituitary hormones take part in learning?

Catecholaminergic antagonists interfere with learning and memory; agonists can stimulate memory function. Amphetamine and also adrenaline, noradrenaline and dopamine cause a favourable effect in behaviour training. On the other hand catecholamines are not essential and act rather as a modulator for learning and memory. The same can be said for acetylcholine. Anti-acetylcholinesterases (e.g. physostigmine) neutralize the learning inhibition caused by scopolamine. These investigations are however of limited significance as amphetamine, scopolamine etc. also affect conscious perception and this effect can not be separated from the actual process of learning and remembering.

ACTH and other pituitary hormones are likewise definitely involved in the learning process [13]. Hypophysectomy causes severe disturbance of learning which can be overcome by injections of ACTH. The question arises as to how a hormone with such a wide action can be specific enough for a special learning function? One would have to postulate that the hormone can act only on the nerve pathways which are directly active. In any case it seems that learning cannot be considered as a simple unified mechanism but a range of overlapping and interacting factors.

Summary

Specificity and plasticity are two basic properties of the nervous system which call for a molecular explanation. The genetic programme must lay down the differentiation of the neuron and the correct coupling of the neural network; however it must be flexible enough and the result must be sufficiently plastic to remain adaptable.

During embryonic development many decisions are made: neuron or glial cell; adrenergic, cholinergic or other transmitter-receptor system; development of a synapse or not. Neurochemistry is searching for the signals of differentiation for this. In autonomic ganglion cells it has been shown that the cholinergic-adrenergic "switch" from non-neural cells, is achieved externally via a trophic factor. Trophic factors are of great significance for the formation and survival of the nervous system. They can be proteins, Ca^{2+}, Na^+, K^+ or other ions, transmitters or hormones. The best researched is the nerve growth factor NGF, a protein which stimulates the growth of neurites in ganglia, induces enzyme activities (e.g. tyrosine hydroxylase) and is essential for the survival of the sympathetic nervous system.

Synaptogenesis has been well investigated in the formation of the neuromuscular synapse. It is independent of the fusion of myoblasts to form myotubes, is neither induced by transmitter, presence of active receptors or Ca^{2+}, nor by electrical activity of the cells involved. The mechanism is still not clear; hypotheses temporarily substitute for facts. A plausible hypothesis postulates a genetically specific growth of axon groups from which by selective stabilization one of the temporarily formed contacts with the target cell is fixed by activity i.e. by the use of the synapse. The path of the axon growth, sometimes many centimetres long, is marked, according to a hypothesis of Sperry, by means of a chemical gradient; according to others, by microfilaments. Special proteins like laminin or celladhesion molecules (N-CAM) may be involved.

Not much more is known about the plasticity of the mature nervous system, about habituation and conditioning and the learning and memory dependent on it. The synapse above all is discussed as the site of this plasticity. Learning is independent of DNA synthesis, but is accompanied by RNA and protein synthesis. Long-term and short-term memory is distinguished by blocking experiments with antibiotics: only the long-term requires protein synthesis. Both antisera against S-100 and certain other brain-specific proteins block the ability to learn in a similar way. Most protein synthesis, however, represents the general growth of a nerve cell or its synapses activated by learning and not specific "memory molecules". Of the transmitters, the catecholamines and acetylcholine (but not serotonin) have a particular relationship to learning. The pituitary hormone ACTH also has a special role.

References

Cited:

[1] Patterson, P.H., "Environmental determination of autonomic neurotransmitter functions", *Ann. Rev. Neurosci.* **1**, 1-17 (1978).
[2] Patrick, J., Heinemann S., and Schubert, D., "Biology of cultured nerve and muscle", *Ann. Rev. Neurosci.* **1**, 417-443 (1978).
[3] Thoenen, H., Schwab, M., and Otten, U., "Nerve growth factor as a mediator of information between effector organs and innervating neurons", *Symp. Soc. Dev. Biol.* **35**, 101-118 (1978).
[4] Spitzer, N.C., "Ion channels in development", *Ann. Rev. Neurosci.* **2**, 363-397 (1979).
[5] Yankner, B., and Shooter, E.M., "The biology and mechanism of action of Nerve Growth Factor", *Ann. Rev. Biochem.* **51**, 845-868 (1982).
[6] Berg, D.K., "New neuronal growth factors". *Ann. Rev. Neurosci.* **7**, 149-170 (1984).
[7] Hamprecht, B., "Neuronenmodelle", *Angew. Chem.* **88**, 211-213 (1976).
[8] Changeux, J.-P., and Danchin, A., "Selective stabilisation of developing synapses as a mechanism for the specification of neuronal networks", *Nature* **264**, 705-712 (1976).
[9] Gottlieb, D.I., and Glaser, L., "Cellular recognition during neural development", *Ann. Rev. Neurosci.* **3**, 303-318 (1980).
[10] Kandel, E.R.: *Cellular Basis of Behavior*. W.H. Freeman and Company, San Francisco 1976.
[11] Thompson, R.F., Berger, T.W., and Madden, J., "Cellular processes of learning and memory in the mammalian CNS", *Ann. Rev. Neurosci.* **6**, 447-491 (1983).
[12] Reinis, S., and Goldman: *The Chemistry of Behavior*. Plenum Press, New York & London 1982.
[13] Dunn, A., "Neurochemistry of learning and memory: An evaluation of recent data", *Ann. Rev. Psychol.* **31**, 343-390 (1980).
[14] Lynch, G., and Baudry, M., "The biochemistry of memory: a new and specific hypothesis", *Science* **224**, 1057–1063 (1984).
[14a] Eccles, J.C.: *The Understanding of the Brain*. 2nd edition. McGraw-Hill, New York 1977.

[15] Barondes, S.H.: "Multiple steps in the biology of memory". In: *The Neurosciences*. Schmitt, F.O. (ed.). The Rockefeller University Press, New York 1970.
[16] Bock, E., "Nervous system specific proteins", *J. Neurochem.* **30**, 7-14 (1978).
[17] Hydén, H., and Lange, P.W.: "Protein changes in nerve cells related to learning and conditioning". In: *The Neurosciences*. Schmitt, F.O. (ed.). The Rockefeller University Press, New York 1970.
[18] Dudel, J., and Kuffler, S.W., "Presynaptic inhibition at the crayfish neuromuscular junction", *J. Physiol.* **155**, p. 514 and p. 531 (1961).
[19] Rosenzweig, M.R., Bennet, E.L., and Diamond, M.C., "Brain changes in response to experience", *Sci. Am.* **226**, 22 (1972).
[20] Shashoua, V.E., "Brain protein metabolism and the acquisition of new patterns of behavior", *Proc. Natl. Acad. Sci. USA* **74**, 1743-1747 (1977).
[21] Paris, C.G., Castelluci, V.F., Kandel, E.R., and Schwartz, J.H., "Protein phosphorylation, presynaptic facilitation, and behavioral sensitization in Aplysia", *Cold Spring Harbor Symp. Quant. Biol.* **48**, 1361-1375 (1983).

Further reading:

Alberts, B., Bray, D., Lewis, J., Raff, M., Roberts, K., and Watson, J.D.: *Molecular Biology of the Cell*. Garland Publishing, Inc., New York and London 1983.
Spitzer, N.C.: *Neuronal Development*. Plenum Press, New York and London 1982.
"Molecular Neurobiology", *Cold Spring Harbor Symp. Quant. Biol.* **48**, 1983.
Edelman, G.M., "Cell adhesion and the molecular processes of morphogenesis", *Ann. Rev. Biochem.* **54**, 135–169 (1985).
Seifert, W. (ed.): *Neurobiology of the Hippocampus*. Academic Press, London 1983.
See also ref. 4, 10, 11, 14.

Chapter 12

Experimental Model Systems

Few would deny that the ultimate aim of neurochemistry is to contribute to our understanding of the human brain in health and disease. For obvious reasons, the opportunities for experimenting with it are extremely limited and so a range of experimental models for various aspects of neural functions must be utilized. Many examples of these have been encountered in preceding chapters. Some of the most important of them will now be reexamined and summarized in this final chapter.

Theoretical models such as cybernetic and mathematical interpretations of brain functions have been excluded. For some of the more biophysical experimental models like the black lipid membrane or the light-driven proton pump of halophilic bacteria the reader is referred to Chapters 3 and 7, respectively. Here some biological model systems will be presented which have certain properties - often in a hypertrophied form - in common with their prototypes, but may differ markedly from them in other ways. The use of an experimental model thus gives information on only one aspect of its prototype and the results must be interpreted with great care. The results from several experimental models can be combined together to produce a representation (though again only a model) of what is thought to be actually taking place. The history of neurobiology, like that of other sciences, is a history of models, proposed, discarded or refined.

Our understanding of the propagated action potential would not have progressed far without the squid giant axon (Fig. 12.1) [1,2]. Similarly, the neuromuscular junction (Fig. 12.2) is the classical experimental model for synaptic transmission [1,2]. The simple nervous systems of the leech (*Hirudo*) and the sea slug (*Aplysia*) (Fig. 12.3) have provided valuable experimental models for the physiology of behaviour [3] (Chapter 11). Specific patterns of behaviour in these animals e.g. swimming in the leech and the flight reflex of *Aplysia* have been traced to identified neurons and have been interpreted as activities of identified neuronal circuits. The identity of these neurons and circuits in many individuals proves that they are genetically determined. Properties such as habituation, sensitization and conditioning which modify the basic behavioural repertoire of *Aplysia* may be used as models for learning and memory and have been studied in the complete biological range from the whole animal through the cellular down to the molecular level (Chapter 11).

In developmental neurobiology a variety of experimental models are available using for example amphibian or chick whole embryos or tissue explants cultured *in vitro*. These have been particularly valuable in identifying factors determining the specificity of neuronal connections. This subject has also been very well investigated in the visual system

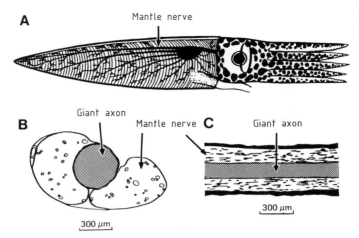

Fig. 12.1. The squid and its giant axon which has a diameter of up to 500 μm (after Eccles [14]). It has been a preferred material for electrophysiological investigations, in particular of the ionic basis of the resting potential and the action potential and its propagation (see also Chapter 5).

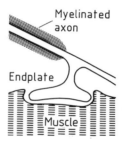

Fig. 12.2. The neuromuscular endplate. Model system for synaptic transmission.

of various organisms. Examples are the classical experiments of Sperry with the reimplantation of eyes after rotation, and the work on cultured *retinal ganglion cells* where the surface markers ensuring the correct interconnections were identified. In Chapter 1 there is another example in the account of Hubel and Wiesel's elegant analysis of the processing of visual information in a much more complex experimental system, the visual cortex of the kitten.

Model systems are being sought for the many diseases of the nervous system. We have already noted in Chapter 9 a very promising model for *myasthenia gravis*. Purified *acetylcholine receptor* injected into an experimental animal produces the typical symptoms, the basis of the disease is thus considered to be the immune response to this receptor protein. *Multiple sclerosis* can be imitated by the injection of the purified myelin basic protein; both diseases are now classified as *auto-immune diseases*. *Scrapie* (Chapter 4) may be a model for several neurological diseases e.g. *Creutzfeldt-Jacob disease* and possibly

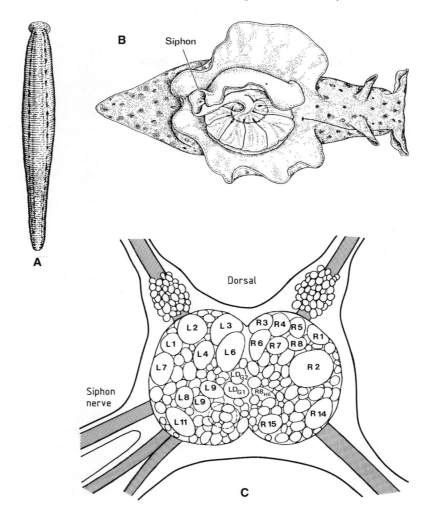

Fig. 12.3. Simple nervous systems with identified neurons. (A) Leech (*Hirudo*) and (B) Sea slug (*Aplysia*) are model systems for the relationship between behaviour and the circuitry of the nervous system. (C) Identified neurons in a ganglion from *Aplysia*. Numbered (and some other ganglion) cells can be recognised in individual specimens by their position and size. L before the number denotes left, R right (after Kandel).

Alzheimer's disease. In psychiatry, it is clearly impossible to find animal models for human consciousness, on the other hand they are vital for the investigation of neuropsychopharmacological drugs. The following is not a comprehensive account of model systems but a selection of a few, mainly those where biochemical analyses are available.

Chemotaxis

"What applies to *E. coli* applies also to elephants." It would be good if this old credo from molecular genetics could be transferred to neurobiology. Although this cannot be done certain analogies can be made between the nervous system of higher forms of life and the mechanism of environmental stimulation in microorganisms. This might make an interesting generalisation possible; there may exist a universal principle of signal transduction and information processing comparable to the universal genetic code. But including unicellular organisms into "neuroscience" would have also a more practical consequence – much simpler experiments could be performed using bacteria instead of nervous tissue. Microorganisms are particularly useful for genetic research as they have a shorter generation time than higher organisms, and mutants can be isolated whose varying behaviour properties can be traced to biochemical differences.

Unicellular organisms have no "nervous system", but information can nevertheless be received by them from their environment. They process information, integrate it and pass it on to molecular structures which produce the reactions or "behaviour" of the organism.

The property of chemotaxis in bacteria is a well investigated and very interesting behavioural phenomenon [4]. In the 1880's Engelmann and shortly after Pfeffer, had already observed that bacteria would swim towards a capillary containing a source of nutrient, thus receiving the information "nutrient" and translating it into action ("behaviour"). In contrast they move away from a toxic substance. This process called chemotaxis has been analysed and many functional elements of the nervous system have been detected in it - learning, memory, decision-making and the ability to form judgements.

Bacteria react chemotactically to a wide spectrum of substances certain molecules of which interact with the cell wall and thus affect the flagellae, the bacteria's organ of locomotion. They move "up" a gradient of a nutrient (or down a gradient of a repellent) by the regulation of their two characteristic phases of movement which is normally smooth forward progression for a while, followed by a stop and tumble and a smooth swimming again in a new direction. When the bacteria sense a gradient to be oriented in the direction of the movement chemotaxis occurs by simply prolonging the smooth swimming phase in this direction.

The sensing of a nutrient occurs through receptors. There are two types of receptor, one consisting of soluble proteins located in the periplasmic space, the other of integral proteins of the inner membrane. Several of them have been isolated and characterized biochemically. Maltose, ribose, and galactose bind to distinct soluble receptors called maltose-binding protein (MBP), ribose-binding protein (RBP), and galactose-binding protein (GBP), respectively. These soluble receptors have a relative molecular mass of 30 000. Aspartate binds to *tar*, serine to *tsr* - both membrane-bound receptors of relative molecular mass 60 000.

Although the receptors for chemotactic signals like galactose or ribose are evidently part of the membrane transport system of the sugar, the transport and metabolism of the molecule is not the cause of the chemical response. After binding to the ligand the receptor seems to undergo a conformational change. This regulates the action of the flagellae either by enzymatic or electrical effects. From the point of view of chemotaxis as a model for conduction and processing of external signals, it is particularly significant that the ribose receptor does not bind galactose; yet galactose inhibits the chemotactic response to ribose (and vice versa). The competition of the two carbohydrates does not apparently take place at the level of the receptor binding sites. From this evidence and that derived from mutants Koshland and others have constructed the following model (Fig. 12.4): various stimuli cause conformational changes in their respective receptors; many of these altered receptors compete at a "central processing system"; this results in a convergence of many stimuli at a few points which integrate the incoming signals; thus RBP and GBP compete for a membrane-bound receptor called *trg*. The situation resembles the receptor-cyclase system (Chapter 9) where there is also a convergence of signals on to a common transducer. The latter, the N-protein (G-protein), corresponds to the *trg* here.

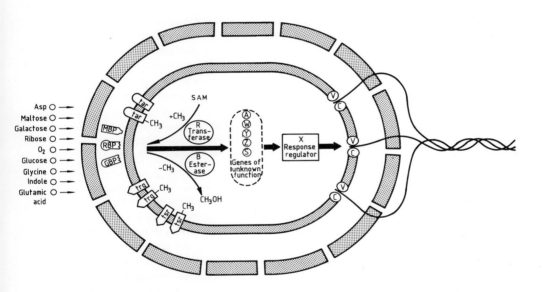

Fig. 12.4. Schematic summary of a bacterial sensing system. A variety of chemoeffectors (O_2, aspartate, glucose, etc.) can pass through pores in the outer membrane and react with receptors in the periplasmic space (GBP, galactose-binding protein; RBP, ribose-binding protein; MBP, maltose-binding protein) and with membrane-bound receptors (*tar* for aspartate, *tsr* for serine). The membrane-bound molecules are methylated by the *cheR* gene product, which utilizes S-adenosyl-methionine (SAM), and demethylated by the *cheB* gene product. The peptides formed by the A, S, W, Z, and Y genes are cytoplasmic and are known to be essential for the signal processing, but have no specifically identified function as yet. They, with R and B, are believed to control the level of the response regulator X. The *cheC* gene in the flagella is a signal detector that records the level of X and transmits the information to the flagella. The *cheV* gene also contributes to this information transfer in an essential but unknown way. Fig. and legend reproduced, with kind permission, from [4a].

This scheme is speculative but it is supported by a large number of genetic and biochemical experiments. Many binding proteins (receptors) have been isolated, the postulated conformational changes have been shown and the molecular identity of the "processing system" clarified. Adler and his coworkers showed that proteins of the bacterial membranes were methylated enzymatically, and that this methylation is an essential part of the chemotactic reaction. Thus in the absence of the methyl donor, methionine, the bacteria are certainly still motile, but their motility is no longer directed towards the nutrient. Methionine is activated by ATP to S-adenosyl methionine (SAM) (Fig. 12.5) and the activated methyl group is transferred to the carboxyl groups of glutamate residues of the membrane bound receptors. The reaction is catalyzed by a methyl transferase. Mutants lacking the enzyme swim smoothly and do not tumble. The methyl groups are removed by a methylesterase. Mutants lacking this enzyme tumble constantly. It has been observed that the methylation pattern of the membrane-bound proteins changes characteristically depending on whether the chemotactic stimulus is an attractant (nutrient) or a repellent.

The above scheme is far from complete. The receptors of many of the stimuli listed on the left side of Fig. 12.4 are not identified or purified, the identity of the gene products A, W, Y, Z, and S is not clear and the mechanism of the signal conversion into flagellum rotation is still under investigation.

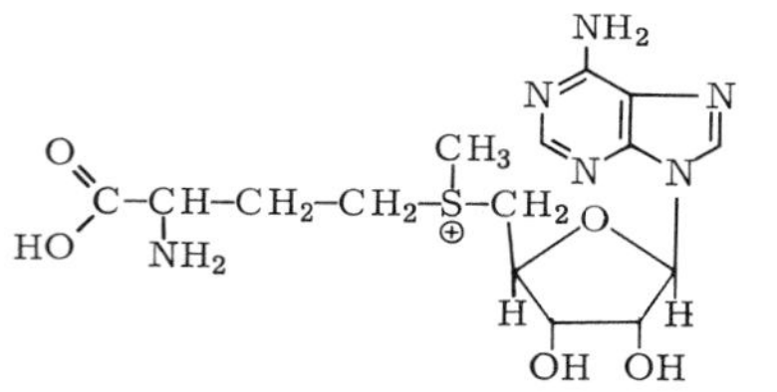

S-Adenosylmethionine (SAM)

Fig. 12.5. The biochemical correlate of signal processing in chemotactic bacteria is the methylation of receptors of the inner membrane. The methyl group transferred by methyltransferase is the active methyl group of S-adenosyl methionine (SAM). In higher organisms protein phosphorylation as well as methylation participate in signal processing at the molecular level.

For further information on this subject and that of flagellal motility specialist literature should be referred to. Here we will return to the analogy to the nervous system of higher animals. Chemotactic bacteria "perceive" (receptors), "distinguish" (different receptors for different stimuli) and "decide" (between opposing stimuli). They are even able to "learn", for example biosynthesis of ribose receptors can be induced by a suitable culture medium and the bacterium thus made sensitive to ribose. Koshland carried the analogy further, when he said that they possess instinct because, with respect to certain receptors, they are constitutive, i.e. they will always react in the same manner. They also have a "memory" for stimuli, whose time span is adapted to the duration of the movement; thus with constant stimulation by a raised but homogenous nutrient concentration, the movement reaction fades away i.e. the bacterium reacts to the signal "concentration increase", but "forgets" it after a certain time (habituation). The analogy between the model "chemotaxis of bacteria" and the mechanism of stimulation in nerve cells is continued at molecular level. Even the connecting link of methylation by means of SAM has been recently demonstrated in neuronal cell membranes.

Chemotaxis also occurs in nematodes [5] - worms with a nervous system consisting of only about 300 nerve cells. In this simply constructed system it is possible to investigate behaviour, again by developing mutants, and identifying their cellular and molecular basis. As evidence that universal principles of stimulus reception and signal processing originate at an early stage of evolution we will mention an example from the plant kingdom: chemotaxis is also shown by the gametes of brown algae [6], which recognize a sex-attractant in the seawater and swim towards it. In this connection one must also mention the slime-mould *Dictyostelium discoideum* whose colony building is regulated by cAMP released in the medium [6].

Behavioural model: Paramecium

Among the unicellular models we will proceed from the prokaryotes to the eukaryotes: *Paramecium* has many properties in common with a nerve cell and is used as a model for the following reasons [5]: it has a membrane which can be stimulated electrically (i.e. an action potential is released); it has a simple "behaviour"; behavioural mutants can be isolated and distinguished as membrane mutants and it can be cloned and cultivated in large quantities.

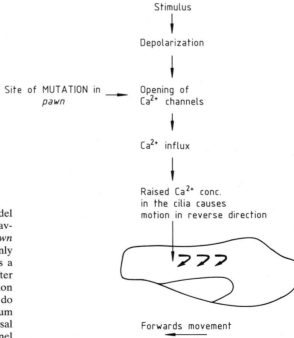

Fig. 12.6. *Paramecium* as a model system for the molecular basis of behaviour. In the behaviour mutant *pawn* which, like the chess piece, can only move forward, the molecular basis is a defect of the calcium channel. After excitation of the membrane by collision with an obstruction these channels do not open as in the wild type. Calcium influx is thus necessary for the reversal of ciliary movement. Other ion channel mutants have since been discovered.

An interesting mutant of *Paramecium* is called "*pawn*" from the chess piece. Whereas the wild type reverses the direction of its swimming when it meets an obstruction, pawn can only swim forwards (see Fig. 12.6). It is now known that the reversing movement is a result of an influx of Ca^{2+} following the disturbing stimulus. This calcium influx interacts with the moving mechanism of the cilia so that for a period these act in the opposite direction. In pawn the calcium channel is mutated and Ca^{2+} ions do not flow in. Another mutant changes swimming direction as in the wild type but in this case continues to swim backwards for minutes. This is due to a mutation of one of the other cell membrane ion channels, more of which have been found there than in the neurons.

Developmental biology model: Hydra

An example of the first life form in evolution to have a nervous system is *Hydra*, a small fresh water polyp. *Hydra* consists of only two cell layers, the ectoderm and the endoderm, and has only five types of cell including nerve cells. Because of this simple construction it is a suitable model system for the investigation of differentiation and development [7]. Molecules have been isolated which will differentiate head cells from undifferentiated cells and others which produce foot cells. "Head activator" and "foot activator" are small peptides found in the nerve cells (under certain conditions also in epithelial cells [8]) which may be early forms of neuropeptides. In addition the nerve cells contain inhibitors – non-peptides of lower molecular mass. These morphogenous substances seem to form a gradient across the organism and regulate the specificity of the various cell areas.

Drosophila

Neurobiologists, with some envy of the geneticists' classical model system of *E. coli*, have hoped for such a model system in the fruit fly *Drosophila*. Its name *Droso-phila* or "dew-loving" indicates a behavioural characteristic: it has a biological clock with a 24 hour rhythm, so it becomes active at dawn. There are point mutants with a 19 hour or 28 hour rhythm, and one without a rhythm at all which is active throughout the day. Many other point mutants have been isolated affecting movement, visual memory and sexual behaviour. More than 2000 species are known of which a hybrid of male and female is particularly interesting.

Point mutants are due to mutation of a single gene, and thus of a single protein, so that the complex behaviour can be traced back to individual proteins. Apart from the well-recorded genetics, *Drosophila* has the following advantages: it has a short generation time, is easy to breed, inexpensive to keep (small, needs little space), harmless, and possesses few but giant-sized chromosomes. An acetylcholine-receptor protein having useful neurochemical properties has already been isolated from it. Its neurons are too small to provide good material for electrophysiological research, but its muscle fibres are accessible for the investigation of neuromuscular synaptic transmission. One mutant, for example, when anaesthetized makes unusual rhythmic leg movements. This behaviour can be traced to a mutation of the voltage-dependent potassium channel, which normally repolarizes the motor neuron after an action potential causing the cessation of impulse transmission. Here too, as in the *pawn* mutant of *Paramecium*, a modification of an ion channel protein of the excitable membrane is the basis of the behavioural change.

Several *Drosophila* mutants are known with well defined memory deficits. With some of them the deficit can be traced back to lesions in cAMP metabolism, others appear to have altered levels of monoamines. With the genetics of *Drosophila* worked out nicely this organism is becoming increasingly important in memory research.

Mouse mutants: genetics as a method for analysis of motor behaviour

Important insights into the control of motor behaviour have been provided by mouse mutants [5,9] with names like *weaver* (wv), *reeler* (rl) *staggerer* (sg) and *nervous* (nr). They all refer to mutations in the circuitry of the cerebellum [9], which controls the animals' movement. They enable the investigation of the relationship between the genetic programme and the cellular environment in the formation of the neuronal network as not only can the exact site of the mutation be identified but also the changes in structure and synaptic circuitry of other cells of the functional unit (Fig. 12.7).

The cerebellum consists of only a few cell types which are connected in a stereotyped fashion. There are two afferent systems, the climbing and mossy fibres, and one efferent system, the Purkinje cells, characterized by their dendritic ramifications. The climbing fibres form direct synaptic contacts with the dendrites of the Purkinje cells, the mossy fibres indirect via the intermediate relay or granule cells.

Weaver mice have no granule cells, as these die early in ontogeny and do not migrate to their scheduled site in the cerebellar cortex. Although the mutation only affects the granule cells, changes also occur in other cells whose environment is altered by the absence of granule cells. Here the significance of the analysis of the behavioural mutants can be seen. The mossy fibres form synapses with the Purkinje cells as their usual target, the granule cells, are not available. The dendritic tree of the Purkinje cell changes its form and it is especially noteworthy that in *weaver* cerebellum it is not one but two climbing fibres which make synaptic contact with it. This change of activity is evidently responsible for the failure of selective stabilization (see Chapter 11) and of regression of redundant fibres. The result of all these changes in the cerebellar system of contacts is the abnormal motor behaviour of the *weaver* mouse.

Analogous causes have been found for other mutant effects. The gene products of the genes concerned with the mutations and their functions, however, have not yet been discovered. We have already mentioned the absence of the cerebellum specific protein P-400 in *staggerer* and *nervous* mice (see Chapter 10). The further significance of the analysis of behaviour mutants is that this kind of model system affords an insight into the genetic causes of "abnormal" behaviour in human diseases of the nervous system. We have already described several inherited diseases of the human nervous system (for example, the lipid storage diseases of Tay-Sachs type in Chapter 2) and we can already predict that a wider range of diseases, whose causes are not yet known (schizophrenia, depression etc.) will have genetic components.

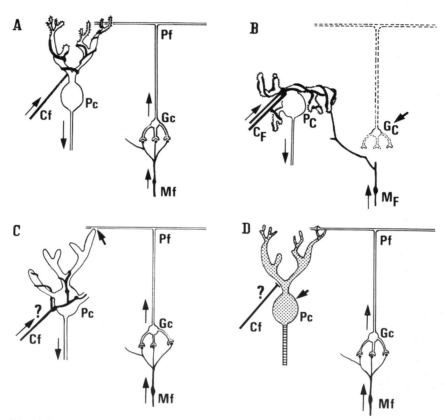

Fig. 12.7. Mouse mutants as model systems for behavioural dysfunction resulting from developmental defects. Four mutations, affecting the neuronal network of the cerebellum and leading to changes in the motor behaviour: Pc, Purkinje cell; Gc, granule cell; Mf, mossy fibre (afferent); Cf, climbing fibre (afferent); Pf, parallel fibre (afferent in relation to Purkinje cell). (A) Normal cerebellar cortex: the three afferents project on to the Purkinje cell. The integrated impulses are conducted through only one efferent. (B) *Weaver* mutant: the mossy fibre projects directly on to the Purkinje cell, as the granule cells have died during development. This leads to a change in the Purkinje cell and its innervation by climbing fibres, evidence that a point mutation (which affects granule cells) by changing the milieu affects genetically normal neurons and synapses. (C) *Staggerer* mutant: here the site of the mutation is the Purkinje cell, which forms no spines on the dendrites. (D) *Nervous* mutant: after normal development the Purkinje cells degenerate. Until now only one gene product concerned with the mutation has been identified, the protein P_{400}, whose function however is as yet unknown. Reproduced, with kind permission, from Sotelo and Elsevier, North Holland [9].

Electroplaques of electric fish: synapse model

The degree to which important scientific progress is dependent on model systems can be shown by the transmitter-receptor problem. "Receptor" was for a long time a functional description without a molecular basis. One of the successes in neurochemistry has been the isolation and chemical characterization of a receptor protein (see Chapter 9). It was achieved chiefly by the choice of an ideal starting material for a model of synaptic transmission of the nerve impulse: the electric tissue of the electric eel (*Electrophorus electricus*) and the various species of electric ray (*Torpedo*) (Fig. 12.8, 12.9).

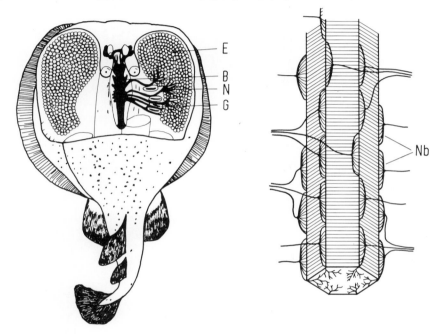

Fig. 12.8. Electric organ of electric ray (*Torpedo*). Model system for the biochemical investigation of synaptic function. From this tissue the nicotinic acetylcholine receptor was the first receptor protein to be isolated in bulk and thus became available for biochemical characterization. Left: *Torpedo* with exposed electric organ (E), nerve (N), brain (B), gill (G). Right: A stack of six-sided electroplaque cells; nerve bundle (Nb), which innervates the cells from the ventral side and synchronizes the discharge.

The electric organ of these fish consists of stacks of cells called electroplaques, which simulate batteries in series. Their voltage across the cell membrane (-70 mV inside negative) can be decreased by a nerve impulse, which, as with the neuromuscular synapse, opens Na^+/K^+ channels of the electroplaque membrane. Similarly the transmitter is acetylcholine and the postsynaptic receptor therefore an acetylcholine receptor. Ontogene-

302   12 *Experimental Model Systems*

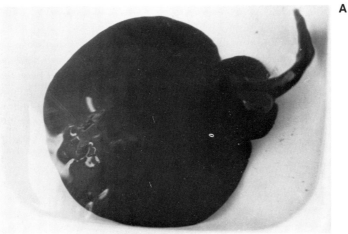

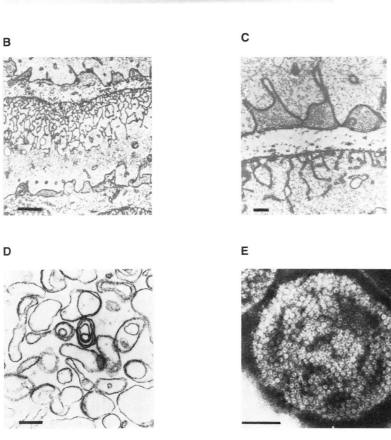

tically the electroplaque is derived from the muscle cell and although it no longer has a contractile mechanism, it is similar in many respects. An electroplaque cell [13] is a 10 x 10 x 0.01 mm large syncitium originating from numerous cells, and contains within it more than 1000 cell nuclei. The potential resulting from 400 to 1000 stacked single cells of the *Torpedo* electric organ in series reaches 40 V in air. Some hundred of these cell columns, connected in parallel, release a temporary electrical current of up to 50 A and can be used as a weapon (a kind of electric club!) for the paralysis of a prey. The electroplaques of eels are not stacked ventro-dorsally but from head to tail. They are 3 x 2 x 0.5 mm in size. Every 5000 form a column and up to 60 columns are connected to form a battery of 600 V and 0.5 A. 2% of the innervated cell surface of the electric eel, and 50% of that of the *Torpedo* consists of synapses.

Thus the electric organ is ideal for neurochemical research because it consists of a large number of similar cells innervated by only one type of synapse. Particularly in the salt water fish *Torpedo* the innervation is so dense that the innervated membrane (only the ventral side of the cell is innervated) consists predominantly of receptor protein. *Torpedo* receptor has been isolated in quantities of 100 mg. The electric eel also contains a good starting material for receptor purification. In this fresh water fish the discharging of the electroplaques is not via the direct depolarizing action of the transmitter as in *Torpedo*, but mainly by action potentials released. The electric tissue of the eel therefore serves also to characterize ion channels which generate action potentials (*Torpedo* electroplaques do not generate action potentials).

Nerve-muscle transmission has been investigated in electric fish at four levels: in the intact organ, in the single electroplaque, in isolated postsynaptic membrane fragments and synaptosomes and in purified receptor protein (see Chapter 9), in the work of the groups of

← Fig. 12.9. "Receptor model" *Torpedo*: From animal to molecule. Electric tissue from electric fish, originally introduced as a rich source of acetylcholinesterase and proved to be a model system for the investigation of neuromuscular transmission. (A) *Torpedo californica*. (B) Cross-section through electric tissue. The central part of the picture is occupied by a single cell, an electroplaque, also referred to as an electrocyte, which is innervated on the ventral surface (below) and whose dorsal surface (above) has many channel-like invaginations of its plasma membrane. While the plasma membrane of the innervated surface is densely packed with nicotinic acetylcholine receptors, the dorsal side contains the ATPase necessary for the removal of the cations entering the cell during excitation (calibration bar 5 µm). (C) The same as (B) at higher magnification. The upper half of the picture shows the ventral side of one electroplaque with the nerve endings (recognizable by their synaptic vesicles containing the transmitter acetylcholine). The lower half shows the dorsal side of the next electroplaque with the ATPase containing invaginations. The light area between is the intercellular space containing some connective tissue (calibration bar 1 µm). (D) Thin section of a membrane preparation containing receptor-rich membranes originating from the innervated side of electroplaques. These membranes form tightly sealed vesicles ("microsacs") on tissue homogenization which resemble the original postsynaptic membrane in their ability to let cations pass only after excitation by an agonist (calibration bar 500 nm). (E) High magnification image of the surface of a receptor-rich membrane vesicle as shown in (D). At this magnification the single receptor molecules become visible. (calibration bar 50 nm). The first isolation of a pure receptor protein from electric tissue was accomplished in 1972 by several laboratories. The electron micrographs (B)-(D) are reproduced, with permission, from S. Reinhardt, *Diplom-thesis*, Berlin 1983, (E) from Schiebler and Hucho, *Eur. J. Biochem.* **85**, 55-63 (1978).

J.-P. Changeux in France, Whittaker in England and Germany, and of Raftery, Karlin and others in the USA who correlated these four levels. The membrane and protein level, i.e. the biochemical part of this success story has already been covered in Chapters 8 and 9.

E. coli of neurobiology: cell cultures

Cultured cells have in common with bacterial cultures the fact that they represent a single cell type. The bacterial culture of microbiology corresponds to the cell line (the clone) of cell biology. Today cell biology plays an ever increasing role as a definitive, reproducible and usually easy-to-handle model system for the investigation of special functions of eukaryotic cells [10–12]. Among many advantages we will mention two which give the basis for the preferred use of cell culture to that of whole animals or their organs. Firstly, like bacteria, some are capable of proliferating so readily that special and rare cell types become available for biochemical investigation. Secondly, they are rapidly available, i.e. an interesting metabolic or developmental stage of a cell type can be fixed and analysed more efficiently in cell culture than if the cell has first to be isolated laboriously from a living animal or organ.

Cells can be cultured, either by dissociation of an organ by means of perfusion with collagenase for instance, or by cloning i.e. the propagation of a single cell. The first method has not yet led to many satisfactory results. In contrast the propagation of cell lines from single nerve cells has been very successful [10]. These and tumours are derived only from cells in an early stage of development since adult differentiated neurons do not divide. In particular the neuroblastoma cell of a single mouse tumour (C 1300) discovered in 1940 and further propagated as a transplantation tumour, was cloned in 1969 and has since become established as a classical model system.

Neuroblastoma cells have important properties typical of nerve cells: they are electrically and chemically excitable, and so have ion channels and receptors; they have the enzymatic apparatus for the synthesis of transmitters, but also the system necessary for their inactivation, i.e. inactivating enzymes like acetylcholinesterase or transport systems for the uptake of the released catecholamines; they form projections similar to nerve fibres. As they grow and divide in culture it has been possible to investigate the biogenesis of these properties and the point of time and regulation of their appearance in the cell cycle (see Chapter 11). Studies have been made of the differentiation of the nerve cell up to the stage when division has ceased; the signals triggering differentiation; the interaction with other cells – glial and muscle cells; synapse formation; the influence of hormones, drugs and also pathogenic bacteria and viruses; problems of nutrition; and the changes due to prolonged stimulation. Some of these applications have already been mentioned in connection with the origin and specificity of synaptic contacts (Chapters 8, 11), the development of excitability (Chapter 11) and the problem of opiates and their properties of tolerance and addiction (Chapter 9). For further information the extensive specialist literature should be consulted.

One other especially useful cell line deserves to be mentioned, the PC 12 cell line cloned from a pheochromocytoma, a tumour of the adrenal medulla. PC 12 cells resemble chromaffin cells in their capacity to synthesize, store and release catecholamines. Like non-neuronal cells they proliferate, but, in response to NGF in the culture medium, they stop dividing, extend neuritic processes and become very much like sympathetic neurons. They develop electrical excitability, respond to acetylcholine and even form functional cholinergic synapses. PC 12 cells are used as model systems to investigate neuronal differentiation, hormonal and trophic factor effects and both the function and metabolism of the hormone receptor (see also the section on NGF, p. 268).

Genetic engineering revolutionizes neurochemistry

This section of my introduction to neurochemistry – and thereby the whole book – would be incomplete if the tremendous potential and some of the early successes of cloning techniques were not outlined briefly. The space occupied by this subject in this text should by no means be taken as a measure of its importance for neurochemistry. On the contrary it is almost inversely related to the role that molecular genetics, recombinant DNA strategies and cloning techniqes (molecular and cellular) are likely to play in the future. The shortness of this section is due to the size of the field which really should have coverage in an extra volume of a neurochemistry textbook. Until such a book is available I have to refer the reader to treatises on molecular biology.

The advent of molecular genetics in the neurosciences has been compared to the introduction of electronics and the electron microscope. This is no exaggeration. Let me briefly outline why this is so. One of the most active fields in neurochemistry was in recent years the identification, isolation and biochemical characterization of neuropeptides from mammalian brains. However one rat hypothalamus for example contains only femtomol quantities of, growth factor-releasing factor (GRF) or corticotropin-releasing factor (CRF). The latter has been recently purified from 60 000 rat hypothalami and sequenced by the most advanced microsequencing techniques. To investigate many of the thousands of known and as yet unknown brain peptides and proteins in this way would be an almost hopeless task. But with recombinant DNA strategies it becomes a real possibility.

In brief, one needs take only a few micrograms of mRNA, the whole mixture present in a tissue producing the required protein. From this mixture of many different mRNA-molecules the one carrying the information for the protein under investigation can be selected or partially purified. A copy-DNA (cDNA) of the various messengers is produced with the help of reverse transcriptase; the cDNA molecules are inserted into a vector, a plasmid, by means of a series of enzymatic steps which are complicated, but already well enough established to be executed in every experienced molecular biology laboratory. The plasmid – actually a population of plasmids each containing one special mRNA inserted – is incorporated into bacteria, a different plasmid (i.e. a different cDNA complementary to a different mRNA) into each bacterium. Now the actual "cloning" takes place: The offspring of one individual bacterium is a clone, and the clone producing the wanted protein has to be selected. This is the most tedious and difficult part, tedious because dozens or hundreds of clones have to be tested; and difficult because if the translation product of the cDNA in the plasmid is not an enzyme or hormone with an activity which is easy to test a more complicated assay may have to be developed. Another difficulty in identifying the correct translation product may arise when the translation product is inactive because it needs posttranslational processing which cannot be performed by the bacterium.

In this case one has to be content with the transcription product, the cDNA. It can be produced now in large quantities by growing the clone and it can be sequenced by the methods of Maxam and Gilbert, or that of Sanger. From the nucleotide sequence of the cDNA the amino acid sequence can be deduced. Of course the primary structure of the wanted protein obtained in this way may be a precursor and it represents just the naked amino acid sequence, without possible carbohydrate, acetyl, methyl or phosphate groups which may be necessary for activity. If the gene coding for the wanted protein is cloned directly (not via the mRNA/cDNA detour) it may even contain intervening sequences (introns) missing in the final gene product.

We have seen in previous sections of this book some applications of recombinant DNA techniques in neurochemistry. One of the most impressive success stories is the elucidation of the origin of the opioid peptides, the identification of three precursors of the enkephalins, of dynorphin and a number of peptide hormones (Chapter 8). Another example is the cloning of the four different polypeptide chains of the nicotinic acetylcholine receptor (Chapter 9). Their sequence as deduced from the respective cDNAs is now known, and active receptor has been expressed from cloned cDNA in oocytes from the *Xenopus* toad, another model system which has proved useful for the neuroscientist.

Hybridoma cloning for monoclonal antibodies

Although it is not truly a neuronal model system another cloned system should be mentioned here because it has also developed into a very useful tool for neurochemistry: the hybridoma cell lines producing *monoclonal antibodies*. Antibody-secreting B-lymphocytes in general secrete only one type of antibody each. The mixture of the many monospecific antibodies forms the normal heterogeneous antiserum. To obtain monospecific antibody-secreting lymphocytes of high productivity Köhler and Milstein fused B cells from an immunized mouse with cells from a B-lymphocyte tumour. Unlike normal lymphocytes the resulting hybrid cells grow and proliferate virtually indefinitely and produce a mixture of antibodies directed against the antigen originally used for immunization. Even if this antigen was a pure protein the antibodies obtained are still a mixture of many individual molecular species each directed against one specific antigenic region of the antigen molecule. To obtain a monospecific serum, i.e. a solution of antibodies directed against only one antigenic region and stemming from one species of hybridoma cells these cells have to be selected and "cloned". The clone now produces "monoclonal antibodies", a homogeneous population of antibodies directed against just one determinant of the antigen. These monoclonal antibodies can be used for a variety of investigations, for example for the mapping of functional areas of a biomolecule. But most important the technique can be used for producing antibodies against a protein which is not pure – out of the mixture of lymphocytes directed against the impure protein fraction the one hybridoma specific for the required protein can be selected and cloned. This monoclonal antibody in turn can be used to isolate the specific mRNA for the protein: the nascent protein protruding from the polysome involved in its translation can be precipitated with the antibodies. The precipitate contains, besides the protein and the ribosome, the mRNA to which the ribosomes were attached. The mRNA finally can be used for "cloning" by the cDNA technique described above.

References

Cited:

[1] Katz, B.: *Nerve, Muscle and Synapse*. McGraw-Hill, Inc., New York 1966.
[2] Kuffler, S.W., Nicholls, J.G., and Martin, A.R.: *From Neuron to Brain*. 2nd edition. Sinauer Assoc., Inc. Publishers, Sunderland 1984.
[3] Kandel, E.R.: *Cellular Basis of Behavior*. W.H. Freeman and Company, San Francisco 1976.
[4] Koshland, D.E.: "Bacterial chemotaxis in relation to neurobiology". *Ann. Rev. Neurosci.* **3**, 43-75 (1980).
[4a] Koshland, D.E.: "The bacterium as a model neuron", *TINS* **6**, 133-137 (1983).
[5] Ehrman, L., and Parsons, P.A.: *The Genetics of Behavior*. Sinauer Assoc., Inc. Publishers, Sunderland 1976.
[6] Jaenicke, L. (ed.): *Biochemistry of Sensory Function*. Springer-Verlag Berlin, Heidelberg, New York 1974.
[7] Gierer, A.: "Biological features and physical concepts of pattern formation exemplified by Hydra", *Curr. Top. Develop. Biol.* **11**, 17-59 (1977).
[8] Schaller, H.C., Rau, T., and Bode, H.: "Epithelial cells in nerve-free hydra produce morphogenetic substances", *Nature* **283**, 589-590 (1980).
[9] Sotelo, C.: "Mutant mice and the formation of cerebellar circuitry", *TINS* **3**, 33-36 (1980).
[10] Haffke, S., and Seeds, N.W.: "The E. coli of neurobiology?" *Life Sci.* **16**, 1649-1658 (1975).
[11] Hamprecht, B.: Glaser, T., Reiser, G., Bayer, E., Propst, F., and Hallermayer, K., "Culture and characterisation of hormone-responsive neuroblastoma x glioma hybrid cells", *Meth. Enzymol.* **109**, 316–341 (1985).
[12] Patrick, J., Heinemann, S., and Schubert, D.: "Biology of cultured nerve and muscle", *Ann. Rev. Neurosci.* **1**, 417-443 (1978).
[13] Changeux, J.-P.: "The acetylcholine receptor: an "allosteric" membrane protein", *Harvey Lect.* **75**, 85-254 (1981).
[14] Eccles, J.C.: *The Understanding of the Brain*. 2nd edition. McGraw-Hill Inc., New York 1977.

Further reading

Spitzer, N.C.: *Neuronal Development*. Plenum Press. New York and London 1982.
"Molecular Neurobiology", *Cold Spring Harbor Symp. Quant. Biol.* **48**, 1983.
Alberts, B., Bray, D., Lewis, J., Raff, M., Roberts, K., and Watson J.D.: *Molecular Biology of the cell*. Garland Publishing Inc., New York and London 1983.
Mc Kay R.D.G., "Molecular approaches to the nervous system", *Ann. Rev. Neurosci.* **6**, 527–546 (1983).
Dudai, Y., "Genes, enzymes and learning in *Drosophila*", *TINS* **8**, 18–21 (1985).
See also ref. 3,4, 5, 12.

Index

absorption spectrum
– rhodopsin of cones   4 f
– rhodopsin of rods   4 f
acetoxycycloheximide   281 ff
acetylcholine   32, 99, 160, 162 f, 181
– and axonal transport   253 f, 255 f
– and cGMP system   228
– and learning   288
– and transmitter differentiation   265
– formula   171 f, 179
– in synaptic vesicles   303
acetylcholine receptor
– muscarinic   170 ff, 181, 222
– muscarinic and nicotinic   186
– nicotinic   62, 68, 100, 170 ff, 181, 208, 214, 266, 298, 303
– and clusters   271
– and myasthenia gravis   220 f, 292
– in *Aplysia* neuron   212
– in synaptogenesis   271 f
– phosphorylation   220
acetylcholinesterase   133, 167, 172 ff
– and synaptogenesis   271 f
– anionic site   174
– binding sites
–– competitive inhibitors   174
–– uncompetitive inhibitors   174
– esteratic site   174
– in fast axonal transport   254
– in myasthenia gravis   221
– in synaptic vesicles   167
– inactivation   173
– reactivation   173
acetylcoenzyme A (acetyl-CoA)   164 ff
ACTH (adrenocorticotropic hormone) (corticotropin)   180, 197 f, 230, 288
actin   134
– in slow axonal transport   254 ff

action potential   20, 95 ff, 105
– analysis   97
– development of   266
– effect of calcium   148
– optical measurements   135
– propagation   98
– threshold   148
active ion transport   18, 95, 139 ff
– models   145
– summary   154
actomyosin   168
adaptation
– as learning paradigm   276
addiction   240 f
adenosine receptor   230
adenylate cyclase   10, 185, 199, 213, 213, 229, 286
– and β adrenergic receptors   224 f
– and opiates   241
– calcium regulation by calmodulin   260
Adler   276, 296
ADP-ribosylation
– of G-protein   231
adrenal cortex   198
adrenal gland   183
adrenaline (epinephrine)   178, 182 f
– and adrenergic receptors   223 f
– and cAMP system   228
– formula   179
adrenergic neurons
– axonal transport   256
adrenergic receptors   208
– α_1-, α_2-, and β_1-, β_2 223
α-adrenergic receptors   229
β-adrenergic receptors   10, 218, 246
adrenocorticotropic hormone (ACTH) (see ACTH)
ADTN (6,7(OH)$_2$-2-aminotetralin)   233
afferents   15
affinity labelling   210 f
afterpotential
– of action potential   97
aggregation-promoting protein factor   272
agonists   169 ff
Ahlquist   223
β-alanine   192, 243
allosteric enzymes   206
alprenolol   186
Alzheimer's disease   293
amacrine cells
– of retina   12
amino acid transport   213
amino acids
– as transmitters   192 ff

γ-aminobutyric acid (see GABA)
aminopyridines 98, 130
amnesia
– anterograde 279
– retrograde 279
amphetamines 189, 236
– and learning 288
amphiphilic
– membrane protein 62 f
amplification (light transduction) 9
anaesthesia 125 ff, 242
– general 127
– local 127
anaesthetics 125 ff, 157
anemonetoxin I, II, III (ATX I, II, III) 123 ff
angiotensin 32, 180
ANS 119
antagonists 169 ff
– noncompetitive 175
anti-acetylcholinesterases
– and learning 288
antibiotics
– translation-blocking 281 f
antidote 173 f
antisera
– against membrane ganglioside G_{M1}
– – and long-term memory 284
– against synaptic membrane fragments
Aplysia californica (sea slug) 22
– and acetylcholine receptors 212
– and actions of acetylcholine 100
– and conditioned reflex 278
– and identified neurons 293
– and neural plasticity 246
– and synaptic transmission 167
– and transmitter differentiation 265
– and transmitter function 181
– as experimental model 291
– sensitization (learning model) 285 ff
apomorphine 233
arabinosylcytosine 281
arachidonic acid 32 f
aromatic amino acid carboxylase 184
aromatic amino acid decarboxylase 190
aspartate 179, 195, 244
– binding to *tar* 294
astrocytes 21, 35
asymmetry
– of biological membranes 60 ff
ATP (adenosine triphosphate) 139, 149, 179
– synthesis 139 ff
ATPase 133
– electrogenic pump 185

– see also sodium/potassium ATPase
atropine
– toxin of deadly nightshade (*Atropa belladonna*) 162 f
ATX I, II, III (sea anemone toxins) 123 ff
autoimmune diseases 85, 220 f, 292
autonomic nervous system 264
autoreceptors 245
axo-axonic synapse 18
axo-dendritic synapse 18
axolemma 75
axon
– of neuron 15
axonal flow 254
axonal ligation 251 f
axonal membrane 34 f, 139 ff
– biochemistry 133 ff
– electron microscopy 133 ff
– ionic permeability & calcium concentration 59
– reconstitution of sodium channel 68
– spectroscopy 133 ff
axonal transport
– anterograde 251 ff
– fast 166, 254
– kinetics 254
– reticular hypothesis 256
– retrograde 251 ff
– slow (axonal flow) 254
axoplasmic transport 19, 163
axo-somatic synapse 18
4-azido-2-nitrobenzyl-triethylammonium-tetrafluoroborate (photoaffinity reagent) 133
A1-protein (basic) 82 ff

bacteria
– chemotactic 276, 294 ff
– halophilic (see halophilic bacteria)
bacterial toxins 39 f
bacteriorhodopsin 142
– and membrane structure 62
– of halophilic bacteria
– – as proton pump 152 f
– reconstitution 68
Bangham, A. 58
barbiturates 242 f
Barondes 282 f
basal lamina 46, 159 ff, 172
batrachotoxin 123 ff
Belleau 208
benzilylcholinemustard 211
benzodiazepine receptor
– endogenous ligand of 243
benzodiazepines 194, 242 f

Bernard, C. 157
Bernstein, J. 91
Berridge & Irvine 32
Betz, H. 243
bicucullin 194, 242
binding studies 208
bipolar cells
– of retina 12
black lipid membrane (BLM) 68 f
bleaching 3
α-blocker 186, 189, 223 f
β-blocker 186, 189
blood groups 46
blood-brain barrier 21
Boll, F. 3
bombesin 181
Bordetella pertussis 231
botulinum toxin 39 f, 101, 165, 175
– and hypersensitization 220
bradykinin 32
brainstem
– presence of serotonin 190
Branton, D. 76
Bretscher, Gordesky & van Deenen 60
Brij 65, 204
bromoacetylcholine 211
bromocryptine 234
α-bungarotoxin 170, 175 ff, 214, 266, 271 f
– and hypersensitization 219 f
β-bungarotoxin 165, 175 ff
butaclamol 234
butyrophenones 234 f

calcineurin 148, 168, 229, 260
calcium channels
– in *Paramecium* 297, 298
calcium ions
– and adrenergic receptor 223
– and facilitation 280
– and muscarinic acetylcholine receptor 222
– and receptor desensitization 218
– and the electrical synapse 159
– as second messenger 227, 229
– as trophic factor 267
– excitation threshold 114
– in chemical synaptic transmission 168
– in ion channels 266
– in ion transport 144
– in visual process 8
– local anaesthetics 129
– possible role in hypersensitization 220
– transport mechanism 142, 148 f
calcium pump 8 f
calcium-ATPase 213

– in visual process 8
– reconstitution 68
calcium-binding sites
– calmodulin 258, 259
calcium-calmodulin-dependent protein kinase
  (see protein kinase)
calcium concentration
– and transmitter release 287
calcium-dependent protease 244, 258
calcium-protein kinase
– phosphatidylserine dependent (protein kinase C) 227
calcium-regulated processes 258
calcium/magnesium activated ATPase 148 f
calmodulin 148, 168, 227 ff, 258 f
calpain (thiol-protease) 244, 281
cAMP (cyclic 3', 5'-AMP) 183, 227 f
– and adrenergic receptors 223
– as second messenger 185, 205, 225, 265
cAMP production
– and serotonin 286
cAMP-dependent protein kinase (see protein
  kinase)
capacitance
– of myelin membrane 75
capping 246
carbamylcholine 217, 220
– formula 171
β-carboline carboxylates 243
cardiac glycosides 143
cardiolipin
– lipid of mitochondrial membranes 80
carnosine 181
carrier 115
Castillo & Katz 167
catecholamine receptors 40, 205, 208, 222 f
catecholamines 182 ff, 232
– and learning 288
– in axonal transport 253 f
– receptors α and β 185 ff
– release by exocytosis 185
– synthesis 227
– uptake and degradation 187
catechol-O-methyltransferase (COMT) 188
caudate nucleus 232
cDNA 198, 305
cell
– surface specificity
– – and ganglioside 34
cell adhesion 46
– and ganglioside 34
cell adhesion molecule (CAM) 263
cell cultures 304
cell differentiation 263 f

– programmes 267
cell line
– (clone) 304 f
cell membrane
– as an electrical circuit 94
– freeze-fractured 55
– permeability barrier for ions 92
cell migration 267
cell recognition
– and ganglioside 34
– intercellular 46
cell survival 267
ceramide (N-acylsphingosine) 34 ff
cerebellar cortex 36
cerebellum 244
– afferent system
– – climbing fibres 299 f
– – mossy fibres 299 f
– and formation of synapse 273
– and S 100 protein 284
– efferent system
– – Purkinje cells 299 f
– Purkinje cells 192
cerebroside 27, 34 ff
– lipid composition of myelin membrane 80
cGMP 8 f, 227 f, 228 f
– and muscarinic acetylcholine receptor 222
Changeux, J.-P. 208, 304
channel, sodium/potassium
– of neuromuscular endplate
– – frequency spectrum of noise 104
– – microelectrode recording 101
CHAPS 66
CHAPSO 66
charge relay system 174
chemical environment
– protein structure and activity 274
chemical synapse 158 ff
chemotaxis 294 ff
chloride channel 194, 212, 242
chloride ions 94
chloroform 127
chlorophenylalanine (pCPA) 191
chlorpromazine 127 ff
– and interaction with dopamine receptor 233
cholecystokinin-like peptide 181
cholera toxin 39 f, 230 f
cholesterol 27, 56
– biosynthesis 33
– lipid composition of myelin membrane 80
– membrane fluidity 33
cholesterol acetyltransferase
– in myelin 81

cholesterolesterase
– in myelin 81
choline 93, 164 ff
choline acetyltransferase (CAT) 166, 266
choline uptake 174
– high affinity 164, 166
– low affinity 164
cholinergic ligands 163
cholinergic synapse
– central 162
– muscarinic 162 f
– nicotinic 162 f
– – and myasthenia gravis 220 f
– peripheral 162
chromaffin granules (storage vesicle of catecholamines) 184 f
chromogranins 185
ciliary movement in $Paramecium$ 297 f
citrate 164
climbing fibres
– of cerebellum 299 f
cloning 305
clorgyline 189
$Clostridium\ botulinum$ 39
$Clostridium\ tetani$ 39
cobra ($Naja\ naja,\ N.\ naja\ siamensis$)
– neurotoxin 176 ff
cocaine 128, 189
codeine 238
colchicine alkaloid ($Colchicum\ autumnale$) 254 f
collagenase 172
collision-coupling hypothesis 213, 225
colour blindness 11
colour vision 11
conditioning
– as learning paradigm 278
conductance of membrane
– potassium (gK) (see potassium conductance)
– sodium (gNa) (see sodium conductance)
conduction
– saltatory
– – unmyelinated fibres 73 ff
– speed of
– – myelinated fibres 74
– – unmyelinated fibres 73 ff
cones 3 ff, 203
– colour vision 11
– function 7
– structure 7
connexins 159
connexons 159
conscious perception
– and dopaminergic pathways 231

convergence 12
convulsants 194
– bicucullin and picrotoxin 242
corpus striatum
– in Parkinson's disease 236
cortex 192, 244
corticotropin-releasing factor (CRF) 305
cortisone 198
Corynebacterium diphtheriae 40 f
crayfish axon 134 f
Creutzfeldt-Jacob disease 86, 292
crystalline phase
– of bio-membrane 56 ff
Cuatrecasas, P. 204
curare (see d-tubocurarine)
cyclic nucleotides
– as second messengers 228 f
cycloheximide 281 f
cytochrome b_5 62
cytoplasm (neuronal) 19
cytoskeleton 268

DAG (diacylglycerol) 30 ff
Dale, H. 99
Dale's Principle 182
Danielli & Davson 61, 76
– membrane model of
– – "sandwich" or "unit membrane" 52 f
dansylchloride (5-dimethylamino-1-naphthalene sulphonylchloride) 120
DDAO (dodecyldimethylamineoxide) 66
de Nó, L. 276
De Robertis & Bennett 167
decamethonium
– formula 171
demyelination 85, 87
dendrites 15
dendritic spine synapses 277
denervation 246
– and hypersensitization 219
DEP (diethyl pyrocarbonate) 120, 133
depolarization 95 ff, 106
deprenyl 189
depression 278, 300
Descartes, R. 277
desensitization
– of acetylcholine receptor 247
detergents
– ionic 63 ff, 204
– non-ionic 63 ff, 204
– zwitterionic 63 ff
development
– of nervous system 263 ff
– activity of the neuron 264

– genetic factors 264
– trophic factors 264
DFP (see diisopropylfluorphosphate)
DHEC (dihydroergocryptine) 234
diacylglycerol (DAG) 30 ff
3,4-diaminopyridine 130 f
Dictyostelium discoideum 297
diethylpyrocarbonate (DEP) 120, 133
digitonine 65, 225
digitoxin 143
digitoxin aglycon 147
dihydroxyacetone phosphate 27 f
dihydroxyphenylalanine (DOPA) 182 ff
diisopropylfluorphosphate (DFP) 165, 173 ff
2,4-dinitrophenol (DNP) 139, 141
diphtheria toxin 40, 87
disc 7 ff
– membrane 4
displacement test 235
5,5'-dithiobis(2-nitrobenzoic acid) (DTNB)
    (Ellman's reagent) 119, 218
DNA 305 ff
– in neuron maturation 264
– recombinant 305
DNA synthesis
– inhibitors 281 f
dolichol 45 f
DOPA (dihydroxyphenylalanine) 237
DOPA-decarboxylase (DDC) 182 ff
dopamine 178, 182 ff, 230 ff
– and cAMP system 228
– formula 179
dopamine receptors 231 ff
– D_1 and D_2 223 ff, 230, 232
dopamine-β-hydroxylase (DBH) 182 f
– and axonal transport 255
– and transsynaptic regulation 265 f
– in fast axonal transport 254
dopaminergic pathways 231 f
dopaminergic systems 231
dorsal horn
– of spinal cord 199
dorsal root ganglion 252
dose-response curve 206, 215, 217
Drosophila (fruit fly) 22, 298 f
Droz, B.
– and model of axonal transport 256
DTNB
– 5,5'-dithiobis(2-nitrobenzoic acid)
    (Ellman's reagent) 119, 218
– – and hypersensitization 219
d-tubocurarine (curare) 157, 162 f, 165, 175 f, 221
– formula 171

dynorphin
- (pituitary polypeptide) 197, 237

E. coli
- galactoside transport 57
e.p.p. (see endplate potential)
e.p.s.p. (see excitatory postsynaptic potential)
E 605 (Parathion) 174
Eccles, J. 182, 281
EC_{50} (effector concentration 50%) 206
Edelman, G. 263
edrophonium (Tensilon) 174, 221
efferents 15
Elapidae 176
electric eel (*Electrophorus electricus*) 117f, 214
- and acetylcholine receptor 170
- as experimental model 301ff
electric organ
- of electric fish 214
- of *Torpedo* 63, 245, 301ff
electric ray (see *Torpedo*)
electrical synapse 158ff
electrocyte (electroplaque) 303
electrogenic pumps 146
electrophoresis
- two-dimensional 267
Electrophorus electricus (see electric eel)
electrophysiology 91ff
- summary of terms 105
electroplaques
- of electric fish 301ff
emotions
- and dopaminergic pathways 231
Emulphogen BC 65, 204
encephalitogenic determinant 84
endoplasmic reticulum 256
β-endorphin 180, 198, 237
endplate 159ff, 292
- crustacean 280
- see also neuromuscular endplate
endplate potential (e.p.p.) 99, 106
engram 277, 279, 281
enkephalins 196ff, 238
- as putative transmitter 178ff
- inhibition of pain conduction 240
- presence with other transmitters 246
epilepsy 140
epileptiform convulsions 20
epinephrine (see adrenaline)
equilibrium potential 93
equivalent circuit 94ff, 106
ergot alkaloids 186, 223f
- derivatives 234

ergotamine 189
erythrocyte membrane 62f
- asymmetry 60
- lipids 26, 60
erythrocytes
- avian 225
- haemolytic effect of snake venom 57
- of frog
-- β adrenergic receptors in 246
- resting potential in 92
eserine (physostigmine) 173f, 175
ethanol 127f
excitability of nerve cells
- chemical 266
- electrical 266
excitatory postsynaptic potential
 (e.p.s.p.) 95, 99, 106, 162
excitatory transmitters 244
exocytosis 163, 227
- and endocytosis 168
- in transmitter release 101, 168
- release of transmitter molecules
-- from chromaffin granules 57
- synaptic vesicles 52
experimental allergic encephalomyelitis
 (EAE) 82, 85f
experimental allergic neuritis (EAN) 85
experimental model systems 291
Eylar, E.H. 83, 85

Fabry syndrome 42f
facilitated diffusion 115
facilitation 169, 279ff
Farber syndrome 43
fat cells
- of rat
-- receptors 213
Fatt, S. & Katz, B. 100
FDNB (1-fluoro-2,2-dinitrobenzene) 60, 133
Flaxedil (see gallamin)
flip flop
- mode of mobility 59
floating receptor hypothesis 212, 241
fluid crystalline phase
- of bio-membrane 56ff
fluid mosaic model
- of membrane 53ff
fluidity of the lipid membrane
- local anaesthetics 127
[^{3}H]fluonitrazepam 211
flupenthixol 234
fluphenazine 233
Fly Agaric fungus (*Amanita muscaria*) 194
- see also muscarine, muscimol

fodrin 244, 281
Folch-Pi, J. 82
foot activator
– of *Hydra* 298
Forbes, A. 276
forskolin 230
freeze-fracture 75
– of cell membrane 55
frog, Costa Rican (*Atepolus chiriquensis*) 121
α-fucosidase 43
fucosidosis 43
Furshpan & Potter 158

GABA (γ-aminobutyric acid) 100, 166, 178, 182, 192 ff
– chloride channels 194
– convulsants 194 f
– formula 179
– in axonal transport 253 f
GABA receptors 242 f
GABA shunt 193
GABA-modulin (CNS-neuropeptide) 243
GABA-transaminase 193
Gaddum, I. 199
galactose-binding protein (GBP) 294 f
α-galactosidase 43
galactosylceramide 34, 38
galactosylcerebroside 35 f
galactosylsulphatide 38
galactosyltransferase 46
gallamin (Flaxedil) 174
– formula 171
ganglion cells
– of retina 12 f
gangliosides 27, 34 ff
– biosynthesis 38
– lipid composition of myelin membrane 80
– metabolism 36
– nomenclature 37
gangliosidoses 43
– G_{M2} 41
gap junction 158 f
gating mechanism 129 ff, 205
– chemical synapse 111 ff
– electrical synapse 111 ff
– of ion channel 103, 112
Gaucher syndrome 42 f
GDP 8 f
genetic control
– and development of nervous system 275
genetic engineering 305
genetics 299 f
Geren, B.B. 76, 78
gigaseal

– of patch clamp method 103 f
gill-withdrawal reflex
– of *Aplysia* 285 ff
glial cell 20 f, 140, 194
– and nerve function 272
– and stimulatory environment 284
glioblast 264
glucagon 230
glucocerebroside 42 f
glucocerebroside-β-glucosidase (glucocerebrosidase) 42 f
glucose
– chemoeffector in bacterial sensing system 295
– transport 213
– utilization
– – and NGF 270
glucosylceramide 35, 38
glutamate 182, 195, 244
– decarboxylase 193
– receptors 244 f
– – in LTP 280 f
– chemoeffector in bacterial sensing system 295
– formula 179
glycerol-3-phosphate 27 f
glycerol-3-phosphate dehydrogenase 27 f
glycine 178, 182, 192 ff
– chemoeffector in bacterial sensing system 295
– formula 179
– inhibitory transmitter 195
– receptors 211, 243
glycogenolysis 148
glycolipids
– in axonal transport 253 f
– in membrane 25 ff
glycolysis 148
glycophorin A 62 f
glycoproteins 27, 44 ff
– function 46 f
– in axonal transport 253 f
– summary 48
glycosidase 37
glycosyltransferase 36 f, 46
glyoxal (1,2-cyclohexanedione) 120
Goldman equation 94
Golgi apparatus 256
Gonyaulax catenella 123
Gorter & Grendel 52
GPPNHP (non-cleavable GTP analogue) 235
G-protein (transducin) 8 f
– see also N-protein
Graham & Gerand 91

gramicidin 68
grayanotoxin 123 ff
Greengard, P. 226 f, 232
grey matter
– of brain 34
growth factor-releasing factor (GRF) 305
growth of developing neuron 263
GTP 8 f, 235 f
GTPase 230
GTP-binding regulator protein (G, N or G/F) 225
– see also G-protein
guinea pig cortex 168

habituation 296
– as learning paradigm 277
Hagins 8 ff
haloperidol 189, 234
halophilic bacteria (Halobacterium halobium) 62, 142, 151 ff
halothane 127
harmane 243
head activator
– of Hydra 298
heart (cardiac) glycosides 147
heart muscle 181, 186
Hebb, D. 276
Hecht 3
α-helix 62 f
hemicholinium (HC-3) 165 f, 175
Henderson, R. & Unwin, P.N.T. 152 f
heroin 238
herpes 256
hexachlorophene 87
hexamethonium
– formula 171
hexosaminidase A 41 ff
high affinity uptake 191
Hill coefficient 215, 222
Hille, B. 115 ff
hippocampus 244, 279 f
– and memory 279
histamine 230
– and cGMP system 228
– formula 179
histone-H1, H3 270
histrionicotoxin (HTX) 165, 175
Hodgkin, A.L. & Huxley, A.F. 91 ff, 111 ff, 115, 129, 133
– analysis of action potential 97 f
– equation 113, 122
Hodgkin & Keynes 139
Hokin & Hokin 32
homosynaptic facilitation 280

homovanillic acid 237
horizontal cells
– of retina 12
Hubel, D. 13
Hubel & Wiesel 273, 292
hybridoma cloning 306
Hydén, H. 281, 283 f
Hydra
– foot activator 298
– head activator 298
– morphogenous substances 298
hydrazinium 115
Hydrophiidae 176
β-hydroxybutyrate dehydrogenase
– mitochondria 63
6-hydroxydopamine 196
hydroxylammonium 115
hydroxylysine 172
hydroxyproline 172
5-hydroxytryptamine (see serotonin, 5-HT)
5-hydroxytryptophan 184, 191
hyperpolarization 96 ff, 106, 194
hypersensitization 219 f
hypophysectomy 288

i.p.s.p. (see inhibitory postsynaptic potential)
IC_{50} 235
imipramine 189
inactivation 106
independence principle 115 f
indole
– chemoeffector in bacterial sensing system 295
induced fit 207, 219
information storage 273 ff, 281
inhibitory postsynaptic potential (i.p.s.p.) 95, 100, 106, 162
inhibitory transmitters 199, 242
inositol trisphosphate 1,4,5 (trisphosphoinositol) ($InsP_3, IP_3$) 31 f
insulin 247
– and β NGF 269
integral proteins
– localization 61
intermediate filaments 258
interneuron
– inhibitory 192
internode 17, 73, 76
intestine
– presence of serotonin 190
inverse agonists 243
invertebrates 75
iodoacetic acid 119
ion channels 104 ff, 109 ff, 205, 208

- acetylcholine regulated 266
- differentiation 266
- in receptor desensitization 219
- noise analysis 105
- opening 215
- potassium 114 ff, 129 ff
- properties 105
- sodium 114 ff
- summary 136
ion channel, single
- conductivity 105
- mean opening time 105
ion permeability 205
- of postsynaptic membrane
-- molecular mechanism 169
-- specificity 169
- potassium channel 130
- sodium channel 116
ion pump 139 ff
ion transport 18
- active (ion pump) 109 ff
- passive (channels) 109 ff
- potassium
-- comparison of passive and active 144
- sodium
-- comparison of passive and active 144
ions
- of axoplasm (intracellular) 92
- of blood of squid (extracellular) 92
isethionate 93
isolated nerve terminals (see synaptosomes)
isoprenaline (isoproterenol) 223 f
- β-sympathomimetic 186
isoprene (isopentenyl-pyrophosphate) 33 f
Iversen, L. 232

kainic acid 195, 244
- chemical lesions 195
Kandel, E. 278, 286 f
Karlin, A. 208, 304
Karrer, P. 3
Köhler & Milstein 306
Korsakoff syndrome 279
Koshland, D. 276, 295 f
- induced fit model 207
Koshland's reagent
- 2-hydroxy-5-nitrobenzyl bromide (HNBB) 120
Kosterlitz & Hughes 196
Krabbe syndrome 43
Krabbe's leucodystrophy 86
krait (*Bungarus multicinctus, B. ceruleus*)
- α-bungarotoxin 178
-- neurotoxin 176 ff

Krebs cycle
- intermediates 193
Krebs, E. 226 ff
Kuffler, S.W. 11
Kühn, H. 8, 10
Kuru 86

lactoperoxidase (LPO) 212
- method for protein localization in cell membranes 61
lactosylceramide 38
Landry-Guillain-Barré syndrome 86
Langmuir trough 67
Larrabee & Bronk 279
Lashley 278
lateral geniculate body 13
L-DOPA (L-dihydroxyphenylalanine) 237
lead 87
learning
- and synapses 34, 275 ff
- biochemical basis of 275 ff
- paradigms 276
learning model
- dynamic 277
- plastic 277
lecithin 57
Lee 175
leech (*Hirudo*) 291, 293
lesions 196, 244
leu-enkephalin 180, 196 ff
Levi-Montalcini 268
Levinthal 275
Librium^R 242
lidocaine 128, 175
ligands for receptor binding sites 210
limbic system 196
lipid
- in membrane 25
lipid bilayer 52 ff
lipid membranes
- artificial 66 ff
-- planar 67
-- vesicular 67
- reconstitution 66 ff
lipid storage diseases 41 ff, 300
lipid-exchange proteins 63
lipid-synthesis
- and NGF 270
lipophilic cations 134
liposome 52, 57 ff, 58, 63
β-lipotropin, β-LPH (pituitary polypeptide) 197 f
lisuride 234
lobe, occipital 2

local anaesthetics 58, 175 f, 218
– calmodulin 259
local potential 95, 105
Loewi, O. 99
long-term memory (see memory)
β-LPH (see β-lipotropin)
LSD (see lysergic acid diethylamide)
LTP (long term potentiation) 279 ff
Lubrol, W. 65
luteinizing-hormone-releasing hormone (LHRH) 181
lysergic acid diethylamide (LSD) 190
– and serotonin receptors 245
lysolecithin 32 f, 57 f
lysophosphatidic acid 28
lysosomes 247, 270

m.e.p.p. (see miniature endplate potential)
magnesium ions 169
magnesium-dependent ATPase
– in vesicle membrane 167
maltose-binding protein (MBP) 294 f
MAO blocker antidepressives 188
MAO (monoamine oxidase) A, B 188
Martin, A.R. 11
mast cells
– presence of serotonin 190
Maxam & Gilbert 306
Meerwein's reagent 121
membrane 51 ff
– asymmetry 60, 62
– central nervous system
– – composition 27
– erythrocyte 26
– fluidity 57, 58
– phase separation 62
– postsynaptic 159 ff, 274
– presynaptic 159 ff
– subsynaptic 159
membrane model
– bilayer 52 ff
– Danielli & Davson "sandwich" 52 ff
– Singer & Nicolson "fluid mosaic" 52 ff
membrane molecules 25 ff
membrane potential 8, 16 ff, 91 ff, 105, 139
– of excitable membranes 142
– recording 92
membrane proteins
– integral 61 f
– peripheral 61 f
memory 241, 244
– and chemotaxis 294
– and dopaminergic pathways 231
– biochemical research methods 275 ff

– deficits in *Drosophila* mutants 299
– localization 281
– long-term 276, 279, 281 f
– short-term 276, 279, 282
– ultra-short-term 282
memory molecules 275
memory-specific proteins 283
merocyanine (dye) 134 f
mesocortical system 232
metachromatic leucodystrophy 43, 86
metarhodopsin I 3, 6
metarhodopsin II 3, 6
metarhodopsin III 6
met-enkephalin 180, 196 ff
methadone 238
methyl transferase 296
methylammonium 115
methylation 188
– of receptors 296
Meyer & Overton 58
micelle 52
– formation from lipid monolayer 54
Michaelis-Menten theory 206
Michell, R.H. 32
microelectrode 91 f
microfilaments 55, 254 ff
microglia 21
microtubule-associated proteins (MAP) 257 f
microtubules 55, 254 ff
migration of developing neuron 263
miniature endplate potential (m.e.p.p.) 100 f, 106, 167 f
Mitchell hypothesis
– chemiosmotic hypothesis 150 ff
mitochondria 148, 161, 164, 168
– calcium accumulation 148
– in axonal transport 253 f
mitochondrial ATP synthetase (proton pump)
– model 152 f
mobile receptors 212
model systems 22, 291 ff
modulation
– of synapse 158
modulatory effects 246
monoamine oxidase (MAO) (see MAO)
monoclonal antibodies 306
Monod, J. 1
Monod-Wyman-Changeux model 207
monolayer 52
Montal-Mueller-method 68
morphine 128, 237 f
morphogenous substances
– in *Hydra* 298
mossy fibres

– of cerebellum 299 f
motor neuron growth factor (MNGF) 271
mouse mutants 260
– jimpy 86
– nervous (nr) 299 f
– quaking 86
– reeler (rl) 299 f
– staggerer (sg) 299 f
– weaver (wv) 299 f
mRNA 305 ff
β-MSH 197
mud puppy, Californian (*Taricha torosa*) 121
Mueller-Rudin-method 68
multiple receptors 186, 223
– opiate receptors 239
multiple sclerosis 82, 85, 292
muscarine
– toxin of Fly Agaric (*Amanita muscaria*) 162 f
muscimol 242
– poison of Fly Agaric fungus 194
muscle receptors 219
myasthenia gravis 18, 85, 158, 220 f, 292
myelin 15, 16 ff, 35, 52 f, 63, 73 ff, 73 ff
– chemical composition 80 f
– cholesterol containing 58
– functions 73 ff
–– proteins & lipids 80 f
– membrane
–– protein composition 80
– protein and lipid analysis in different organisms 81
– structure 74, 77 f
–– enzymes 80 f
– summary 87
myelination 38, 74
– at birth 87
myosin 134
– in slow axonal transport 254 ff
myosin light chain kinase 227 f

N-acetylgalactosamine 36
N-acetylglucosamine 36, 44
N-acetylneuraminic acid (NANA) (sialic acid) 37
– in axonal transport 254
Nachmansohn, D. 133, 166, 172, 208
Naja naja
– α-toxin 214
N-allylnormetazocine 239
nalorphine 238
naloxone 238
NANA (see N-acetylneuraminic acid)
Na^+,K^+-ATPase (see sodium/potassium-ATPase)
N-CAM (see neuronal cell adhesion molecule)
Neher & Sakmann 101
nematodes 22, 297
β-neodynorphin 198
neostigmine 173 f, 221
neostriatum 192
Nernst potential 93 f, 142
nerve
– functions 15
nerve cell
– electrical properties 91 ff
– differentiation 304
– see also neuron
nerve fibres
– afferent 15
– efferent 15
nerve growth factor (NGF) 247, 256, 266 f, 268 ff, 305 f
nerve impulse
– modulation 157
nerve membrane 51 ff
– as circuits in parallel 96
– electrical model 94
– ion distribution 92
– properties-summary 70
nervous system
– development 263 ff
–– activity of neuron 264
–– genetic factors 264
–– trophic factors 264
– plasticity 263 ff
– stabilization 263 ff
nervous system antigens (NS1, NS2, L1, L2) 261
N-ethylmaleimide (NEM) 62, 118 f
neural plate 263
neuraminidase 37, 118
neuroactive drugs and toxins
– in synaptic transmission 165
neuroblast 264
neuroblastoma cells 20
– nerve cell properties of 304
– of single mouse tumour (C 1300) 304
neuroblastoma x glioma-cells 20, 240
– and formation of chemical synapse 272
neurochemistry
– definition
–– as integrating science 14
–– exemplified by visual process 1 ff
neurofilament subunits
– in slow axonal transport 254 ff
neurofilaments 254 ff
neuroglia 20

neuroglia adhesion molecule (Ng-CAM) 263
neuroleptic drugs 231, 236
neuromodulators 196 ff
neuromuscular endplate 159, 162, 169, 301
- potential 169 ff
neuromuscular junction
- as experimental model 291 f
- see also neuromuscular endplate
neuron (nerve cell)
- adrenergic 255
- catecholaminergic 255
- cholinergic 255
- definition 1 ff
- excitation
-- action potential 95 ff
-- local potential (electrotonic potential) 95 ff
- functional elements 15 ff
-- axon and dendrites 18
-- cytoplasm 19
-- glial cell 20
-- myelin 16 ff
-- plasma membrane 16 ff
-- synapse 16 ff
- functions 2
- inhibitory 192
- ion transport 18
- structure
-- axon 15
-- dendrites 15
-- nucleus (cell) 15
-- soma (cell body) 15
neuron maturation 264
neuronal cell adhesion molecule (N-CAM) 261, 263
neuronal cell death 272
neuronal cytoplasm 19, 251 ff
neurons, identified 293
neuron-specific proteins 251 ff
- summary 261
neuropeptides 180, 196 ff, 246, 298
neurotensin 180
neurotubules 228
neurotoxin
- cobra 176 ff
- krait 176 ff
- sea snakes 176 ff
α-neurotoxin
- from snake venom 165
neurotoxins 39 f, 58, 117, 195 f
- as tools 121 ff
- binding proteins 260
- binding sites 126
- sodium channel 126

NGF (see nerve growth factor)
Nicholls, J.G. 11
nicotine 163
Niemann-Pick syndrome 42 f
nigrostriatal (extrapyramidal) system 232
N,N'-dicyclohexylcarbodiimide (DCCD) 151
node of Ranvier 15, 73, 75
- and ATX II 123
- definition 17
- of frog sciatic nerve
-- model of potassium channel 131
noise 101 ff, 103, 106
non-suppressible-insulin-like-activity (NSILA)
- and β NGF 269
noradrenaline (norepinephrine) 178, 182 f
- and adrenergic receptors 223 f
- and transmitter differentiation 265
- formula 179
- in schizophrenia 236
norepinephrine (see noradrenaline)
N-protein 10, 31 f, 229 ff, 295
- and cholera toxin 40

occupancy theory 206
octopamine 184
- and cAMP system 228
β-octylglycoside 66
olfactory nerve 252
oligodendroglial cells (oligodendroglia) 21, 27, 34
oligosaccharide 55
opiate peptides 196 f, 230
opiate receptors 186, 196, 204, 237 ff
opiates 196, 237 ff
- endogenous 196
- presynaptic inhibition 240
opsin 3 ff
- formula 5
optic chiasma 13
optic nerve 13, 75
optic tract
- birds 252
- fish 252
organophosphates 175
ouabain (strophanthin) 109, 143 f, 146 f
- and short-term memory 282
oxaloacetate 164
oxidative deamination 188
oxytocin 180

palmitoyl-CoA 35
PAM 173 f
Paramecium 22, 297
- behaviour mutant pawn 297 ff

- calcium channels 297 f
- ciliary movement 298 f
- resting potential in 92
parasympathetic system 162
Parathion (E 605) 173
Parkinson's disease 18, 158, 231, 236 f
patch clamp method 67, 101 ff, 117
patching 246
pawn (behaviour mutant of *Paramecium*) 297 ff
PC 12 270, 304 f
p-chloromercuribenzoate (pCMB) 118 f
- as ionic SH-reagent 61
PcI Duarte protein 261
pCMB (see p-chloromercuribenzoate)
peptide transmitters
- and cAMP system 228
perception 3
perikaryon
- of neuron cell body 19
peripheral nervous system 36
pertussis toxin
- from *Bordetella pertussis* 230 f
Pfeffer 294
phase transition
- phospholipid membrane 56
phenothiazine 189, 232
phenoxybenzamine 186
phenylalanine 86, 184
phenylalanine hydroxylase 86
phenylalanine-rich acidic protein (PAP Ib-protein) 260
phenylephrine (α-sympathomimetic) 186
- and adrenergic receptors 223 f
phenylethanolamine-N-methyltransferase (PNMT) 182
phenylethylamine 184
phenylketonuria 86
phenylpyruvic acid 86
pheochromocytoma
- cell line PC 12 270, 304 f
phosphatidic acid 28
phosphatidylcholine (PC) 63, 166
- distribution in lipid membranes 60
-- lecithin 26 ff
phosphatidylethanolamine (PE) 6 f, 26 ff, 166
- distribution in lipid membranes 60
phosphatidylinositol (PI) 26 ff, 31
- in myelin 81
phosphatidylserine (PS) 26 ff
- distribution in lipid membranes 60
- in myelin 81
phosphodiesterase (PDE) 213, 226, 229
- calcium regulation by calmodulin 260

- in light transduction 8 ff
phosphoglycerides
- biosynthesis 30
phospholipase 32 f, 118
- A_1, A_2 57
- C 31
phospholipid membrane
- phase transition 56
phospholipids
- amphipathic 54
- in axonal transport 253 f
- in membrane 25 ff
- lipid composition of myelin membrane 80
- structure 27 ff
- summary 47
phosphoprotein phosphatase 260
phosphorylase b-kinase 148, 227
- calcium regulation by calmodulin 260
phosphorylation 183
- acetylcholine receptors 220
- and *Aplysia* sensitization 286 f
- and modulation 246
- in sodium/potassium pump 145
- of rhodopsin 10
- protein 226 ff
- proteins of postsynaptic membrane 172
- sites in rods 4
photoaffinity labelling 211 f, 243
photoaffinity reagents 211 ff
photoreception 3, 4
photoreceptors 3, 203
physostigmine (eserine) 165
- and learning 288
picrotoxin 194, 242
pimozide 234
pindolol 186
pinocytosis 266
Pinto da Silva & Miller 76
pituitary
- polypeptide 197
pituitary hormones
- and learning 288
plasma membrane 16 ff, 51
- composition
-- bacterium 25
-- chloroplast 25
-- erythrocyte 25 f
-- mitochondrion 25
-- myelin 25
- summary 47
plasmalogen 27 ff
- in myelin 81
plasticity 246, 274 ff
- of nervous system 263 ff

- of synapse   157 f
platelets, blood
- blood presence of serotonin   190
polar lipid heads
- of myelin membrane   79
polarity
- of synapse   158
poliomyelitis   256
polysomes
- formation and NGF   270
pons
- presence of serotonin   190
pore   115 ff
postganglionic synapse   162
postsynaptic density   162
postsynaptic membrane   46, 169
- and chemical synapse   160 ff
postsynaptic potential   169
postsynaptic potentiation   169, 244
post-tetanic depression   281
post-tetanic facilitation (potentiation)   169, 278, 279 ff
posttranslational modification   219
potassium channel   129 ff
- biochemical characterization   132 f
- model   132
- modulation in Aplysia   246
- physical properties   132
potassium conductance gK 97, 110, 129
potassium equilibrium potential   93 f
potassium permeability   213
- excitable membrane   130
potassium potential   20, 93 f
potentiation   278
- long term potentiation LTP   279 ff
- postsynaptic   244
- post-tetanic   279 ff
prazosin
- and adrenergic receptors   223
precursor proteins   198
preformed equilibrium   207
pre-pro-opiomelanocortin   198
presynaptic inhibition   100
presynaptic membrane   46
- and chemical synapse   160 ff
procaine (Novocaine^R)   128 f, 165, 175
prodynorphin   197 f
proenkephalin   197 f
proline   172, 179
proopiomelanocortin   197
propanolol   186
- as antagonist   246
propylbenzilylcholine mustard   222
prostaglandins   230

protein kinase   205 f
- calcium-calmodulin-dependent I & II   227 f, 257, 259 f
- cAMP-dependent   183, 226 ff, 227, 257 f, 265, 270, 286 f
- cGMP dependent   227
- inhibitor   287
- Mg^{2+} requiring
-- cAMP-independent   10
protein kinase C   31
- and facilitation   280
protein of myelin
- acidic (Wolfgram protein)   80, 82, 87
- basic (A1)   80, 82 ff, 87
- proteolipid   87
protein phosphorylation
- and NGF   270
protein synthesis   253
14-3-2 protein   260
protein S6
- of ribosomes   270
protein-lipid interactions   62
protein-synthesis
- and NGF   270
proteolipid (PLP)   82
- of myelin membrane   80
proteolytic degradation   198
proton pump   62, 142, 149 ff
pseudo-cholinesterases   174
P-site   230
psychopharmacological drugs   157
pteridine   184
Purkinje cells
- of cerebellum   192, 299 f
puromycin   281 f
putative transmitter   178 ff, 184
pyridoxal phosphate (Vitamin B_6)   184
pyruvate dehydrogenase   164
P-400 (cerebellum specific protein)   260, 299 f

quantum   106, 167
- of transmitter   100
quinacrine   176

Racker, E.   149, 151
radioimmune assay (RIA)   221
Raftery, M.   304
Ramón y Cajal   157, 276
raphé nuclei
- presence of serotonin   190
receptive field   12 f
receptor   106, 169, 303
- and regulation mechanism   246 f
- α_1-adrenergic   32

- criteria 203
- definition 203
- dopamin 223
- for peptide hormones 218
- GABA 242
- muscarinic cholinergic 32
- nicotinic cholinergic 265
- opiate 237
- peptidergic 32
- presynaptic 245
- properties 204
- serotonergic 32
- specific binding 209
- specificity of 203 f
- see also individual receptors
α-receptor 186 f
β-receptor 186
- ($β_1$, $β_2$) 186 ff
receptor cells
- of retina 12 f
receptor ligands
- irreversible 211
receptor models 204
- *Torpedo* 301, 303 f
receptor potentials 8, 95
- integration 13
receptor turnover 221
receptor-desensitization 218
reconstitution
- methods 67 f
- sodium channel 117
refractory period 105
- of action potential 97
regulatory protein (N, G) 229
- see also N-protein
rehabilitation 278
relaxin
- and β NGF 269
Renex 30 65
repolarization 96 f
reserpine 189
resting potential 91 f, 105, 194
reticular theory 158
retina 2, 3 ff
- cellular construction 12
retinal all-*trans* 3 ff
- formula 5
retinal ganglion cells 292
retinal 11-*cis* 3 ff
- formula 5
retinal$_1$ 4 f
- formula 5
retinal$_2$ 4 f
- formula 5

retinol 3
- all-*trans* 7
- 11-*cis* 7
rhodopsin 3 ff
- absorption spectrum
-- cones 4 f
-- rods 4 f
- phosphorylation of 10
- reconstitution 68
- transducin 8 f, 225, 230
ribose 294
ribose-binding protein (RBP) 294 f
RNA
- in nerve tissue during learning 281
- synthesis and NGF, *270*
rods 3 ff, 203
- inner segment 4
- outer segment 4
-- disc membrane 4
- structure 4
-- function 7 f
Rojas, E. 113

S-adenosylmethionine (SAM) 29, 183, 188
- and bacterial sensing system 295 f
saltatory conduction 73
Sandhoff-Jatzkewitz disease 41 ff
sarcoplasmic reticulum 148 f
sarin 173
saxitoxin (STX) 98, 109, 117 f, 122 ff
Scatchard diagram 209
Schiff's base 4
schizophrenia 158, 231, 261, 300
- dopamine hypothesis 236
Schoffeniels & Nachmansohn 214 f
Schwann cell 17 ff, 20, 75 ff
sciatic nerve 135
scopolamine
- and learning 288
scorpion toxins (ScTX) 125 f
scotophobin 275
scrapie 86, 292
SDS (see sodium dodecylsulfate)
sea anemone (*Anemonia sulcata*) 123
- toxins 123 ff
sea slug (see *Aplysia californica*)
sea snakes
- (e.g. *Laticauda semifasciata*)
-- neurotoxin 176 ff
second messenger 31, 227 f, 267
- in brain
-- calcium 148, 227 f
-- cAMP 227 f, 265
-- cGMP 227 f

– – DAG 31
– – IP$_3$ 31
selective stabilization
– of synapse 273 f, 299
selectivity filter 112 ff, 129 ff, 205
– of ion channel 103
sensitization 277, 280
sensory ganglion 268
serine 34 f
– binding to tsr 294
serotonin (5-hydroxytryptamine, 5-HT) 178, 182, 190 f
– and axonal transport 255
– and cAMP system 228
– degradation by MAO 190
– formula 179
– synthesis from tryptophan 190
serotonin receptors 86, 244 f
– and adenylate cyclase 286
Sherrington, A. 157
short-term memory (see memory, short term)
sialogalactosylceramide 36
signal perception
– in chemotactic bacteria 276
signal processing 296
signal transduction
– and energy coupling 142
Singer & Nicolson 52 ff, 61
single channel recording 101 ff, 103
Skou, J. 143
sleep regulation 190
slow virus infection 86
smooth muscle 187
snake venoms 176 f
– and NGF 269
– haemolytic effect 57
Snyder, S. 204
sodium channel
– biochemical characterization 117 f
– cardiac muscle 134
– cDNA 118
– gating mechanism 112 f
– ion permeability 116
– physical properties 117
– primary structure 118
– selectivity filter 114 ff
sodium cholate 65
sodium conductance gNa 97, 110, 112
sodium dodecylsulphate (SDS) 65, 204
sodium ions
– in ion channels 266
sodium permeability
– excitable membrane 124
sodium taurodesoxycholate 65

sodium/potassium-ATPase 109, 140, 143 f, 213
– and memory 282
– electrogenic character 146
– inhibition by cardiac glycosides 147
– reconstitution 68
– transport system 61
soma
– of neuron 15
somatostatin 181
spare receptors
– theory of 206
specificity
– of synapse 158
Sperry, R. 273, 292
sphingolipidoses 41 ff, 86
– summary 48
sphingolipids 34 ff
– summary 48
sphingomyelin 26 f, 36
– distribution in lipid membranes 60
– in myelin 81
sphingomyelinase 42 f
– in myelin 81
sphingosine 34 ff, 43
spinal cord 192, 199
spiperone (spiroperidol) 234
squalene 33
squid giant axon 91 ff, 111, 291 f
– and neurofilaments 258
– and sodium channel 118
– comparison with electric cable 96
– electrode insertion 91 ff, 96
– model of potassium channel 131 f
stabilization
– of nervous system 263 ff
sterol
– in membrane 25 ff
Sterox, A.J. 65
Stoeckenius, W. 151
strychnine 211, 243
Stryer, L. 8 f
styrylpyridine derivatives 175
subcellular fractionation 252
suberyldicholine
– formula 171
submaxillary salivary gland 269
substance P 32, 180, 196, 199, 240
– as putative transmitter 178 ff
– presence with other transmitters 246
substantia nigra 232
– in Parkinson's disease 236
sulphatide 27, 34 ff
sulpiride 234

Sutherland, E.   225
Svennerholm   35
sympathetic nervous system   162, 265
sympathetic neurons   305
– survival of   268
sympathomimetic substances   185
synapse   15 ff, 157 ff
– and memory   279
– axo-axonic   18
– axo-dendritic   18
– axo-somatic   18
– chemical   99, 158 ff
–– summary of functions   200
– cholinergic   162 ff
–– see also cholinergic synapse   162 ff
– definition   18
– electrical   99, 158 ff
– excitatory   157
– formation   264
– formation and stabilization   267, 273
– in neural membrane   16
– inhibitory   157, 181
– model   301 ff
– modulation   158, 218
– neuromuscular
–– and stabilization   274
– nicotinic cholinergic   160, 164
– plasticity   157 f
– polarity   158
– receptors   203 ff
– selective stabilization   273
– specificity   158
synapses
– and learning   34, 279
synapsin   228, 260
synaptic cleft   99, 160, 167 f, 172
synaptic complexes
– and chemical synapse   159 ff
synaptic modulator   180
synaptic transmission   99, 291
– and sphingolipids   39
– chemical
–– individual stages of   163 ff
– inhibitors   175
– neuroactive drugs and toxins   165
synaptic vesicles   101, 167 f, 303
– and chemical synapse   159 ff
– and transmitter differentiation   265
synaptogenesis   267, 271 f
synaptosomes (isolated nerve terminals)   35, 166
– and axonal transport   252
– and chemical synapse   159 ff
– and choline uptake   164

– and cholinergic autoreceptors   245
– and enkephalins   199
– and glycosyltransferase   36
S-100 protein   260, 283 f

tabun   173
Tanzi & Lugaro   276
tar   294 f
taurine   192, 243
– formula   179
Tay-Sachs syndrome   41 ff, 300
TEA (see tetraethylammonium)
Terenius & Wahlström   196
tetanus toxin   39 f
tetrabenazine   189
tetracaine   128, 176
tetraethylammonium (TEA)   98, 111, 129 ff, 144
tetranitromethane (TNM)   121
tetraphenylphosphonium   136
tetrodotoxin (TTX)   98, 109, 111, 117 f, 121 f, 126, 144
– and facilitation   280
thalamus   13
thioridazine   233
thioxanthene   234 f
third messenger   227
threshold   105
thyrotropin-releasing hormone (TRH)   181
TNBS (2,4,6-trinitrobenzyl sulphonic acid)   60
tolerance   240 f
Torpedo
– and synaptic vesicles   167 f
– and acetylcholine receptor   170, 214 ff
– as experimental model   301 ff
– electric organ   63, 245, 301 ff
– electroplaque
–– and acetylcholine receptor   219
toxins as tools   231
toxogonin   174
tranquillizers   194
– diazepines (benzodiazepines)   242
transducin (G-protein)   8 f, 225
transduction   7
translocase   25
transmitter   106
– criteria and classification   178 ff
– differentiation   265
– inactivation   163, 172, 174 f
– inhibitory   195
– release   163, 167
–– calcium-dependent   160
– reuptake   163

transsynaptic regulation 184
- orthograde 265 ff
- retrograde 265 ff
trg 295
trifluoperazine 233
trimethyllysine 258
trisphosphoinositol (inositol trisphosphate 1,4,5) (InsP$_3$,IP$_3$) 31 f
Triton X-100 65 f, 204
trophic factors 267
trophism 264
tropolon 189
tropomyosin 134
troponin C 260
trypsin 172
tryptophan 191
tsr 294 f
TTX (see tetrodotoxin)
tuberoinfundibular system 232
tubulin 134, 257 ff
- and microtubules 255
- in slow axonal transport 254 ff
tubulin assembly protein (TAP) 257
turkey (avian) erythrocytes 225
Tween 66
two state model
- for transmitter and hormone receptors 207
tyramine 184
tyrosine 182 ff
tyrosine hydroxylase (TH) 182 ff, 227, 270
- and transsynaptic regulation 265

UDP-glucose 45
uptake 198
uridine
- uptake and NGF 270
uridine diphosphate (UDP) 36

vagus nerve 181
Vagusstoff 99
valinomycin 68
Valium^R 194, 242
van Deenen 63

van Heyningen, S. 39
varicosity 245
- of neuron 188
vasoactive intestinal polypeptide (VIP) 181
vasopressin 32, 181
veratridine 123 ff
vesicles
- formation from lipid bilayer 54
- large
-- catecholamine storage 185
- small dense core
-- catecholamine storage 185
Vibrio cholerae 39
vinblastine 255
vincristine 255
VIP (vasoactive intestinal polypeptide) 181
Virchow, R. 20
visual cortex 2, 13, 35
visual field 13
visual purple 3
visual transduction 230
vitamin A aldehyde (retinal) 3 f
vitamin A (retinol) 3
voltage clamp 98
von Euler, U. 199

Wald, G. 3
- cycle 22
Waldeyer 157 f
Weiss, P. 251
white matter
- of brain 34, 36
Whittaker, V.P. 160, 167, 182, 245, 304
Wiesel, T. 13
Wolfgram protein (see protein of myelin)

Xenopus (toad) 272
- oocyte 242, 306
X-ray diffraction
- myelin 78

yohimbine 223
Young, J.Z. 91

Notizen

Notizen

Notizen

Notizen

Notizen

Notizen